Cholera and the Ecology of *Vibrio cholerae*

Cholera and the Ecology of *Vibrio cholerae*

Edited by

B.S. Drasar

Department of Clinical Sciences, London School of Hygiene and Tropical Medicine, UK

and

B.D. Forrest

Biocine Clinical Research, Chiron Biocine, Emeryville, California, USA

CHAPMAN & HALL

London · Weinheim · New York · Tokyo · Melbourne · Madras

Published by
Chapman & Hall, 2–6 Boundary Row, London SE1 8HN, UK

Chapman & Hall, 2–6 Boundary Row, London SE1 8HN, UK

Chapman & Hall GmbH, Pappelallee 3, 69469 Weinheim, Germany

Chapman & Hall USA, 115 Fifth Avenue, New York NY 10003, USA

Chapman & Hall Japan, ITP Japan, Kyowa Building, 3F, 2-2-1, Hirakawacho, Chiyoda-ku, Tokyo 102, Japan

Chapman & Hall Australia, 102 Dodds Street, South Melbourne, Victoria 3205, Australia

Chapman & Hall India, R. Seshadri, 32 Second Main Road, CIT East, Madras 600 035, India

First edition 1996

Typeset in 10/12pt Palatino by Acorn Bookwork, Salisbury

Printed in Great Britain by St Edmundsbury Press, Bury St Edmunds, Suffolk

ISBN 0 412 61220 8

A catalogue record for this book is available from the British Library
Library of Congress Catalog Card Number: 96–83671

∞ Printed on permanent acid-free text paper, manufactured in accordance with ANSI/NISO Z39.48-1992 and ANSI/NISO Z39.48-1984 (Permanence of Paper).

Contents

List of contributors		vii
Preface		ix
1	***Vibrio cholerae*** B. Said and B.S. Drasar	1
2	**The Asiatic cholera: an historical determinant of human genomic and social structure** G.C. Cook	18
3	**Management of cholera: the vital role of rehydration** G.C. Cook	54
4	**Intestinal immunity to *Vibrio cholerae*** B.D. Forrest	95
5	**Cholera: pathogenesis and vaccine development** M.M. Levine and J.B. Kaper	125
6	**Ecology of *Vibrio cholerae*: role of aquatic flora and fauna** M.S. Islam, B.S. Drasar and R.B. Sack	187
7	**Zoological microhabitats of *Vibrio cholerae*** S. M. Lewin	228
8	**The geography of cholera** A.E. Collins	255
9	**The fate of *Vibrio cholerae* in wastewater treatment systems** T. Curtis	295
10	**Problems of the epidemiology of cholera and the ecology of *Vibrio cholerae*** B.S. Drasar	333
Index		349

Contributors

Dr A.E. Collins
Department of Geography
King's College London
Strand
London WC2R 2LS, UK

Dr G.C. Cook
Hospital for Tropical Diseases
St Pancras Way
London NW1 0PE, UK
and
Department of Clinical Sciences
London School of Hygiene
and Tropical Medicine
Keppel Street
London WC1 7HT, UK

Dr T. Curtis
Department of Civil Engineering
University of Newcastle upon
Tyne
Newcastle upon Tyne
NE1 7RU, UK

Professor B.S. Drasar
Department of Clinical Sciences
London School of Hygiene and
Tropical Medicine
Keppel Street
London WC1 7HT, UK

Dr B.D. Forrest
Biocine Clinical Research
Chiron Biocine
4560 Horton St
Emeryville
California 94608
USA

Dr M.S. Islam
Environmental Microbiology
Laboratory
Laboratory Sciences Division
ICDDR, B
GPO Box 128
Dhaka 1000
Bangladesh

Dr J.B. Kaper
Centre for Vaccine Development
University of Maryland
School of Medicine
Department of Paediatrics
685 West Baltimore St
Baltimore
Maryland 21201
USA

Professor M.M. Levine
Centre for Vaccine Development
University of Maryland
School of Medicine
Department of Paediatrics
685 West Baltimore St
Baltimore
Maryland 21201
USA

Dr S.M. Lewin née Thom
143 Wentworth Drive
Lansdale
Pennsylvania 19446
USA

Professor R.B. Sack
Environmental Microbiology
Laboratory
Laboratory Sciences Division
ICDDR, B
GPO Box 128
Dhaka 1000
Bangladesh

Dr B. Said
Laboratory of Enteric
Pathogens
Central Public Health
Laboratory
61 Colindale Avenue
London NW9 5HT, UK

Preface

As these words are written, the world peers into the abyss of an emerging eighth cholera pandemic. Not since the devastation of the fifth pandemic has cholera posed such a significant threat to both the developed and developing worlds. As the number of reported cases of clinical cholera due to *Vibrio cholerae* O1 decline, outbreaks of disease caused by *V. cholerae* O139 have appeared in Europe but, as yet, *V. cholerae* O139 cholera has not established a foothold there. However, at least 19 countries are experiencing the impact of this new threat. *V. cholerae* O139 represents a threat of considerable magnitude due to its novelty. This serotype produces more cholera toxin than either Classical or El Tor *V. cholerae* O1 strains, and shares very few surface antigens. As a result, everyone is immunologically naïve, reflected in observations that even those convalescent individuals recovering from a clinical attack of *V. cholerae* O1 possess no protection against *V. cholerae* O139. The contributors to this book have examined the collective knowledge of *V. cholerae,* from which a better understanding of the issues surrounding the emergence, management, and environmental control of current and future outbreaks of cholera can be derived. A key area of interest is in the inter-epidemic survival of *V. cholerae,* and the factors that permit the emergence of new, more virulent serotypes of *V. cholerae* O1.

Said and Drasar describe the complex taxonomic relationships of the various species comprising the genus *Vibrio,* and offer a much clearer understanding of what defines the species *V. cholerae.* In addition, they address the nature of clinical disease caused by non-O1 *V. cholerae* and the 10 other species of *Vibrio* that have been associated with human disease, providing insight into the striking difference in the behaviour of *V. cholerae* O139 from the other 138 serotypes. The spread of cholera and the nature and severity of disease appears to be a complex interaction between the organism, the host, and the intertwining social structure.

Cook provides a thorough historical perspective into this symbiotic relationship, showing how each epidemic and pandemic has provided in a stepwise manner, an accumulation of knowledge on this relationship that has led to the implementation of social control measures to restrict the spread of this disease. It is clear that when the infrastructure providing this control collapses, then cholera progresses. These periodic outbreaks and pandemics and the modern-day factors responsible for them are described by Collins, who examines the issues surrounding the geography of cholera. Cholera is a disease controlled by sanitation. Collins links the ecology of cholera, with global finance, human migration, and introduces some new thoughts on the seasonality of cholera, that together provide great insight into the reasons for the failure of environmental control in an era that possesses considerable resources. Curtis picks up the issues of environmental management by reviewing the issues surrounding wastewater management, and the design and implementation of wastewater treatment systems that provide an effective process for the elimination of *V. cholerae*. A variety of wastewater management systems are described and their relative efficacy in the removal of *V. cholerae* are compared. Their suitability for a particular social or economic situation is considered, and their impact on *V. cholerae* in the general environment is assessed. Most importantly, Curtis brings to our attention the relative ignorance that exists on the fate of *V. cholerae* in water management systems, and the role that many inefficient systems may have in propagating the conditions necessary for the emergence of new clinically relevant serotypes.

It is the survival of *V. cholerae* in the environment that presents the most difficulties in designing control and management systems. *V. cholerae* has been generally accepted as an autochthonthous aquatic resident. It is also recognized that terrestrial animals, including humans, have no important role in the interepidemic survival of *V. cholerae*, which has been demonstrated to be able to survive in a variety of marine habitats, possibly making use of viable but non-culturable forms. Islam and Sack provide a review of the environmental factors associated with the survival of *V. cholerae* in aquatic environments, providing an insight into the role of aquatic flora and fauna in the survival of *V. cholerae* in the environment. They provide a detailed examination of the relationship between both freshwater and marine aquatic microphytes and macrophytes, and algae, with *V. cholerae*. Their review clearly describes the importance of these relationships in permitting the survival of *V. cholerae* in environmental waters. Lewin examines the role of zoological microhabitats in the propagation of *V. cholerae*, and focuses on the inter-epidemic survival of *V. cholerae* O1 in association with aquatic fauna. She also corrects some of common misconceptions surrounding the role of specific aquatic life in the long-term survival of *V. cholerae* in the aquatic environment.

The transition of *V. cholerae* from aquatic microhabitat to humans is made with terrible consequences for the new host. A severe and prolific diarrhoea represents the well-known picture of cholera, but it is frequently forgotten that most clinical cholera is associated with mild diarrhoea that is rarely life-threatening. However, untreated severe diarrhoea is frequently fatal, and although the management of clinical cholera is simple fluid replacement, this can be problematic in many areas where cholera is endemic, or in a new epidemic. The clinical management of cholera is reviewed by Cook, with special emphasis placed on the role of oral rehydration therapy, and the history of its development. Fluid replacement as an effective therapy for cholera has been well-recognized for more than 160 years. Cook provides a detailed history of the problems associated with the development of an effective form of rehydration suitable for field application during an epidemic, concluding with a practical review of the established recommendations, and the uses and cautions associated with rehydration therapy.

However, although practical and effective management is strongly desirable, the development and efficient delivery of a highly efficacious vaccine would be of considerable benefit to most individuals living in communities at risk of epidemic cholera. In addition, even a vaccine with as little as 50% protection, would provide considerable relief to the overwhelmed health care systems in many of these developing communities through a reduction in mortality and morbidity. It is now generally accepted that such a vaccine should be administered orally in order to stimulate local intestinal immunity that is known to be associated with protection against clinical disease. Since cholera toxin is the principal mediator of the clinical illness, Forrest provides a review of the role of local intestinal immunity to the cholera toxin of *V. cholerae* O1 and O139. A brief review of the mechanism of induction of local intestinal immunity, is followed by a more detailed investigation of the role of anti-cholera toxin secretory IgA and local cellular immunity in the inhibition of cholera toxin binding to the intestinal epithelium, and in the mediation of protection against subsequent clinical disease. The translation of findings in key animal experiments to humans, and eventually to the field is reviewed, in the context of possible application of anti-toxin immunity in vaccine development. Kaper and Levine describe the development of vaccines against cholera with special emphasis on the development of live, attenuated, non-toxic mutants of *V. cholerae* O1 and O139. The development of such vaccines uses advanced techniques of molecular biology to identify key toxin genes responsible for a significant contribution to the clinical disease. The highly successful nature of several of these candidate vaccines in healthy adult volunteers, as well as in several field trials for the study of clinical efficacy, has provided hope for the future. Such vaccines may provide useful tools in the control of epidemic or endemic cholera, but do not provide an alternative to the much-needed efficient wastewater management systems that may provide a means to control *V. cholerae* in the environment.

Finally, Drasar provides an erudite assessment of the problems relating to the epidemiology and ecology of *V. cholerae*. In particular, the issues surrounding inter-epidemic survival of *V. cholerae* are critically reviewed, and the mechanism of this survival, ecological models for evaluating this, and the possible relationship between the postulated genetically unstable, viable but non-cultural forms and the development of pandemic strains of *V. cholerae* provide an integrated picture of how *V. cholerae* remains the threat today that it has been for centuries.

We wish to thank the contributors to this book, and a special thanks to Sue Horrill for administrative support.

B.D. Forrest
B.S. Drasar

Vibrio cholerae 1

B. Said and B. Drasar

1.1 INTRODUCTION

The genus *Vibrio* is among the oldest still recognized by bacteriology. Taxonomy was reviewed by Baumann *et al.* (1984) who listed 20 species; more recently, Farmer and Hickman-Brenner (1992) listed 34 species. However, not all of these have been isolated from clinical material. These studies illustrate the way in which our knowledge of the genus *Vibrio* has advanced. Until recently, *Vibrio* consisted of a heterogenous and poorly described group of organisms. However, a more strict definition of the genus, coupled with the transfer of some members to other well-understood groups, has allowed the taxonomy to be clarified. Some, such as *Campylobacter*, have been transferred to their own genera.

1.2 TAXONOMY OF VIBRIOS

The *Vibrio* genus is part of the family Vibrionaceae. The family also includes the genera *Photobacterium*, *Aeromonas* and *Plesiosomonas*. It is not certain that this family is a natural grouping in any evolutionary sense. It arose from the wish to separate these organisms from the Enterobacteriaceae. DNA homology studies suggest that *Vibrio* and *Photobacterium* may be closely related. However, *Aeromonas*, is more distantly related and may in fact be closer to *Pasteurella multocida* and Enterobacteriaceae than to vibrios. Members of the family are usually oxidase-positive and have polar flagella for motility. Peritrichous flagella may occasionally be produced.

Cholera and the Ecology of Vibrio cholerae. Edited by B.S. Drasar and B.D. Forrest.
Published in 1996 by Chapman & Hall, London. ISBN 0 412 61220 8.

Family definition: rigid Gram-negative rods, motile by polar flagella in liquid media but peritrichous flagella may be produced on solid media in some species. Chemo-organotrophs; metabolism is both respiratory (use oxygen) and fermentative. Utilize glucose fermentatively, catalase-positive, most species are oxidase-positive and reduce nitrate to nitrite. Facultative anaerobes without exacting nutritional requirements.

The genus *Vibrio,* as currently understood, is comparatively well characterized, and consists mostly of oxidase-positive fermentative organisms. All members are facultative anaerobes and grow well on culture media. It is usual to speak of these organisms as belonging to two groups: those isolated from clinical specimens, and the marine vibrios. However, this distinction is by no means clear.

All vibrios benefit by addition of sodium chloride to their media used for identification. An absolute requirement for sodium is not restricted to vibrios of marine origin. Indeed, many of those isolated from clinical material are thought to originate from marine sources. Of the 11 vibrios isolated from clinical materials, most have also been associated with marine environments. They have been isolated from food poisoning associated with consumption of seafoods, and from wound infections resulting from contamination with a variety of marine materials.

1.3 THE GENUS *VIBRIO*

Historically, taxonomically, and medically the genus *Vibrio* focuses on the organisms that cause cholera. These are largely members of the species *Vibrio cholerae;* further species may occasionally cause clinically identical illness. Advances in molecular biology, immunology and bacterial taxonomy mean that *V. cholerae* is now among the best understood of bacterial species. The taxonomy of the organisms, the virulence mechanisms by means of which it causes disease, and the means of treating apparent cholera have all been thoroughly investigated. So we have greatly enhanced our understanding of the vibrios; hence, the question must be asked as to the extent to which cholera patients benefit.

Genus definition: short asporogenous rods, axis curved or straight 0.5 × 3 μm single or occasionally united in S-shapes or spirals. Motile by single polar flagellum which is sheathed in certain if not all species. Some species produce lateral (peritrichous) flagella on solid media. Carbohydrates are fermented with the production of acid and no gas. A wide range of extracellular enzymes is produced, including amylase, chitinase, DNAse, gelatinase, lecithinase and lipase. Grow on simple mineral media. NaCl stimulates growth with an optimum of 1–3%. Some strains require NaCl for growth. Temperature optima range from 18–37°C and pH range 6.0–9.0. Usually sensitive to 0/129 (2,4-diamino-6,7-diisopropyl pteridine). G + C content of DNA 40–50 mol %.

1.3.1 IDENTIFICATION, NOMENCLATURE AND CLASSIFICATION OF *V. CHOLERAE* AND *V. MIMICUS*

The genus *Vibrio* includes several species which are pathogenic to humans, and infections are usually food-borne. Of more than 30 species, 11 are associated with human diseases; *V. cholerae, V. mimicus, V. parahaemolyticus, V. alginolyticus, V. hollisiae, V. fluvialis, V. furnisii, V. vulnificus, V. damsela, V. metschnikovii* and *V. cincinnatienisis* (Bryant, 1986). These divide biochemically into three groups (Table 1.1); *V. cholerae* and *V. mimicus* belong to group 1 (Furniss *et al.*, 1978; Brayton *et al.*, 1986; Janda *et al.*, 1988).

V. cholera and *V. mimicus* are short, rod-shaped (straight or slightly curved) Gram-negative bacteria with a single polar flagellum. These organisms are facultatively anaerobic, oxidase-positive and ferment glucose without forming gas. They grow readily on an agar medium at an optimal pH between 7.6 and 8.6. Colonies on agar are usually translucent and amorphous but sometimes wrinkled or rugose. Strains of *V. cholerae* and *V. mimicus* may produce a pellicle when grown in liquid media. Strains of *V. cholerae* and *V. mimicus* can grow in the absence of NaCl, a feature which distinguishes these bacteria from other vibrios. Other useful biochemical characteristics (Table 1.1) are: lysine decarboxylase and ornithine decarboxylase-positive, but arginine dihydrolase-negative, acid from mannitol but not from inositol or arabinose.

The different serogroups of *V. cholerae* are very similar biochemically. However, all strains of *V. cholerae* serogroup O1 ferment mannose, whereas the reaction is variable with strains of *V. cholerae* belonging to non-O1 serogroups. *V. mimicus* is biochemically distinguished from *V. cholerae* by the inability to ferment sucrose, a negative reaction in the Vogues–Proskauer (VP) test, a lack of amylase production and failure to haemagglutinate chick red blood cells (Sakazaki and Donovan, 1984).

1.3.2 SEROTYPING OF *V. CHOLERAE* AND *V. MIMICUS*

Serology has enabled strains of *V. cholerae* and *V. mimicus* to be divided into 140 groups based on epitopes located on *V. cholerae* lipopolysaccharide, also termed the somatic or 'O' antigens. Strains belonging to *V. cholerae* serogroup O1 regardless of biotype are agglutinated by anti-O1 antisera, while all other subgroups, which are not agglutinated by

Table 1.1 Biochemical characteristics of pathogenic *Vibrio* spp.

	NaCl -requiring	Lysine	Ornithine	Arginine	Vogues–Proskauer reaction	Sucrose (gas)	Glucose
Group 1*							
V. cholerae	–	+	+	–	v	+	–
V. mimicus	–	+	+	–	–	–	–
Group 2†							
V. vulnificus	+	+	v	–	–	–	–
V. parahaemolyticus	+	+	+	–	–	–	–
V. alginolyticus	+	+	+	–	+	+	–
V. hollisae	+	–	–	–	–	+	–
V. cincinnatiensis	+	+	–	–	+	+	–
Group 3‡							
V. fluvialis	+	–	–	+	–	+	–
V. furnisii	+	–	–	+	–	+	+
V. damsela	+	v	+	v	v	–	–
V. metschnikovii§	+	v	–	v	+	+	–

* Non-halophilic, arginine dihydrolase-negative group.
† Halophilic, arginine dihydrolase-negative group.
‡ Halophilic, arginine dihydrolase-positive group.
§ *V. metschnikovii* is unusual in that it is the only member of the family which is oxidase-negative.
+, positive; –, negative; v, variable reaction.

Table 1.2 Serotyping of *Vibrio cholerae**

	O antigen determinants
V. cholera O1	
Subtype	
Inaba	A, B
Ogawa	A, C
Hikojima	A, B, C
V. cholerae non-O1	
{non-cholera vibrios (NCVs)}	O2 – O140
{non-agglutinable vibrios (NAGs)}	
{non-O1 *V. cholerae*}	
V. cholerae O139 Bengal	O139

* H antigen is common to all groups and *V. mimicus*

anti-O1 antisera, are designated *V. cholerae* non-O1. Strains of *V. mimicus*, once thought to be a variant of *V. cholerae* unable to utilize sucrose, have now been classified as a distinct species (Davis *et al.*, 1981).

All strains of *V. cholerae* and *V. mimicus* share a common 'H' antigen. Apart from the O1 and O139 there is no link between the O-group and pathogenicity. The O1 antiserum is available commercially as *V. cholerae* 'polyvalent' antiserum. Subtyping of O1 strains may be done by testing for agglutination with absorbed Inaba and Ogawa antisera (Table 1.2). Since strains of *V. mimicus* and *V. cholerae* can belong to the same serogroups, a single scheme is applicable to both species.

Individual strains of both species (*V. cholerae* and *V. mimicus*) can be differentiated by the structure of cell wall lipopolysaccharide, as detected by specific antisera prepared to the somatic ('O') antigen. Flagellar antigens are of little value for strain differentiation since flagellar or 'H' antigens are seemingly identical for all strains of *V. cholerae* and *V. mimicus*. Early attempts to differentiate strains of *V. cholerae* by serology were performed by Gardner and Venkatraman (1935), who successfully divided a panel of strains of *V. cholerae* into six O-subgroups, in which O1 was assigned to 'cholera vibrios' (Gardner and Venkatraman, 1935). In addition to the category 'cholera vibrios', a large number of vibrios, which are biochemically similar but which are not agglutinated by O1 antisera, have been recognized for many years in cholera-epidemic areas. These have been called non-agglutinable (NAG) or non-cholera vibrios (NCV). Since 1972, when the International Subcommittee on Taxonomy of Vibrios recommended that the species *V. cholerae* no longer be restricted to the cholera vibrio, they have been included in the species definition and are known collectively as

V. cholerae non-O1 (Sakazaki and Donovan, 1984; West and Colwell, 1984). The antigenic scheme of Sakazaki and Donovan was extended throughout the 1970s, incorporating the scheme of Smith, to 83 serogroups (Smith, 1979; Sakazaki and Donovan, 1984). Recent studies have further extended the scheme to show that the species *V. cholerae* contains over 140 serogroups (Shimada *et al.*, 1994).

1.4 *VIBRIO CHOLERAE*

The definition of the species *V. cholerae* and the relationship of the species to the cholera vibrios that cause cholera has long been a source of confusion. Rather like *Escherichia coli*, *V. cholerae* is a large and diverse species, not all members of which are pathogens. In the past it was customary to speak of cholera vibrios, NCVs and NAGs. It is now clear that these organisms are all members of the same species. Taxonomic studies including DNA–DNA homology have shown that *V. cholerae* is a single, relatively homogenous, closely related species (Citarella and Colwell, 1970). Like other Gram-negative bacteria, V. cholerae has a complex cell wall which includes lipopolysaccharides (LPS), the carbohydrate components of which are designated as O antigens. Most of the strains of *V. cholerae* isolated from cases of cholera belong to the O1 serogroup. Within this serogroup are further subdivisions, three serotypes Inaba, Ogawa and Hijokima, and two biotypes, Classical and El Tor. Strains of *V. cholerae* can be divided into the two biotypes on the basis of the tests set out in Table 1.3. Each biotype may include members of all serotypes.

It should be remembered that not all O1 strains cause cholera. Members of groups O2 to O138 and O140 may cause diarrhoea but are

Table 1.3 The biotypes of cholera vibrios (*V. cholerae* O1)

	Biotype	
	Classical	El Tor
O1 antiserum	+	+
Vogues–Proskauer reaction	−	+
*Acetoin production		
Haemolysis of sheep erythrocytes	−	+
Chick red cell agglutination	−	+
Polymyxin, 50 i.u.	S	R
Classical phage IV	S	R
El Tor phage V	R	S

*This difference is quantitative. Many Classical strains produce trace amounts of acetoin not detected by the standard test. R, rough; S, smooth.

not thought to cause epidemic or endemic cholera. Although groups O2–O138 and O140 are generally recognized to be less pathogenic than the Classical or El Tor biotypes, they can cause symptoms very similar to or identical to those of cholera and may cause epidemics of enteritis. Indeed, they are more commonly isolated in Britain from cases of enteritis than the El Tor or Classical biotypes. Only *V. cholerae* O1 is reportable to the World Health Organization (WHO) as cholera. The Classical biotype is seldom isolated and the current seventh cholera pandemic is caused by the El Tor biotype. Members of the recently isolated O139 group cause a disease clinically indistinguishable from cholera and have been responsible for epidemics (Albert *et al.*, 1993; Bhattacharya *et al.*, 1993; Ramamurthy *et al.*, 1993). It is likely that these O139 strains will be responsible for the eighth pandemic of cholera.

1.5 *VIBRIO MIMICUS*

Until recently *V. mimicus* has been identified as *V. cholerae*. Organisms that are unable to ferment sucrose form green colonies on thiosulphate-citrate bile salt (TCBS) agar but which are otherwise phenotypically similar to *V. cholerae* have been isolated from shellfish, brackish water and stools of human beings with diarrhoea in many parts of the world. Recent studies on more than 50 such organisms show that they have at species level DNA relatedness to each other but are related only distantly to *V. cholerae* (DNA–DNA homology, 20–50%). Thus, these bacteria would appear to be distinct species (Davis *et al.*, 1981). V. mimicus shares O group serotypes with *V. cholerae*. Only some strains produce enterotoxin.

1.6 EPIDEMIOLOGY OF CHOLERA

Diarrhoeal diseases are a significant cause of mortality and morbidity worldwide, particularly in developing countries. *V. cholerae* is an important cause of diarrhoea in both children and adults, particularly in Southern Asia and the Indian subcontinent where cholera and cholera-like diseases are endemic. Cholera is often associated with poverty, overcrowding and low socioeconomic status.

Worldwide the number of cholera cases reported to the WHO, in 1993, was 376 845 in 78 countries with 6781 deaths (WHO, 1994a). This represents a substantial decrease in the figure for 1991, when due to the spread of the seventh pandemic to South America, 595 000 cases were reported (WHO, 1994b).

In Asia alone, there were 18 007 cases reported by 13 countries in 1989; this increased sharply in 1990, due to an outbreak in Nepal, to 30 979 cases reported by 12 countries (Anonymous, 1991). In 1991, there were 49 791 cases and 1286 deaths; this declined to 1992 to 16 229 cases and 372 deaths. Again, the difference in numbers was due to the continuing outbreak in Nepal which accounted for 30 000 cases in 1991 (Anonymous, 1993).

Cholera is a pandemic disease *par excellence*. Although it has undoubtedly been common in India since the beginning of records, and has spread to neighbouring countries on frequent occasions, the great days of cholera came after the resumption of world trade following the Napoleonic wars. Whether widespread cholera had occurred previously is not known. Writings of Europeans in India from the earliest days have made reference to this disease. Epidemics undoubtedly occurred among Europeans and European troops stationed in India (Orton, 1831; MacPherson, 1872; MacNamara, 1876). Until 1817 spread was probably limited, but then the first pandemic started and reached Europe. This was rapidly followed by the second pandemic which reached England through the port of Sunderland in 1831. It is notable that Snow, who later became famous for the elucidation of the waterborne transmission of cholera (Snow, 1849, 1855) was working as a doctor's apprentice in Newcastle at about this time. Early 19th century Europe was ravaged by a series of epidemics of cholera. We do not know for certain what caused these epidemics, although it is assumed to be *V. cholerae*. It was not until the fifth pandemic that Robert Koch (1884) succeeded in isolating and identifying the organism. Even when the organism had been isolated, there were many who did not believe that it was the cause of the disease, though few went so far as Pettenkoffer (1892) in their wish to demonstrate that *V. cholerae* was non-pathogenic. He drank a culture and did not get cholera. As we now know this probably reflects the relative resistance of some people to cholera vibrios. Comparable studies in healthy volunteers have not always been successful in inducing disease.

When the El Tor biotype was discovered it was regarded as a harmless variant. However, *V. cholerae* El Tor strains rapidly displaced the Classical biotype in endemic areas, and proved capable of causing epidemic cholera. The seventh and current pandemic started in 1961, and has spread from South-East Asia to the Middle East and parts of Europe, reaching Africa in 1970 and South America in 1991 (Janda *et al.*, 1988; Gangarosa and Tauxe, 1992).

We may now be about to embark on the eighth pandemic caused by *V. cholerae* O139 (Table 1.4).

For most practical purposes it is the local epidemics in endemic areas that are of the greatest importance. Historically it is likely that endemic cholera has been restricted to the Ganges delta. However, during the seventh pandemic, many other endemic areas have been detected. How an area becomes an endemic focus is unclear; the answers probably relate to aspects of the ecology of *V. cholerae* that we do not fully

Table 1.4 The chronology of cholera

At home	To 1817	Endemic on the Indian sub-continent	
The Historic Pandemics	1817–1823	First Pandemic	John Snow and the Broad Street Pump (1854)
	1829–1850	Second Pandemic	
	1852–1860	Third Pandemic	
	1863–1879	Fourth Pandemic	
The Classical Pandemics	1881–1896	Fifth Pandemic	Robert Koch and the comma bacillus (1884)
	1899–1923		
The El Tor Pandemic	1961 to date	Seventh Pandemic	De and Chatterjee Cholera toxin (1953)
O139 Bengal	1992 to date	Eighth Pandemic?	

understand. In endemic areas successive, usually seasonally related, epidemics of cholera occur (Feacham, 1981).

Historically, cholera has been considered to be a water-borne disease (Feacham, 1982; Feachem *et al.*, 1981). Indeed, there is considerable evidence that water-borne transmission of cholera has been and is very important. However, in an epidemic situation, food-borne transmission is also very significant. *V. cholerae* has often been isolated from seafood and shellfish. The vibrios are able to grow well in cooked rice and on other foods. However, external contamination of foods may be equally important. In the recent outbreak in Chile, cholera was transmitted by vegetables irrigated with untreated sewage. Similar transmission, related to melons, was seen in Peru (Crowcroft, 1994). Thus, both food and water can act as vehicles for transmission (Anonymous, 1991; Glass *et al.*, 1991). Poor sanitation is undoubtedly important for maintaining transmission during epidemics. This probably explains why secondary transmission is rare when cases are imported into countries with good sanitation. Investigations of other outbreaks in an epidemic situation may demonstrate what vehicle is important.

1.6.1 RECENT DEVELOPMENTS: THE SPREAD OF *V. CHOLERAE* O1 TO SOUTH AMERICA AND THE EMERGENCE OF *V. CHOLERAE* O139

The seventh pandemic of cholera spread and increased steadily from 1961 until 1966, and in 1970 a major upsurge occurred as countries in Africa were infected. Reviewing the global cholera situation in 1980, the WHO working group expressed 'concern that if cholera should reach any of the countries of South and Central America which are considered to be "receptive", there could be another dramatic increase in the number of affected countries' (WHO Scientific Working Group, 1980). In January 1991, strains of *V. cholerae* O1 were isolated in Peru, the first time it had been identified in the Americas. By 1992, 391 220 cases and 4002 deaths had been recorded (Anonymous, 1992) in 14 of the 21 countries of South and Central America.

Untreated drinking water was thought to be a potential vehicle of infection with *V. cholerae* and the organism was also isolated from fish and seawater. The consumption of vegetables irrigated with raw waste also contributed to the spread of this organism. Genetic analysis revealed that strains isolated from the outbreak in South America were similar to strains of *V. cholerae* isolated from the pandemic occurring in Asia and Africa, and clearly distinguishable from strains causing sporadic cases of cholera along the Gulf coast of the USA (Wachsmuth *et al.*, 1991, 1993).

As the seventh pandemic spread through South and Central America an apparently 'new' strain of *V. cholerae* emerged in India and Bangladesh. The first reported outbreak of cholera caused by this organism occurred in October 1992 in Madras, India (Ramamurthy *et al.*, 1993). The illness was clinically indistinguishable from epidemic cholera, except for the frequency of abdominal cramps (Bhattacharya *et al.*, 1993). The outbreak spread to Calcutta (November, 1992) with over 15 000 cases and 230 deaths. The number of cases of diarrhoea caused by *V. cholerae* O139 was increasing whereas the number of cases of cholera caused by *V. cholerae* O1 was decreasing; by December 1992, 95% of cases were due to *V. cholerae* serogroup O139. There followed outbreaks in Southern Bangladesh (December, 1992) (Albert *et al.*, 1993), Dhaka (January and February, 1993 – 10 000 cases, 500 deaths) and a further major outbreak in Calcutta (February, 1993) (Anonymous, 1993). The outbreaks, both in India and Bangladesh, were apparently caused by the same clone of *V. cholerae* O139 (Shimada *et al.*, 1993). The severity of the illness, the high attack rate in the adult population, and the lack of immunological protection in individuals with a history of clinical cholera suggested that this was a new strain, one to which the population had not been previously exposed. This was supported by the lack of protection conferred by vaccination with heat-inactivated whole-cell *V. cholerae* O1 antigen. None of the strains reacted with a panel of 138 anti-lipopolysaccharide antisera O1–O138 and all were positive for cholera toxin (CT). The outbreak strain was designated as *V. cholerae* serogroup O139, synonym Bengal, a novel non-O1 clone which appeared to be spreading and displacing the *V. cholerae* O1 El Tor strain (Swerdlow and Ries, 1993) just as the El Tor strain once superseded the Classical *V. cholerae* O1. Sporadic indigenous cases were also confirmed from Thailand, China, Malaysia and Nepal (WHO, 1994); imported cases were reported in Estonia, Germany, USA and UK (Anonymous, 1993; Cheasty *et al.*, 1993). This strain of *V. cholerae* belonging to serogroup O139 may have the potential to cause a new pandemic (Swerdlow and Ries, 1993).

Despite microbiological and clinical advances, cholera remains a persistent problem for the developing and to a lesser extent, the developed world. The emergence of strains of *V. cholerae* belonging to serogroup O139, the appearance of multiply resistant strains of *V. cholerae* and the spread of the seventh pandemic to South America has lead to renewed interest in *V. cholerae*.

1.6.2 *V. CHOLERAE* NON-O1, *V. MIMICUS* AND ASSOCIATED CLINICAL DISEASE

Strains of *V. cholerae* belonging to serogroups other than O1 have been implicated in sporadic cases and outbreaks of gastrointestinal disease. The symptoms vary from bloody diarrhoea with fever to a severe watery diarrhoea indistinguishable from cholera. In 1986 it was reported that strains of *V. cholerae* belonging to serotypes other than O1 were a significant cause of mortality in Bangladesh with 25.8% fatality rate (Islam and Shahid, 1986). Volunteer studies in 1990 (Morris, 1990) confirmed that strains of *V. cholerae* non-O1 can cause diarrhoea with a severity comparable with that seen in cholera, even by strains unable to make CT, a key virulence property expressed by *V. cholerae* O1. If CT is produced by non-O1 serogroups, as with the epidemic *V. cholerae* O139 strain, the presentation is typical of clinical cholera. Reports of outbreaks possibly caused by strains of *V. cholerae* other than O1 date back to 1954 (Yajnik and Prasad, 1954); since then there have been several published reports of outbreaks of gastroenteritis attributable to strains of *V. cholerae* non-O1 and some of these have, retrospectively, been linked to particular O-serogroups (Donovan, 1984). For example, during an outbreak of cholera in Czechoslovakia in Aldova *et al.* (1968) there were 56 cases, and *V. cholerae* was isolated from 42 of the patients. Strains of *V. cholerae* belonging to serogroup O5 were implicated, with potato salad as the probable vehicle of transmission (Aldova *et al.*, 1968) (Table 1.5). Another outbreak occurred among passengers onboard an aeroplane destined for Australia; the outbreak was linked to the consumption of chopped egg but the serogroup of *V. cholerae* involved was not identified (Dakin *et al.*, 1974). In a waterborne outbreak, reported in 1971, there were 544 cases and 31 fatalities. The outbreak was suspected to have resulted from faecal contamination of drinking water and caused by a strain of *V. cholerae* serogroup O37 (Kamal, 1971).

Sporadic cases associated with *V. cholerae* non-O1 diarrhoea have also been reported. McIntyre reported 19 persons with sporadic illness from Bangladesh; many had severe dehydrating diarrhoea and 50% required intravenous fluid therapy (McIntyre *et al.*, 1965). Spira *et al.* reported on 14 adults with cholera in Bangladesh; symptoms included vomiting (100%), abdominal pain (71%), pyrexia (43%) and muscle cramps (21%). The mean duration of illness was 42 hours (Spira *et al.*, 1978). Another study from the USA reported on 13 patients whose symptoms included abdominal cramps (92%), nausea (77%), vomiting (69%), chills (64%), fever (58%), mucus in stools (30%), and bloody stools (8%); the duration was approximately 7 days (Hughes *et al.*, 1978). In summary, a range of symptoms has been associated with cases of *V. cholerae* non-O1 gastroenteritis, including fever and bloody diarrhoea which are encountered only rarely in 'typical' cases of cholera.

V. cholerae has been established as the aetiologic agent of cholera and until recently it was generally accepted that only strains of *V. cholerae* belonging to serogroup O1 were responsible for causing epidemics of cholera. However, in 1992/3 a non-O1 strain of *V. cholerae* emerged

Table 1.5 *V. cholerae* non-O1 outbreaks

Year	Place	Clinical details	Suspected mode of transmission	Serogroup	Virulence factor	Reference
1965	Czechoslovakia	Diarrhoea 1–2 days, vomiting (25%), fever (11%)	Potato salad	O5	?	Aldova *et al.* (1968)
1968	Sudan	Rice-water stools	Water	O37	?CT	Kamal (1971)
1974	Australia	Diarrhoea and vomiting abdominal cramps 18–24 h	Chopped egg	?	?	Dakin *et al.* (1974)
1992	India	Acute cholera-like? diarrhoea	?Water	O139	CT	Ramamurthy *et al.* (1993)
1992	Bangladesh	Acute cholera-like	?Water	O139	CT	Albert *et al.* (1993)

CT, cholera toxin.

with epidemic, and possibly pandemic, potential. This strain of *V. cholerae* belonging to O139 appeared to be displacing strains of *V. cholerae* belonging to serotype O1 El Tor in endemic areas, and may prove to be the cause of an eighth pandemic, the first caused by a non-O1 strain of *V. cholerae* (Swerdlow and Ries, 1993).

Strains of *V. cholerae* belonging to serogroups O1 and O139 produced cholera toxin; however, toxin production is not a prerequisite for infection and non-toxigenic strains of *V. cholerae* O1 are able to cause illness. This usually involved diarrhoea, and symptoms vary from the mild and transitory to an illness resembling *cholera gravis*. Infections with non-toxigenic *V. cholerae* O1 may be of increasing prevalence (Honda *et al.*, 1988). Most *V. cholerae* non-O1 do not produce CT, although other enterotoxigenic mechanisms may or may not be present.

V. mimicus has also been reported as a causative agent of diarrhoea (Shandera *et al.*, 1983); however, as with *V. cholerae* non-O1, production of CT is rare and most strains produce toxins other than CT (Chowdhury *et al.*, 1987; Ramamurthy *et al.*, 1994). The wider range of presenting symptoms seen with *V. mimicus* and *V. cholerae* non-O1 as opposed to *V. cholerae* O1 infections is probably related to the presence or absence of a number of virulence factors.

Strains of both *V. cholerae* non-O1 and *V. mimicus* are also associated with non-gastrointestinal infections and are capable of causing extra-intestinal infections. Septicaemia may occur, with the causative organism being present in bile, gallbladder, urine, blood, sputum, appendix, peritoneal fluid and cerebrospinal fluid (Safrin *et al.*, 1988; Clark *et al.*, 1989). In general, patients which develop septicaemia have underlying disease, such as cirrhosis of the liver, or have wound infections contaminated with sea or estuary water. Extraintestinal infections, due to *V. cholerae* O1, have also been reported, although these are rare (Johnston *et al.*, 1993). Where extraintestinal infections have occurred the strains of *V. cholerae* were found not to produce CT, indicating that other factors may be involved. The pathogenic mechanisms possessed by strains of *V. cholerae* causing extraintestinal infections remain to be elucidated.

1.7 REFERENCES

Albert, M.J., Siddique, A.K., Islam, M.S., Faruque, A.S.G., Ansaruzzaman, M., Faraque, S.M. and Sack, R.B. (1993) Large outbreak of clinical cholera due to *Vibrio cholerae* non-O1 in Bangladesh. *Lancet*, **341**, 704.

Aldova, E., Laznickova, K., Stepankova, E. and Lietava, J. (1968) Isolation of non-agglutinable vibrios from an enteritis outbreak in Czechoslovakia. *Journal of Infectious Diseases*, **118**, 25–31.

Anonymous. (1991) Cholera in 1990. *Weekly Epidemiological Record*, **66**, 133-7.

Anonymous. (1992) Cholera update. *Communicable Disease Report*, **2**, 181.

Anonymous. (1993) Cholera in 1992. *Weekly Epidemiological Record*, **68**, 149-56.

Baumann, P., Furniss, A.L. and Lee, J.V. (1984) Section 5. Facultatively anaerobic gram-negative rods, in *Bergey's Manual of Systematic Bacteriology*, vol. 1, (eds J.G. Hot and N.R. Krieg), Williams and Wilkins, Baltimore, USA, pp. 518–38.

Bhattacharya, S.K., Bhattacharya, M.K., Nair, G.B. *et al.* (1993) Clinical profile of acute diarrhoea cases infected with the new epidemic strain of *Vibrio cholerae* O139: designation of the disease as cholera. *Journal of Infection*, **27**, 11–15.

Brayton, P.R., Bode, R.B., Colwell, R.R., MacDonell, M.T., Hall, H.L., Grimes, D.J., West, P.A. and Bryant, T.N. (1986) *Vibrio cincinnatiensis* sp nov, a new human pathogen. *Journal of Clinical Microbiology*, **23**, 104–8.

Bryant, T.N. (1986) *Vibrio cincinnatiensis* sp. nov., a new human pathogen. *Journal of Microbiology*, **23**, 104–8.

Cheasty, T., Said, B., Rowe, B. and Frost, J. (1993) *Vibrio cholerae* serogroup O139 in England and Wales. *British Medical Journal*, **307**, 1007.

Chowdhury, M.A.R., Aziz, K.M.S., Kay, B.A. and Rahim, Z. (1987) Toxin production by *Vibrio mimicus* strains isolated from human and environmental sources in Bangladesh. *Journal of Clinical Microbiology*, **25**, 2200–3.

Citarella, R.V. and Colwell, R.R. (1970) Polyphasic taxonomy of the genus *Vibrio*. *Journal of Bacteriology*, **104**, 434–42.

Clark, R., Bracy, W., Hanna, B. and Love, G.L. (1989) Case report: *Vibrio cholerae* non-O1 infection presenting as localised cellulitis. *American Journal of the Medical Sciences*, **298**, 328–30.

Crowcroft, N.S. (1994) Cholera: current epidemiology. *Communicable Disease Report Review*. Public Health Laboratory Service, London, **4**: Review No. 13: R157–64.

Dakin, W.P.H., Howell, D.J., Sutton, R.G.A., O'Keefe, M.F. and Thomas, P. (1974) Gastroenteritis due to non-agglutinable (non-cholera) vibrios. *Medical Journal of Australia*, **2**, 487–90.

Davis, B.R., Fanning, R., Madden, J.M., Steigerwalt, A.G., Bradford, H.B., Smith, H.L. and Brenner, D.J. (1981) Characterization of biochemically atypical *Vibrio cholerae* strains and designation of a new pathogenic species *Vibrio mimicus*. *Journal of Clinical Microbiology*, **14**, 631–9.

Donovan, T. (1984) Serology and serotyping of *Vibrio cholerae*, in *Vibrios in the Environment*, Wiley-Interscience, New York, pp. 83–101.

Farmer, J.J. and Hickman-Brenner, F.W. (1992) The General *Vibrio* and *Photobacterium*, in *The Prokaryotes*, vol. III, (eds A. Barlows, H.G. Trüper, M. Dworkin, W. Harder and K.-H. Schleifer), Springer-Verlag, New York, pp. 2952–3011.

Feachem, R.G. (1981) Environmental aspects of cholera epidemiology. I. A review of selected reports of endemic and epidemic situations during 1961–1980. *Tropical Diseases Bulletin of the Bureau of Hygiene and Tropical Diseases*, **78**, 675–98.

Feachem, R.G. (1982) Environmental aspects of cholera epidemiology. III. Transmission and control. *Tropical Diseases Bulletin of the Bureau of Hygiene and Tropical Diseases*, **79**, 1–47.

Feachem, R., Miller, C. and Drasar, B. (1981) Environmental aspects of cholera epidemiology. II. Occurrence and survival of *Vibrio cholerae* in the environment. *Tropical Diseases Bulletin of the Bureau of Hygiene and Tropical Diseases*, **78**, 865–880.

Furniss, A.L., Lee, J.V. and Donovan, T.J. (1978) *The Vibrios*, Public Health Laboratory Service Monograph Series, **11**, HMSO, London.

Gangarosa, E.J. and Tauxe, R.V. (1992) The Latin American cholera epidemic, in *Cholera*, Plenum Medical Book Company, New York, pp. 351–8.

Gardner, A.D. and Venkatraman, K.V. (1935) The antigens of the cholera group of vibrios. *Journal of Hygiene*, **35**, 262-82.

Glass, R.I., Claeson, M., Blake, P.A., Waldman, R.J. and Pierce, N.F. (1991) Cholera in Africa: lessons on transmission and control for Latin America. *Lancet*, **338**, 791–5.

Honda, T., Kasemsuksakul, K., Oguchi, T., Kohda, M. and Miwatani, T. (1988) Production and partial characterization of pili on non-O1 *Vibrio cholerae*. *Journal of Infectious Diseases*, **157**, 217–18.

Hughes, J.M., Hollis, D.G., Gangarosa, E.J. and Weaver, R.E. (1978) Non-cholera vibrio infections in the United States: clinical, epidemiological and laboratory features. *Annals of Internal Medicine*, **88**, 602–6.

Islam, S.S. and Shahid, N.S. (1986) Morbidity and mortality in a diarrhoeal diseases hospital in Bangladesh. *Royal Society of Tropical Medicine and Hygiene*, **80**, 748–52.

Janda, J.M., Powers, C., Bryant, R. and Abbott, S. (1988) Current perspectives on the epidemiology and pathogenesis of clinically significant *Vibrio* species. *Clinical Microbiology Reviews*, **1**, 245–67.

Johnston, J.A., Morris, J.G. and Kaper, J.B. (1993) Gene encoding zonula occludens toxin (*zot*) does not occur independently from cholera enterotoxin genes (*ctx*) in *Vibrio cholerae*. *Journal of Clinical Microbiology*, **31**, 732–3.

Kamal, A.M. (1971) Outbreak of gastroenteritis by non-agglutinable (NAG) vibrios in the republic of the Sudan. *Journal of the Egyptian Public Health Association*, **46**, 125–73.

Koch, R. (1884) Ueber die Cholerabacterian. *Deutsche Medizinische Wochenschrift*, **10**, 725–8.

MacNamara, C. (1876) *A History of Asiatic Cholera*, Macmillan, London.

MacPherson, J. (1872) *Annals of cholera from the earliest periods to the year 1817*, Ranken and Co., London.

McIntyre, O.R., Feeley, C., Greenough, W.B., Benenson, A.S., Hassan, S.I. and Saad, A. (1965) Diarrhoea caused by non-cholera vibrios. *American Journal of Tropical Medicine and Hygiene*, **14**, 412–18.

Morris, J.G. (1990) Non-O group 1 *Vibrio cholerae*: a look at the epidemiology of an occasional pathogen. *Epidemiological Reviews*, **12**, 179–91.

Orton, R. (1831) *The Epidemic Cholera of India: Madras*. Burgess and Hill, London.

Pettenkofer, von M. (1892) Ueber Cholera nut Berücksichtigung der jüngsten Cholera – epidemic in Hamburg. *Munchener Medizische Wochenschrift*, **46**, 345–8.

Ramamurthy, T., Garg, S., Sharma, R., Bhattacharya, S.K., Nair, G.B., Shimada, T., Takeda, T., Karasawa, T., Kurazano, H., Pal, A. and Takeda, Y. (1993) Emergence of novel strain of *Vibrio cholerae* with epidemic potential in southern and eastern India. *Lancet*, **341**, 703–4.

Ramamurthy, T., Albert, M.J., Huq, A., Colwell, R.R., Takeda, Y., Takeda, T., Shimada, T., Mandal, B.K. and Nair, G.B. (1994) *Vibrio mimicus* with multiple toxin types isolated from human and environmental sources. *Journal of Medical Microbiology*, **40**, 194–6.

Safrin, S., Morris, J.G., Adams, M., Pons, V., Jacobs, R. and Conte, J.E. (1988)

Non-O1 *Vibrio cholerae* bacteraemia: case report and review. *Reviews of Infectious Diseases*, **10**, 1012–17.

Sakazaki, R. and Donovan, T.J. (1984) Serology and epidemiology of *Vibrio cholerae* and *Vibrio mimicus*. *Methods in Microbiology*, **16**, 271–89.

Shandera, W.X., Johnston, J.M., Davis, B.R. and Blake, P.A. (1983) Disease from infection with *Vibrio mimicus* a newly recognised species: clinical characteristics and epidemiology. *Annals of Internal Medicine*, **99**, 169–71.

Shimada, T., Nair, G.B., Deb, D.C., Albert, M.J., Sack, R.B. and Takeda, Y. (1993) Outbreak of *Vibrio cholerae* non-O1 in India and Bangladesh. *Lancet*, **341**, 1347.

Shimada, T., Arakawa, E., Itoh, K., Okitsu, T., Matsushima, A., Asai, Y., Yama, S., Nakazato, T., Nair, G.B., Albert, M.J. and Takeda, Y. (1994) Extended serotyping scheme for *Vibrio cholerae*. *Current Microbiology*, **28**, 175–8.

Smith, H.L. (1979) Serotyping of non-cholera vibrios. *Journal of Clinical Microbiology*, **10**, 85–90.

Snow, J. (1849) *On the Mode of Communication of Cholera*. John Churchill, London.

Snow, J. (1855) *On the Mode of Communication of Cholera*, 2nd edn, John Churchill, London.

Spira, W.M., Daniel, R.R., Ahmed, Q.S., Huq, A. and Sack, D.A. (1978) Clinical features and pathogenicity of O group/non-agglutinating *Vibrio cholerae* and other vibrios isolated from cases of diarrhea in Dacca, Bangladesh. *Proceedings of the 14th Joint Conference US–Japan Cooperative Medical Science Program Cholera Panel, Medical Science Program Cholera Panel Tokyo University*. Tokyo, pp. 137–53.

Swerdlow, D.L. and Ries, A.A. (1993) *Vibrio cholerae* non-O1 – the eighth pandemic? *Lancet*, **342**, 392–3.

Wachsmuth, I.K., Bopp, C.A., Fields, P.I. and Carrillo, C. (1991) Difference between toxigenic *Vibrio cholerae* O1 from South America and US Gulf Coast. *Lancet*, **337**, 1097–8.

Wachsmuth, I.K., Evins, G.M., Fields, P.I., Olsvik, O., Popovic, T., Bopp, C.A., Wells, J.G., Carrillo, C. and Blake, P.A. (1993) The molecular epidemiology of cholera in Latin America. *Journal of Infectious Diseases*, **167**, 621–6.

West, P.A. and Colwell, R.R. (1984) Identification and classification of *Vibrionaceae* – an overview, in *Vibrios in the Environment*, John Wiley, New York, pp. 285–363.

World Health Organization Scientific Working Group. (1980) Cholera and other vibrio-associated diseases. *Bulletin of the World Health Organization*, **58**, 353–74.

World Health Organization. (1994a) Cholera in 1993: part 1. *Weekly Epidemiological Record*, **69**, 205–12.

World Health Organization. (1994b) Cholera – update, end 1993. *Weekly Epidemiological Record*, **69**, 13–17.

Yajnik, B.S. and Prasad, B.G. (1954) A note on vibrios in Kumbh fair, Allahabad, 1954. *Indian Medical Gazette*, **89**, 341–9.

The Asiatic cholera: an historical determinant of human genomic and social structure

2

G.C. Cook

Historically, the world has encountered numerous major pestilences, but the Asiatic cholera has arguably been the most dramatic and awe-inspiring, and has hence formed the basis of numerous novels (Scott, 1939; Singer and Underwood, 1962; McNeill, 1976; Barua and Greenough, 1992). The clinical course from onset of the first symptom to death may be alarmingly – spectacularly – rapid. Pandemics have possibly swept the globe since time immemorial (see below) (McNeill, 1976); it is, however, impossible to ascertain whether these 'historical' disease(s) were identical with cholera as seen today. Survivors of epidemics (pandemics) would have embodied a genotype possessing survival qualities; this infection has therefore exerted a profound effect in the moulding of the human genome. Time and again, fear of cholera has been instrumental in creating the origin(s) of sanitary wisdom (Scott, 1939; Singer and Underwood, 1962; Cook, 1994); public health legislation in mid-19th century Britain owed much to *Vibrio cholerae* infection – *'the great sanitary reformer'* (Singer and Underwood, 1962; Cook, 1994).

The reason(s) why the disease suddenly became worldwide in its distribution in the mid-19th century remains largely obscure (Van Heyningen and Seal, 1983); several theories have been advanced. A gradual change in the course of river beds affecting decomposition processes in the soil, and/or an increase in cultivators – per square mile – under British rule in India have been incriminated (Van Heyningen

Cholera and the Ecology of Vibrio cholerae. Edited by B.S. Drasar and B.D. Forrest.
Published in 1996 by Chapman & Hall, London. ISBN 0 412 61220 8.

and Seal, 1983); the famines in Bengal following the eruption of Mount Tambora in the Dutch East Indies (Indonesia) in 1815 have also attracted attention in this context.

2.1 ANCIENT HISTORY OF CHOLERA

Did the *χολέρα* of Hippocrates (470–400 B.C.), the disease(s) described by Celsus (in the reign of Tiberius) – and some 80 years later by Coelius Aurelianus, that recorded by Aretaeus of Cappadocia (AD 81–?138) (Barua and Greenough, 1992), or that of other ancient writers constitute the same entity as that of the present day (Scott, 1939)? We shall never know! Nor shall we ever be able to identify the nature of the disease(s) producing dehydrating diarrhoea described in the *Sushruta Samhita* (Sanskrit) (500–400 B.C.) (Barua and Greenough, 1992), Galen (129–99 B.C.), and Wang-Shooho. Severe diarrhoeal disease accompanied by dehydration, was also recorded in Chinese and Hindu writings (Scott, 1939). Thomas Sydenham (1624–1689) is given credit for coining the term *cholera morbis* to distinguish the disease from cholera – the 'state of anger' (Barua and Greenough, 1992). There is good evidence that many early accounts recorded a dysenteric-type illness (Barua and Greenough, 1992).

A great deal of conjecture therefore remains concerning the true antiquity of cholera (Van Heyningen and Seal, 1983); much evidence indicates that although it had swept intermittently throughout India and neighbouring countries (ones characterized by a dense population, poor sanitation, extensive surface-water area, temperatures exceeding 17°C, and humidity of $\geqslant$40% – for much of the year) (see above), it had not, until the early 19th century, 'broken loose' outside this geographical location.

2.2 THE DISEASE IN ASIA

The first European writer of modern times to describe a choleric disease was Bontius (or Bonsseus) – a physician to the Dutch East India Company who practised in Batavia in 1629 (Scott, 1939). However, the earliest account of the disease in epidemic form, probably related to the outbreak which devastated Ahmed Shad's military forces in 1438 (Scott, 1939).

There were several severe outbreaks in India in the 18th century (Van Heyningen and Seal, 1983). In 1770 it was endemic in the Madras and Travencore regions. In March 1781, it attacked British troops on the march at Ganjam (between Madras and Calcutta); and from Calcutta it shifted to the holy place Hardwar on the banks of the Ganges – approximately 200 miles north of Delhi – killing 20 000 pilgrims within

8 days. Further devastating outbreaks occurred at Arcot and Vellore in 1787 and 1794 respectively, at Ganjam in 1790, and at Malabar and Coromandel in 1796 (Scott, 1939; Barua and Greenough, 1992). Therefore, when the disease first reached England in 1831 (see below), some British physicians and also administrators already had considerable knowledge and experience of the disease – in India (Van Heyningen and Seal, 1983). An historical account of the disease at Calcutta has been provided by De (1961).

2.3 ESCAPE FROM ASIA: THE EARLY 19TH CENTURY

Medical records with an accurate historical value begin with what has been widely termed the First Pandemic (1817–1823) – in the early years of the 19th century (Scott, 1939; Pollitzer, 1959; McNeill, 1976; Van Heyningen and Seal, 1983; Barua and Greenough, 1992). This started in 1817. Reliable dates for its diffusion indicate that it began at Kishnagur (on the Hooghly) in May of that year; there can be little doubt, however, that cholera had already become endemic in Behar by 1816 (Scott, 1939). In India, a marked difference in case-fatality was recorded in European and 'native' troops – 21 and 10 per 1000, respectively (Barua and Greenough, 1992). By 1818, the disease crossed the border(s) to invade Ceylon (in 1819), Mauritius – and subsequently East Africa (1820), the Philippines, China, and Japan (1822). In a north-westerly direction it 'travelled' to Arabia (1821), Persia and Syria (1822), and Russia (1823) (Scott, 1939; McGrew, 1965). During this period the disease had therefore traversed territory almost 100° in longitude – from Nagasaki (147°E) to the Syrian coast (52°E), and more than 67° in latitude – from Bourbon (now Réunion) (21°S) to Astrakhan (46° 21′N). Although the epidemic did not cross into Europe it came extremely close to doing so (Scott, 1939)! The observed route(s) of extension – along rivers – was ascribed by contemporary observers to commerce. Removal of inhabitants to a 'new' site frequently ended an outbreak; therefore, the disease was believed not to be *contagious* and in consequence quarantine measures were regarded as being useless (Scott, 1939).

The Second Pandemic (1826–1837) also started in India (Scott, 1939; McNeil, 1976; Van Heyningen and Seal, 1983; Barua and Greenough, 1992); there had been a lapse of only 3 years since the first had ended. This one extended over an even larger area, and progressed in several directions, from India to China, India to Russia (in 1830) (Barua and Greenough, 1992), and then to Poland, Germany (Evans, 1987), Great Britain and France, and thence to Belgium, Holland, and Norway. On 9 April 1832, Heinrich Heine, in a letter written in Paris, described how on 29 March a masked ball was in progress when one of the harlequins collapsed with the cholera (Scott, 1939); shortly afterwards carriage-loads of affected individuals were hurried to the Hôtel Dieu where they died; in order to prevent a panic they were 'thrust into rude graves – without even removal of their dominoes' (Scott, 1939). The public halls were soon filled with the dead, sewed up in sacks for lack of coffins; long lines of hearses queued outside Père Lachaise. The rich fled from the town and over 120 000 passports were issued at the Hôtel de Ville. Meanwhile, other parts of the Continent seem to have been affected via Great Britain, rather than contiguously; these countries included Spain, Italy, Sicily and Switzerland (Scott, 1939). It also seems probable that Britain harboured the focus from which Canada and hence the USA (via Irish immigrants; McNeill, 1976), Cuba, and South and Central America were afflicted. The first official case in Montreal was recorded on 10 June 1832 (Barua and Greenough, 1992), and by 23 June the disease had moved down to New York (Van Heyningen and Seal, 1983); historical evidence suggests, however, that the New York case(s) occurred before those in Canada – being brought in via an infected ship (at the time the New York Board of Health suppressed these facts) (Van Heyningen and Seal, 1983). By 1834 it had reached most major cities of the USA. This pandemic died down in 1837, but was not extinguished until 1851 (Barua and Greenough, 1992); and there are no satisfactory records of the disease in Europe, Africa, or America for almost a decade.

During the Second Pandemic, quarantine and 'a triple cordon of troops' were used unsuccessfully in an attempt(s) to contain the disease (Scott, 1939); in Spain the death penalty was imposed for infringement of the rule(s), but this strategy proved futile. A contemporary viewpoint (see below) was that a poison within the soil (a *miasma*), which 'checked perspiration, congested the bowels and led to inflammation', was aetiologically important (Scott, 1939). Numerous observations failed to establish the transfer of disease to those who cared for the sick; therefore, contagion was not considered aetiologically important. It was during this pandemic that intravenous injections of saline were first used in management (Chapter 3); copious bleeding (popular at the time for numerous illnesses) was beginning to find few advocates.

The Third (1846–1863) and Fourth (1865–1875) (Scott, 1939; Van Heyningen and Seal, 1983; Barua and Greenough, 1992) Pandemics behaved similarly to those which had gone before; during the third, the disease spread over the whole of the northern hemisphere – to 25°S in the Old, and 30°S in the New World; the fourth passed to southern Europe (by sea) from the coast of Arabia, in contrast to the routes pursued by the former pandemics – via Afghanistan, Persia, and Russia; the shorter route led to its spread in Europe within a few weeks. The third pandemic killed 23 000 people in England and Wales in 1854 alone; the disease was even more virulent in southern Europe. In the fourth pandemic the death toll in Britain was a mere 14 000 (Van

Heyningen and Seal, 1983). No pandemics of the magnitude of these four have occurred in succeeding years. A Fifth (minor) Pandemic (1881–1896) produced an epidemic in Toulon and Italy in 1884 (Pollitzer, 1959; Van Heyningen and Seal, 1983; Barua and Greenough, 1992); perhaps the major significance of this one lay in the fact that Robert Koch (1843–1910) established – in outbreaks rampant in Egypt and Calcutta – that the disease was caused by a specific gastrointestinal pathogen (Pollitzer, 1959). The Sixth Pandemic (1899–1923) caused many deaths in Mecca, Egypt, and Russia (McGrew, 1965); however, it had minimal consequences in western Europe and failed to exert a major impact in the USA, where the last indigenous cases occurred in Massachusetts and New York in 1911 (Barua and Greenough, 1992).

Scott (1939) and Pollitzer (1959) have both stressed the fact that several countries 'escaped' from these massive 19th century pandemics; numerous potentially causative factors [including elevation, soil saturation, rainfall, etc. (see below)] have undergone scrutiny, but no theory seems entirely satisfactory. Mortality rate varied in the different pandemics; in India, for example, it amounted to 20–40% among European troops, depending on the locality, whereas in the Bombay Presidency in 1831–1838, it ranged from 33% to 50%. Between 1847 and 1849, >1 million, and in England alone 53 293, deaths were attributed to the disease (Scott, 1939).

In summary, cholera was probably unknown in Europe before the first half of the nineteenth century (Scott, 1939; Barua and Greenough, 1992; Tramont and Gangarosa, 1995). It spread beyond its original boundaries (from India to Russia) on three occasions – overland to Europe – via Mecca pilgrims to Egypt (see below), and thence to Marseilles from where a woman introduced the disease into Paris in 1865. The major outbreaks of 1865, 1873 and 1884 also entered France via the Mediterranean; in the last of these, the route was: Alexandria to Naples, Marseilles, Toulon, Nantes, Paris, and Spain; the disease reappeared the following year in Toulon.

2.4 THE PART PLAYED BY PILGRIMAGES IN WORLD DISTRIBUTION

Scott (1939) has summarized available data for spread of the disease by pilgrims in the late 18th and 19th centuries. The earliest reference refers to a pilgrimage to Bengal in 1783; 20 000 died of the disease, but fortunately this was confined to bathing-places and died out when the multitude(s) dispersed. In a classical text, *The Influence of Tropical Climates on European Constitutions* (1812), James Johnson (1777–1845) recorded the scenario on the Ganges (Scott, 1939):

'The banks of this river present, particularly about the rising and

> setting of the sun, a motley group of all classes, and sometimes both sexes, sacrificing to the Goddess Cloacina, in colloquial association, not, indeed, offering their gifts in temples, but committing them freely to the passing current.'

An account of the outbreak at the pilgrimage of 1867 at the Hardwar Fair, Lower Bengal, has also been recorded (Scott, 1939); on the ninth day of the event one case of cholera occurred and on the 13th there were eight; by the time the pilgrims dispersed on the 15th, many were infected and the disease had been transmitted widely. A similar situation occurred in 1891.

At Mecca, cholera established itself during the Moslem pilgrimage in 1831 (McNeill, 1976); it subsequently afflicted this pilgrimage around 40 times between 1831 and 1912 (Barua and Greenough, 1992). Cases continue to occur even today!

2.5 THE ASIATIC CHOLERA REACHES ENGLAND

The background to the first English epidemic (1831–1832) is admirably portrayed in the opening lines of T. Shapter's text, *The History of the Cholera in Exeter in 1832*, published in 1849:

> 'THE Asiatic or pestilential Cholera, which showed itself in 1817 in the Delta of the Ganges, confined its ravages almost exclusively to the territories of the East Indies, until 1830, when, bursting forth in Russia, it evinced something like a regular progress in a westerly direction. Its occurrence in St. Petersburgh, during the early part of 1831, forcibly arrested public attention, and the government of this country was induced to send thither Dr. Russell and Dr. Barry to watch the progress of the disease, and acquaint themselves with its details; while at home, by an order in council, dated June 21, a "Board of Health" was established [by the King]. As the disease approached that portion of Europe immediately contiguous to the seas which wash our own shores, it became the subject of a somewhat remarkable controversy; for while all felt that a new and deadly pestilence was about to overwhelm the country with its terrors, fierce and unseemly disputes ensued, as to the very existence of any such disease; and its occurrence in a severe form, at the neighbouring port of Hamburgh, having induced the Lords of the Privy Council to issue an "order" in relation thereto, a panic arose, and with it a new storm of denial of its existence; nay more, the absolute occurrence of the disease at Sunderland, and afterwards in the metropolis, did not, for some period at least, serve to dispel this curious phenomenon of human incredulity.'

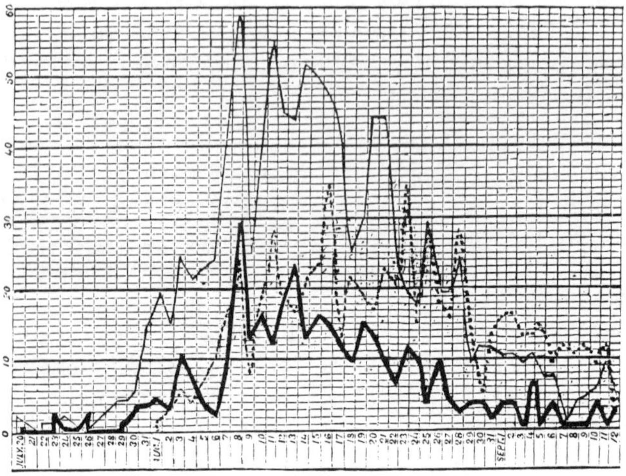

Fig. 2.1 Chart showing the progress of the Asiatic cholera in Exeter from 19 July to 12 September, 1832. Symbols: ——, new cases; ▬▬, deaths; ...., recoveries.

Figure 2.1 summarizes numbers of cases, deaths, and recoveries during the Exeter epidemic of 1832 (Shapter, 1849). The outbreak had in fact reached Britain in October 1831, via the port of Sunderland (Van Heyningen and Seal, 1983; Barua and Greenough, 1992). The first recorded case occurred on 9 October 1831; the first death took place on 12 October, but it was not until 26 October that Britain's 'first officially acknowledged victim of cholera' – William Sproat – died at his home at Fish Quay, Scotland (Van Heyningen and Seal, 1983). By the spring of 1832, cholera had assumed a major importance in London, and >7000 had died (Howell and Ford, 1985); early in 1833 the epidemic burnt itself out! Fifteen years later (1848) it again entered Britain, also via London, spreading out through the towns and cities in a nationwide epidemic; this epidemic was not extinguished until 1849. In London alone, nearly 7000 individuals died in one month. Local boards of health were established, and more significantly, in 1848, Parliament established a General Board of Health (Pelling, 1978) – but only one week before cholera appeared in England for the second time (McNeill, 1976); this Board instituted far-reaching programmes to improve public sanitation. A new strategy of the 1840s – championed very largely by the social reformer Edwin Chadwick (1800–1890) (an earnest disciple of Jeremy Bentham [1748–1832]) – lay in constructing narrow sewers out of smooth ceramic pipe; sufficient water was passed through them to flush the waste matter to a distant depository, far removed from

Fig. 2.2 The "Silent Highway"-Man. "Your MONEY or your LIFE!" (Reproduced from *Punch*, 10 July, 1858.)

human habitation (McNeill, 1976). Figure 2.2 gives an insight into the grossly insanitary state of the River Thames even as late as 1858; much of the drinking water for the metropolis originated here! Clearly, the installation of newer systems involving water and sewer pipes, the development of powerful pumps to deliver water (under pressure) to individual houses, and compulsory elimination of older sewage systems was essential. However, it took the 'lively fear of cholera – the great sanitary reformer', to achieve these goals!

2.5.1 CHOLERA IN A LONDON HOSPITAL

Experience of the disease at a recently established London hospital has been recorded: The Free (later Royal Free) Hospital had been founded in Greville Street, EC1, in 1828 (Sandwith, 1960). With the onset of the cholera outbreak in London preparation was made for the reception of afflicted individuals; on 20 February 1832, two wards (one male and the other female) were opened and the first cases admitted. The London Free was the only major London hospital to admit cholera patients, 300 during the first few months. William Marsden (1796–1867), founder of the Hospital, claimed (and published in a treatise on cholera) that large doses of calomel – if necessary followed by castor oil and tincture of rhubarb – 'removed the poisons from the system' (Sandwith, 1960). He did not advocate opium and/or brandy (both widely used); indeed, he was of the opinion that those with liver disease (resulting from excess alcohol intake, a common problem of the time) were unduly susceptible to the disease. Marsden's initial management consisted of an unlimited oral water intake; he also administered repeated doses of concentrated saline solution to stimulate 'the repulsive action of the stomach' (Sandwith, 1960). When this technique failed he resorted to the intravenous route (see above); saline of approximately the equivalent concentration to that of serum was given, to 'restore the pulse'. He also gave 'ten drops of laudanum with one ounce of compound tincture of bark' – undiluted, every hour. Marsden recorded that 10 patients 'who were restored', drank 225 gallons of water in 78 hours (Sandwith, 1960). Some 580 out of 700 cholera patients (many of them moribund at arrival) admitted during the 1832–1833 outbreak recovered; the London Free Hospital's wards were fully occupied for 5 months (Sandwith, 1960). While the emergency lasted, board meetings had to be held in the homes of the members of the hospital committee, the boardroom being given over to the care of convalescents.

2.5.2 CHOLERA ON THE THAMES

Occurrence of cholera on the river was obviously a major potential problem with serious ramifications; the Seamen's Hospital Society (SHS) was at the centre of preventive activities (Cook, 1992). Figure 2.3 shows a notice which was distributed by the SHS to all captains of ships in the Port of London, and many others also. In July 1853, the *Lancet* recorded (Anonymous, 1853):

> 'SEAMEN'S HOSPITAL SOCIETY. – CHOLERA ON THE RIVER.
> On Tuesday, a quarterly general court of the governors of this charity was held at the offices, 74, King William-street, City. It was announced that the Government had been applied to with respect to "a cholera ship," for the purpose of receiving cases of cholera occurring amongst the sailors on the river. To this judicious request the Government has acceded; and her Majesty's ship the *Devonshire* has been granted for that purpose. It is moored near the *Dreadnought* [the second SHS hospital-ship] off Greenwich, and patients are

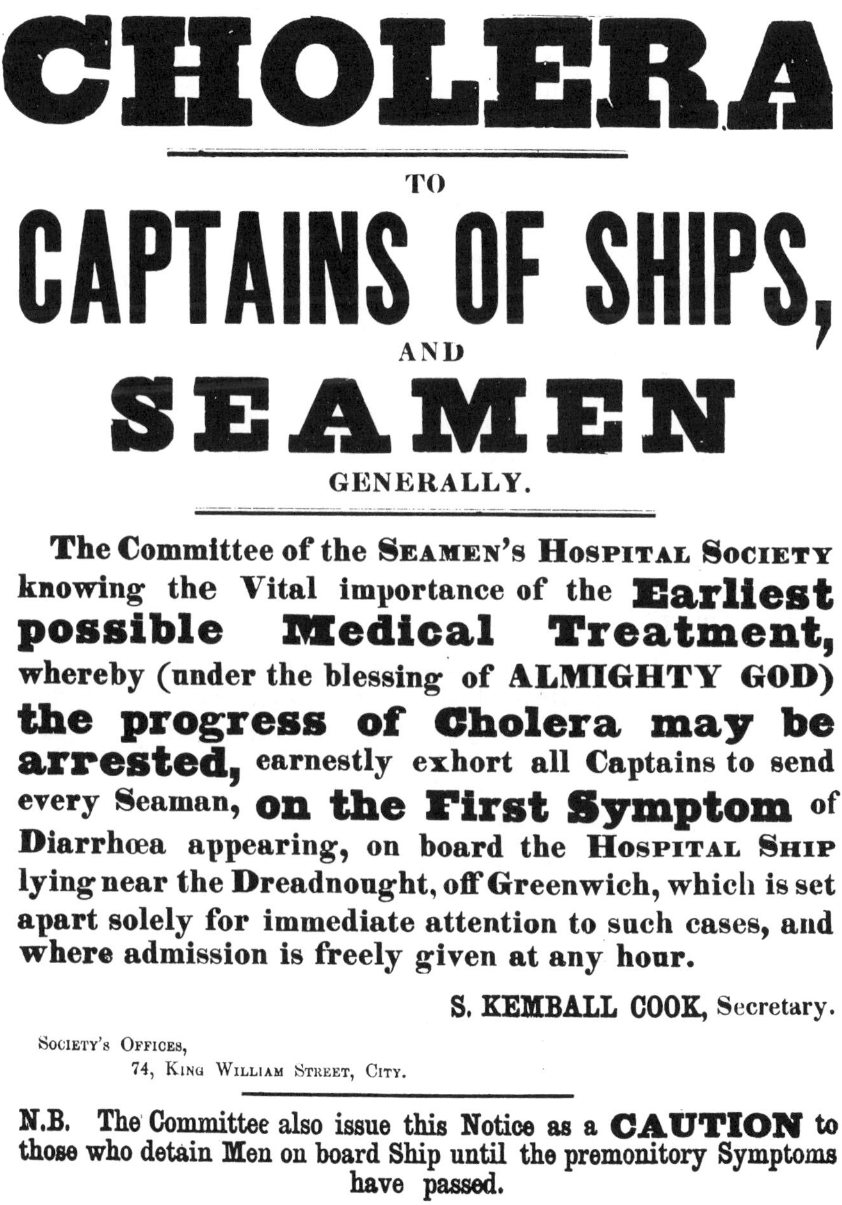

CHOLERA

TO

CAPTAINS OF SHIPS,

AND

SEAMEN

GENERALLY.

The Committee of the SEAMEN'S HOSPITAL SOCIETY knowing the Vital importance of the **Earliest possible Medical Treatment,** whereby (under the blessing of ALMIGHTY GOD) **the progress of Cholera may be arrested,** earnestly exhort all Captains to send every Seaman, **on the First Symptom** of Diarrhœa appearing, on board the HOSPITAL SHIP lying near the Dreadnought, off Greenwich, which is set apart solely for immediate attention to such cases, and where admission is freely given at any hour.

S. KEMBALL COOK, Secretary.

SOCIETY'S OFFICES,
74, KING WILLIAM STREET, CITY.

N.B. The Committee also issue this Notice as a **CAUTION** to those who detain Men on board Ship until the premonitory Symptoms have passed.

Fig. 2.3 Notice distributed by the Seamen's Hospital Society to ships' captains in the Port of London, 1853, warning of the dangers of cholera (Anonymous,1853).

received at all hours of the day and night. As yet there have been but 20 cases, as we are informed, on the river, and the prevalence of the disease is decreasing rather than otherwise, so far as the Thames is concerned. The *Devonshire*, the "Cholera ship," has accommodation for two hundred. There are ample provisions in the way of nurses, physicians, &c. It will be remembered that the same course was adopted by this charity when the cholera last visited our shores. On that occasion the ship *Iphigenia* was granted for that purpose, and 250 cases were treated therein. The agents of the society are now also engaged in distributing a medicine found useful and adapted to sailors, gratuitously, at the sailors' homes, Thames church-ship, and other places, that all may have recourse to it when the preliminary symptoms appear.'

In 1859 and 1868, the *Lancet* carried reports of 'cholera cases' admitted to the third of the *Dreadnought* hospital-ships (Anonymous, 1859, 1868); while the five cases in the first article probably did suffer from a *Vibrio cholerae* infection (other causes of acute gastrointestinal disease cannot be excluded), the two in the latter report (both recovered) probably did not.

Cholera in the Port of London remained a topical subject (although not one of any great practical importance) for some time, and in 1871 the *Lancet* recorded (Anonymous, 1871):

'...Two meetings of the sub-committee appointed by the local authorities [bordering the Thames] have recently been held, at which three important resolutions were proposed and adopted: (1) To place a floating hospital at Gravesend for the reception of cases of cholera; (2) to obtain permission from the Metropolitan Asylums Board to send cases of cholera to the *Dreadnought* hospital ship; (3) to appoint a medical adviser and superintendent, to conduct an inspection of the shipping lying in the river and the docks, and to inaugurate and carry out such hygiene arrangements as may be required. The Mayor of Gravesend and Dr. Letheby have made arrangements with the Government for the conveyance of the *Rhin*, now lying in the Medway, to moorings at Gravesend when required, and it is stated that she can be sent round in about twelve hours. The committee, in accordance with their resolution, have obtained leave from the Metropolitan Asylums Board to make use of the *Dreadnought* if required. There are still two districts (those of Rotherhithe and West Ham) the authorities of which object to unity of action, so that these districts will be compelled to provide for their own sick at their own cost. We hope to record next week that the port of London is at all events as well protected against the advance of cholera as the outposts, and hence that, practically, the piecemeal system of sanitary *surveillance,* against which we have so long protested, has ceased to exist.'

2.6 19TH CENTURY VIEWS ON THE CAUSATION OF CHOLERA

In his *History of Contagious Cholera* (1832), James Kennedy described five 'Laws of Cholera' (Scott, 1939):

1. Climatic influence: 'The contagion of cholera may spread in every climate...',
2. Predisposition: 'certain states of bodily health' e.g. fatigue and undernutrition, render individuals susceptible,
3. Latent infection: 'The period of time during which the contagion lies dormant in the system rarely exceeds three days',
4. Increase and Decline: 'When the cholera appears in a town it extends rapidly, and in general, runs its course in the space of a few weeks',
5. Contagion: 'Cholera is contagious and its contagion is of a highly diffusible nature'.

Scott has also summarized various 'hypotheses' regarding the cause(s) and mode of propagation of cholera held rather later, in the mid-19th century (1850–1860) (Scott, 1939); these – based on observations of (James) Ranald Martin (1793–1874) – effect a compromise between the two directly opposing viewpoints which were currently held, namely contagion versus miasmas:

(a) 'a special poison of eastern or foreign origin' – in the presence of certain local conditions and predisposition of the inhabitants – was responsible,
(b) a 'true epidemic' was preceded by diarrhoea, dysentery, and other form(s) of flux; these 'prepare the [population] for its inception',
(c) after the subsidence of an epidemic, mortality from fluxes never entirely 'recedes within its former limits' [this suggests a carrier-state],
(d) individuals at home were more 'ready for attack by a disease of a flux character' [this implies overcrowding and bad sanitation],
(e) although it might be 'smouldering' in England, the 'flame causing conflagration might be brought from the East'; when 'the "terrene" and "atmosphere" meet they become the "shears of Fate" and true cholera resulted,
(f) certain meteorological conditions, e.g., high barometric pressure, low temperature (60°F), low rainfall, still air, and an overcast sky 'assist the spread of cholera',
(g) the disease tended to linger in certain districts; certain 'local circumstances have great influence in determining its seat',

(h) local causes of insalubrity – by lowering the general state of health – were necessary for the evolution of an epidemic,
(i) special poisons were capable of 'exciting ... fermentation in the blood'; one of these was cholera,
(j) the 'exciting cause was more likely to be simply toxical...',
(k) the 'nidus was not in the air or water, but in the soil itself',
(l) 'impure water alone' would not constitute a necessary causative factor for cholera, nor would mere odour – 'privy odour'.

A later conference held in Constantinople in 1866 decided that (Scott, 1939):

(a) India was the permanent home of cholera,
(b) the disease was propagated by man as a result of the rapidity of his own movements,
(c) water and food might serve as vehicles for transmission,
(d) the most potent of all causes was the movement of pilgrims who 'tended to develop and propagate it', and
(e) no existent evidence indicated propagation by animals.

The conference (there was still no unanimity regarding miasmatic or contagious theories) supported the conclusions of Doctors W. Baly (1814–1861) and W.W. Gull (1816–1890) (see below) and recommended: measures of general cleanliness, fresh air, avoidance of overcrowding, disinfection of buildings – where cases occurred on shore, and of merchandise and ships – when cases occurred on board.

An International Sanitary Conference was held at Vienna in 1874 with the object of drawing up generally acceptable quarantine regulations (Scott, 1939); most conclusions formulated at the 1866 conference were ratified. In 1875, a further conference was held at Brussels to consider 'Prevention of cholera'; conclusions on that occasion were:

(a) 'the cholerigenic miasma spontaneously developed in certain conditions' in India – notably in the Delta of the Ganges and in the low-lying districts surrounding Madras and Bombay; from there it had migrated to Europe, Africa, and America,
(b) cholera was contagious, and 'its poison could be dissolved in water and diffused in air',
(c) sources of contagion were: the 'ejecta', corpses, linen and clothes, ships, rooms, carriages, latrines, contaminated water, air, animals and merchandise,
(d) the 'miasma' penetrated the affected individual via the respiratory and gastrointestinal systems.

Yet another Sanitary Conference held in Rome, in 1885, recommended disinfection to guard against the introduction and spread of cholera, referring especially, on this occasion, to shipping.

Despite extensive investigations, the origin(s) of cholera remained a total mystery; for many years to come, attempts were made to relate outbreaks to rainfall, ambient temperature, change(s) in wind force and direction, temperature of the Thames water, and presence of ozone and 'electricity' in the atmosphere (see below).

'Miasmas' (see above) were generally considered (certainly within the medical profession) more plausible than 'contagion' as being relevant in the aetiology of Asiatic cholera. However, Dr John Snow (1813–1858) considered that a 'specific poison' in the excreta of patients suffering from cholera when ingested by others set up the disease (see below) (Snow, 1849; Scott, 1939); this multiplied in the affected individual, who in turn passed the disease to others – many of them, when the 'poison gained [access] to a drinking-water supply'. Dr William Budd (1811–1880) (see below), working at Bristol suggested that a living organism, perhaps fungal (Pelling, 1978), multiplied within the intestinal lumen, and gave rise to the symptoms of the disease; it was disseminated in food and water. However, Baly and Gull reported to the Royal College of Physicians in 1854 that spread of cholera was independent of the sick and healthy, and that infection was probably wind-borne (Figure 2.4) (Scott, 1939; Pelling, 1978)!

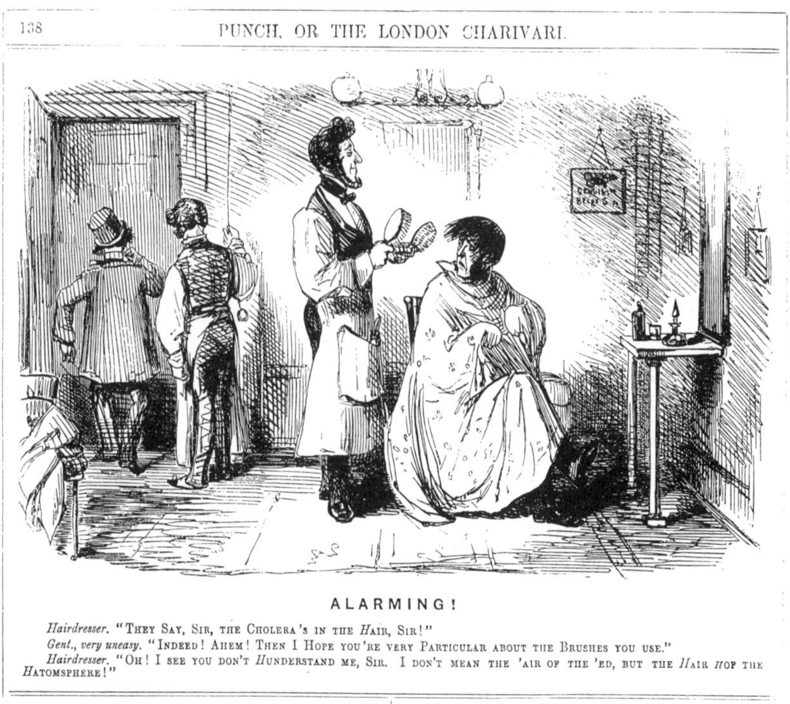

Fig. 2.4 'Alarming!' First suggestions that the cholera infection was wind-borne. (Reproduced from *Punch*, 3 April, 1852.)

Pelling (1978) has listed numerous contemporary 'theories of cholera': the telluric, electrical and ozonic theories; the animacular and fungoid; the zymotic and humoral; and the theories incriminating ingesta, putrid effluvia, or even a specific poison as in smallpox (Anonymous, 1849a; Rosenberg, 1960). Budd (see above) was an exponent of the 'fungoid theory'; Snow seems not to have appeared in this context at all. To this list may be added some other by-products of the contemporary influence of *chemistry*. The 'deficiency' theories were due to Tunstall and Blacklock. Tunstall's was based on an observation of Davy that air expired by cholera patients contained an abnormally low proportion of carbon; he considered that the excretion of carbon had an antiseptic action in the body, and deduced that 'available' carbon was deficient (rather than unexcreted) – from reports of the therapeutic value of petroleum (Tunstall, 1848). 'Carbon theories' were certainly not uncommon, and could usually be traced to von Liebig's *Animal Chemistry*. Blacklock, writing in India, attributed cholera to the 'continued use of a diet deficient in sulphur'; sulphur was necessary to maintain the body in its normal electrical condition, and its absence gave rise to 'morbid ganglionic excitability' (Pelling, 1978).

While the 'animacular and fungoid' theory was dominant, the electrical and ozonic ones also drew much attention. Many found it difficult to believe in a cause which could not be detected by a known chemical means. Electricity was a little-known 'force', which varied in the earth's atmosphere, and yet seemed to be in some way among the basic properties not only of matter itself, but also of living beings, since it appeared to be analogous (if not identical) with nervous energy, or the vital principle itself! Most importantly, it could be experimented with, and even measured. It was therefore a popular component in cholera theories, either as a cause in itself or, more frequently, as one taken in conjunction with others. The leading authority on electricity and cholera seems to have been Sir James Murray, whose work, *Electricity as a cause of cholera & other epidemics, & the relation of galvanism to the action of remedies* (1849) was reviewed in the *London Medical Gazette* and elsewhere (Pelling, 1978). Blacklock combined electrical with chemical factors; others envisaged swarms of organisms which were attracted, or even generated, by electrical force(s).

The 'ozonic' theory was sponsored in England by Hunt (then active in the British Association) who was for many years chief scientific writer for the *Athenaeum* (Hunt, 1849). C.F. Schönbein of Basle, the discoverer of ozone, had remarked that its presence in the atmosphere varied in an exact ratio with electrical intensity, and that it was formed during all combustion processes. Hunt drew attention to its great powers of oxidation and disinfection, and also to some observations of Quetelet indicating that during a cholera outbreak the electrical intensity of the atmosphere was abnormally low. He concluded that under normal circumstances ozone was designed by nature to prevent poisons forming during decomposition, and that cholera epidemics were the result of a deficiency of ozone resulting from a low level of electrical activity. In support of his theory he gave the Birmingham example, whose puzzling exemption from cholera could be explained by postulating that the great industrial fires of the city had maintained the level of atmospheric ozone. Hunt added that Moffat (another member of the British Association) had, on the basis of similar correlations, suggested that influenza epidemics might be caused by an ozone excess. These theories were easily criticized, but remained attractive, in part at least, because they suggested feasible lines for research and possible verification (Pelling, 1978).

The 'ozonic' and 'fungoid' theories offered two entirely different kinds of explanation; they were however, comparable in certain other respects. Each named a discoverable entity as the causative agent; the fungoid theory reflected the influence of the 'new microscopy', whereas the ozonic one was derived from recent advances in chemistry. To some extent rival interests were involved. Contemporary views of the limits and peculiar virtues of microscopy were often critically compared with the capabilities of chemical analysis (Hunt, 1849). Organic chemists had shown animal and vegetable substances to be composed of the same limited range of elements; to a microscopist this was an argument for making them the major basis of classification. Knowledge of the chemical constitution (or properties) of the lower forms of life was minimal. Some of the earliest workers to investigate adulteration and pollution regarded the microscope as the finer tool, in part because smaller 'quantities' could be analysed. The proponents of the 'cholera-fungus' theory at first depended almost entirely on microscopical evidence; later in the debate they were criticized on this count, and some conclusions were based on chemical grounds. Chemists and microscopists joined forces in defining the conditions in which the 'cholera-bodies' were found (Pelling, 1978). Some writers – arguing from analogy – claimed that certain of these conditions provided an adequate explanation for the presence of these bodies.

In 1849, Brittan produced evidence for the existence of microscopical bodies in 'cholera atmosphere'; Quekett concurred that these so-called bodies were of 'fungoid' origin (Anonymous, 1849b). Several other investigators also, became involved in the 'fungal (fungoid) controversy'. Late on the scene was George Busk (1807–1886) (Letter, 1849), now a distinguished microscopist; he had dealt with cholera while a surgeon with the SHS, serving on the first two hospital-ships, HMS *Grampus* and HMS *Dreadnought*, at Greenwich.

Sachs (1890) has given an account of the state of mycology at this time and of its relevance to broader issues. In 1849, after little more than 10 intensive years of study, the field was expanding, but was far from being well-defined (Pelling, 1978); this was almost enough to account for the occurrence of the 'cholera-fungus' controversy, which raged throughout 1849. At the outset, all the recent developments (products of French and German rather than British research) were rehearsed, and it was claimed that no hypothesis was supported by a 'stronger body of plausible analogies'. However, the question of whether fungi possessed a cause or effect relationship was much debated; it was generally considered, even in the context of fermentation, that their role was merely secondary. [In its review of Starr's work, the *Lancet* termed the process of fermentation 'the great "cheval de bataille" of our theorists'. By the end of the 1840s, it was generally accepted that yeast consisted wholly or in part of fungi, but the nature of the relationship between the organism and the fermentation process remained unclear.] As the 'appointed executioners and nimble scavengers of nature', the presence of fungi was generally seen as a retribution for some personal or sanitary neglect (Pelling, 1978). Even to an enthusiast like Wilkinson, the 'essential part of the malady' was the 'conditions producing the soil'. Carpenter had already indicated that where fungi appeared unequivocally on living tissues, there was reason to believe that they were 'generally the indications of a state of previous disease' (Anonymous, 1849b; Wilkinson, 1849).

2.6.1 THE ROLE OF CLIMATE

The role of climate, especially humidity, in the causation of cholera remained enigmatic (Scott, 1939). Much later (in 1926), Sir Leonard Rogers (1868–1962) gave an account of the epidemiology of cholera in India over a 50-year period (Rogers, 1926, 1933, 1950); spread of the disease from the west of Bengal was related to a rise of absolute humidity to >1 cm in the 'invaded areas' in the months of increase; however, in lower Bengal, south-east Madras, and the west coast of Bombay, absolute humidity was never <1 cm and cholera was present throughout the year. Rogers also noted a temporary decline in the incidence of cholera in Bengal in January/February when absolute humidity fell almost to the critical point, i.e. 1 cm.

Russell (1926) also developed his 'periodogram analysis' showing that rainfall and temperature had little association with cholera, whereas humidity had a high positive correlation. It is, he wrote, 'fairly conclusive that high humidity is favourable to [a] high incidence of cholera, while pressure is associated in exactly the opposite manner' (Scott, 1939).

Rogers was able to show later that during a 45-year period, there had been fewer cholera epidemics in Lower Bengal than in the United

Provinces; in 40 out of 41 outbreaks, the rains of the preceding monsoon, the winter rains, or both, had failed. Furthermore, an unusually early rise in absolute humidity seemed to favour an early recrudescence, or epidemic spread of the disease. By studying these factors, Rogers considered that he was able to forecast outbreaks some months ahead (Scott, 1939).

2.6.2 THE ROLE OF ALTITUDE

Evidence that altitude was an important factor in the aetiology of cholera was succinctly summarized by the *Lancet* in 1854 (Anonymous, 1854):

> 'A VERY important and, in our opinion, a fairly arrived at law connected with the natural history of the choleraic pestilence is that first clearly and scientifically established by Mr. Farr [William Farr (1807–1883) was the statistician at the Office of the Register-General; see Howell and Ford, 1985], demonstrated in his masterly report on the cholera in 1849, and since confirmed by the experience of the past year, as shown by the weekly reports of the Registrar-General. This law may be thus expressed – the increments of mortality in individual districts of a given spot bear a certain constant relation to the order of the altitudes of these districts. It is true that this law has been established chiefly upon data afforded by observation in the London Basin; but since the first calculations of Mr. Farr, Dr. Duncan, of Liverpool, has, at the request of the former gentleman, examined the influence of the elevation of soil on the increments of mortality from cholera in that city, and confirms Mr. Farr's law so far as to show that "when the districts of approximating elevations are grouped together, and the groups so formed contracted, the results distinctly point to a relation between the elevation of the soil and the mortality from cholera." More general, though far less exact (mathematically speaking) observation fully confirms the truth of Mr. Farr's law. In whatever part of the globe the history of the cholera there raging is investigated, it will be found as the rule that the disease affects most fatally the low-lying seaports and deltas of rivers, sparing the high grounds, even round river sources. In India we know it began in the delta of the Ganges, ravaged all low-lying places; while sepoy lines, placed from sixty to a hundred feet above the general level of the country, had scarcely any cases, excepting such as occurred in persons on guard at the different outposts. The higher classes of natives and Europeans generally inhabiting the better raised and more airy parts of the towns suffered proportionally less than the lower ranks, and indeed the general result was such as

to lead Mr. Jameson, in his report of the Bengal cholera in 1817–18–19, to say, that "there is abundant proof that in high, dry, and generally salubrious spots it was both less frequent in its appearance and less general and fatal in its attacks than in those that were low and manifestly unwholesome." In illustration of the law, (established at any rate for the great "London Basin,") that *the mortality from cholera is in the inverse ratio of the elevation*, we may remark, that if the districts of London are classified according to their elevation above the level of the Thames, in those whose elevation is not on an average twenty feet above its water-line, it will be found that "on this bottom of the London Basin the mortality was at the average rate of 102 in 10,000; while in a district, 100 feet high, the mortality was 17 in 10,000; and in another (Hampstead), of an altitude of 350 feet, it was 8, or, deducting a stranger infected at Wandsworth, but who died there, 7 in 10,000." Now if the mortality from cholera, according to the data here involved, and as occurring in relation to the elevation and mortality of other groups of districts included between those of less than 20 and above 350 feet of altitude be closely examined, the relation between elevation and death will be expressed as in the following table [see Table 2.1] on the one hand, as the facts actually occurred, and, on the other, as they should occur

Table 2.1 The relationship between cholera mortality and altitude according to Mr. Farr

Elevation in districts in feet above high-water mark	Deaths in 10 000. Mortality observed	Calculated series
20	102	$\frac{102}{1} = 102$
20–40	65	$\frac{102}{2} = 51$
40–60	34	$\frac{102}{3} = 34$
60–80	27	$\frac{102}{4} = 26$
80–100	22	$\frac{102}{5} = 20$
100–120	17	$\frac{102}{6} = 17$
340–360	7	$\frac{102}{18} = 6$

according to a series calculated from an equation worked out by Mr. Farr in his cholera report.'

2.6.3 THE ROLE OF SOIL

The part played by soil also remained a dominant theme (Scott, 1939). Workers in India had recorded that whereas hill dwellers (in their native surroundings) did not suffer from cholera, when they came down to the plains they no longer remained immune to the disease (Scott, 1939). But, the soil theory certainly did not fit all cases; outbreaks had certainly occurred on board ship – although it was widely believed that the best way of avoiding infection was to put out to sea (Scott, 1939)! Outbreaks at sea were generally uncommon and there were good grounds for believing that when they occurred the infection had initially been acquired in port.

2.6.4 THE WATER-BORNE THEORY

The suggestion that the 'cholera poison' was conveyed by drinking-water had, in the mid-19th century, ardent advocates and equally vocal opponents (Scott, 1939). It was widely appreciated at the time that water obtained from the River Thames (from which London's drinking-water was derived) was very heavily contaminated (Figure 2.5). The 'soil theory' (see above) and the 'drinking-water theory' existed side-by-side; the possibility that 'well-water' might be infected via soil contamination seems to have been given little credence; neither did the possibility that food contaminated by soiled hands, flies, etc., receive serious consideration. It was well recognized that sick attendants often escaped the disease; therefore, it was assumed that the cholera patient eliminated an infective agent which in its primary form was not a potent 'cholera poison', and only attained its specific power after undergoing change/maturation outside the body. Alternatively, multiplication or reproduction of the 'cholera poison' proceeded independently of the patient, and clung to him/her (whether sick or healthy) or to various objects carried by them – thus giving rise to an outbreak (even an epidemic) when it met conditions conducive to its reproduction.

2.7 'ON THE PATHOLOGY AND MODE OF COMMUNICATION OF CHOLERA'

Cholera remained a major health hazard, and continued to receive wide publicity in the press and in the satirical magazine *Punch* (Figure 2.6). The investigations of Dr John Snow (1813–1858) Sheppard, 1995) (Figure 2.7) – an eminent London anaesthetist – into the Broad Street

188 PUNCH, OR THE LONDON CHARIVARI.

THE WONDERS OF A LONDON WATER DROP.

The freshest fruits of microscopical research are the wonders which have been revealed in a drop of London water through the Molecular Magnifier, illuminated by the Intellectual Electric Light. For the ability to behold these astounding marvels, a certain preparation is necessary, bearing, superficially considered, some resemblance to Mesmerism. The person intended to be the Seer is placed on a seat. Any competent individual then takes him in hand, and explains to him the composition of water, showing him how the pure fluid differs from the liquid constituting the Thames, and from that which exists in the metropolitan wells, when the former has received the contents of the sewers, and the latter the oozings of intramural graveyards. Some delicate subjects, even of the male sex, cannot endure this process, it affecting them with faintness and nausea.

Having been subjected to the above preliminaries, most people are in a sufficient state of enlightenment to discover, by the aid of the Molecular Magnifier, the curiosities contained in

A DROP OF LONDON WATER.

The drop to be magnified is taken from a mixture of the common well-water of London with that supplied by the various Companies. MR. HASSELL, it is already known, has enabled philosophers to discriminate between these waters, by the verminous and other peculiarities which he has demonstrated in each particular form of beverage.

The Molecular Magnifier differs from all other microscopes, in displaying the ultimate constitution of objects; a spectacle not only defying the naked eye, but all vision which is not in a measure psychical.

And wondrous indeed is the scene disclosed within the sphere of a little drop of water—of that water which Londoners drink, swallowing daily, myriads and myriads of worlds, whole universes instinct with life, or life in death! It transcends all that has hitherto been deemed astonishing. America herself will confess that it stumps the revelations of ANDREW JACKSON DAVIS.

Creatures—who shall name them? things in human shape—in all appearance London citizens—aldermen, deputies, common councilmen,—are seen disporting in the liquid dirt as in their native element. Behold them, fiercely hustling each other in competition for atomic garbage. What pushing, poking, fighting, kicking, scrambling! There goes an unfortunate wretch fast as if for dear life, with a hook-nosed homunculus—evidently a genuine water-bailiff—darting after him. Here a cheap slop-seller has caught a smaller individual of the same species by the head, and is trying to bolt him. There again, as plainly as possible, you see a funeral procession with an undertaker at the head

Fig. 2.5 'The wonders of a London water drop'. First realizations of the level of contamination of London's drinking-water. (Reproduced from *Punch*, 11 May,1850.)

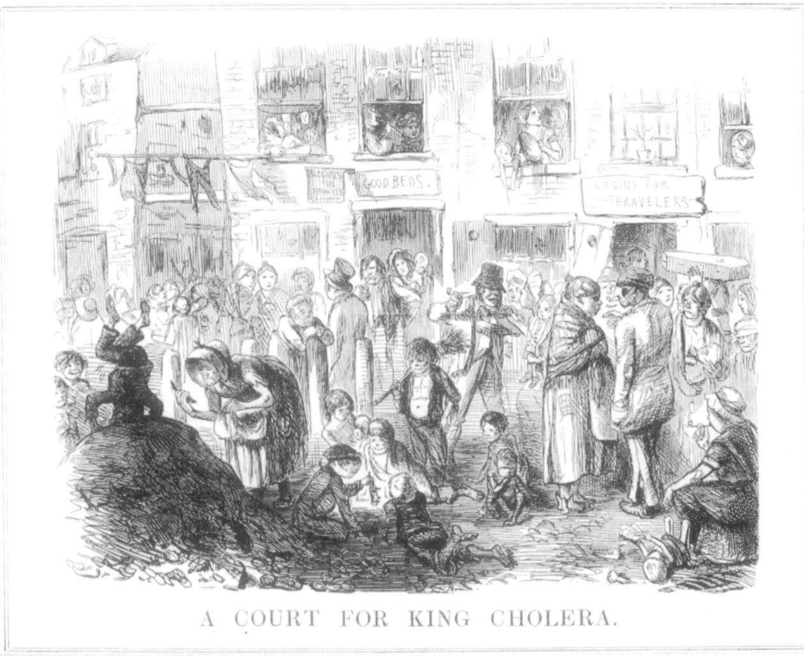

Fig. 2.6 'A Court for King Cholera'. An attempt to increase public awareness of the health hazards of cholera. (Reproduced from *Punch*, 25 September, 1852.)

pump water, and its relation to the cholera outbreak in Westminster in 1854 (see below) clearly indicated (although his views were not readily accepted, see below) that the fouling of drinking-water by ejecta actually caused the disease; these observations were made some 30 years before the specific infecting organism was demonstrated. Snow initially summarized his views at a meeting of the Westminster Medical Society, held on 13 October 1849:

'DR. SNOW said that he had been led, contrary to the usual opinion, to consider cholera as being, in the first instance at least, a local affection of the mucous membranes of the alimentary canal, and not an affection of the whole system, depending on a poison in the blood. In diseases caused by a special poison absorbed into the blood, general illness of a febrile character preceded any local symptoms which might arise; but this was not the case in cholera. The loss of the watery part of the blood, in all the cases he had seen, was sufficient to account for the collapse and all the general symptoms; and when cholera commenced gradually, it could often be checked, and prevented from proceeding beyond the stage of

Fig. 2.7 Dr John Snow (1813–1858).

diarrhoea, by remedies adapted to act merely on the mucous membrane. The recent discovery of peculiar microscopic cells, believed to be of a vegetable character, in great abundance, in the cholera discharges, tends to confirm this view of the nature of cholera. An attentive examination of the history of cholera, as an epidemic, showed that it was communicable by human intercourse; and although there were many facts opposed to the theory of cholera being contagious, in the same way that the eruptive fevers were believed to be – yet, in the sequel, these facts would afford the strongest evidence of the communication of the disease. If the alimentary canal were the seat of cholera, and the disease were communicable, it was clear that it must be conveyed by something which passed from the mucous membrane of the intestines of one patient to that of another, which it could only do by being swallowed; and as the disease grows in a community by what it feeds upon, attacking a few people in a town first, and then becoming more prevalent, the cholera poison must multiply itself by a kind of growth, like every other morbid poison. The instances in which minute quantities of the ejections and dejections of cholera patients must be swallowed were sufficiently numerous to account for the spread of the disease; and it spread most where the facilities for this mode of communication were greatest. The bed-linen nearly always became wetted by the cholera evacuations, which were almost without colour and odour; and the hands of persons waiting on the patient became soiled without their noticing it; consequently, unless they were very cleanly in their habits, and had an opportunity of washing their hands, they would be liable to swallow some of the excretion, and leave portions on the food they handled or prepared, which had to be eaten by the rest of the family, who, amongst the working classes, had to eat their meals generally in the sick room; hence the thousands of instances in which, in this class of the population, a case of cholera in one member of the family was followed by other cases; whilst medical men and others who merely visited the patient, without taking food, generally escaped. The mining population of this country had suffered more from cholera than any other; there were no privies in the mines, and as the workmen usually stayed down for eight hours at a time, they took food with them, which they ate with unwashed hands; hence, in the event of one workman getting the cholera, the others were very liable to contract the disease, and take it to their families. One important medium of the conveyance of the cholera poison from one patient to another was the drinking water, when it became contaminated by their evacuations, either by permeating the ground and getting into wells, or by their being conveyed by sewers into a river. Since he (Dr. Snow) first published his views on this point, Dr. William Budd has found the microscopic bodies before alluded to in such drinking-water of cholera districts as received the contents of sewers. Dr. Snow then related a number of instances where, as in Albion Terrace, Wandsworth Road, there was a very high mortality from cholera in connection with the contamination of the water with discharges of the patients. He attributed the high mortality on the south side of the Thames to three causes – viz., the drinking from certain tidal ditches, which prevails to so great an extent in Rotherhithe and Bermondsey; the water supplied by the waterworks in this part of London, which is generally procured from the Thames in the midst of sewers; and to the contamination of a number of the wells by the contents of the cesspools, which are often as deep as the wells. The low elevation of the ground had no connection with cholera, except when combined with infected water. Westminster had the same elevation as the Borough, and yet the mortality from cholera was not half so great. Bethlem Hospital and the Queen's Prison had all but escaped cholera, situated in a low level, but having pure water of their own. The Brixton district was chiefly on elevated ground, and yet the mortality there had been high, as a great part of the district was supplied with water got out of the Thames near the Hungerford Suspension Bridge, by the Lambeth Waterworks. In Exeter, in 1832, the greater part of the people were supplied with water, into which, as he had learnt from Dr. Shapter, one of the chief sewers emptied itself. This sewer brought the sewage from the street in which two of the first cases of cholera introduced from Plymouth occurred; the disease was in a few days scattered through the town, and was very fatal. Since that time Exeter had been copiously supplied with water quite free from the sewage of the town, and this year there have been very few cases of cholera, and those chiefly of strangers just arrived from places in which it was prevailing. Hull had also got a new and more plentiful supply of water since 1832, at which time it was scantily supplied with water conveyed from some springs. The new supply was from the river Hull, which receives half the sewage of the town, and which is washed up by the tide past the water works, as he had learnt from Dr. Horner. In 1832, cholera in Hull was confined to the poor, of whom it carried off 300; this year it had assailed all classes, and about six times that number had died, although eight or ten thousand people had left the town to escape the disease. Certain towns, as Birmingham, Bath, Cheltenham, and Leicester, had almost escaped cholera, both in 1832 and the present year. These towns were supplied with water quite uncontaminated by the sewers. He gave some recommendations for the prevention of cholera; the chief of which were – extreme cleanliness in those waiting on the sick, and avoidance of tainted water, or at least having it well boiled before it was used. Dr. Brittan had found microscopic bodies in the atmosphere, which he considered to be the same as those existing in the alimentary canal. He (Dr. Snow) understood that others had not succeeded in finding them in the air, and he was of opinion, that if they should be generally found in the atmosphere, even of infected districts, they could not be the real cause of cholera, for all the evidence he had collected was opposed to the idea that the cause of cholera existed in the air.'

It is clear therefore that Snow's views had been consolidated *before* he produced his pamphlet (1849) and subsequently the definitive text (1854).

2.7.1 THE BROAD STREET SAGA

The major outbreaks in England occurred in 1831–1832, 1848–1849, and 1853–1854, and the minor ones in 1868 and 1893 (see above) (Longmate, 1966). It was towards the close of the 1848–1849 epidemic (in London) that Snow followed up his hypothesis that excreta from cholera patients could – when accidentally ingested – cause the disease.

The inhabitants of St James, Westminster, escaped lightly in each of the first two epidemics (Howell and Ford, 1985); in 1831–1832 there were 13 deaths per 10 000 population, and in 1848–1849, 15 per 10 000, while the corresponding overall figures for London were 56 and 76. In the third epidemic, the disease had first appeared in London in July 1853; 2 months later the death toll in Newcastle had reached 1500. In July 1854, a sudden peak in cholera deaths occurred in London; during the last week of that month 133 people died. However, the relative 'immunity' of the population of St James came abruptly to an end in August; on 29 August a 5-month-old girl at 40 Broad Street fell ill with 'summer diarrhoea'. On 30 August, a major epidemic broke out in Soho; in Broad Street alone, 56 individuals died within 48 hours, and by 2 September the deadhouse at the workhouse (in Poland Street) contained 82 bodies. The Middlesex Hospital admitted 120 cases in 3 days; 80% were from the parish of St James. A similar situation prevailed at University College Hospital, Charing Cross Hospital, and the Workhouse Infirmary; however, most patients died at home.

By 3 September, there was general agreement that the epidemic was abating. On 15 September, *The Times* published an eyewitness account 'by a reporter at the scene' (Editorial, 1854; Howell and Ford, 1985):

'The outbreak of cholera in the vicinity of Golden Square is now subsiding but the passenger through the streets which compose that district will see many evidences of the alarming severity of the

> attack. Men and women in mourning are to be found in great numbers, and the chief topic of conversation is the recent epidemic. ... An oil shop puts forth a large cask at its door, labelled in gigantic capitals 'Chloride of Lime'. The most remarkable evidence of all, however, and the most important, consists in the continual presence of lime in the roadways. The puddles are white and milky with it; ... great splashes of it lie about in the gutters, and the air is redolent with its strong and not very agreeable odour ... the parish authorities have very wisely determined to wash all the streets of the tainted district with this powerful disinfectant; and accordingly the purification takes place regularly every evening. The shop keepers have dismal stories to tell – how they would hear in the evening that one of their neighbours whom they had been talking with in the morning had expired after a few hours of agony and terror. It has even been asserted that the number of corpses was so great that they were removed wholesale in dead-carts for want of sufficient hearses to convey them; but let us hope this is incorrect.'

In less than 2 weeks, 700 individuals within the Golden Square vicinity died of cholera; a report in the *Medical Times and Gazette* confirmed that it had been necessary for the dead to be transported in carts (Howell and Ford, 1985). The Board of Guardians met frequently during the crisis, and on 7 September was addressed by Snow who outlined his views on the 'cholera poison' – in his opinion, the causative factor (see above) (Chave, 1958); his experience of the disease had been gathered during the 1831 Newcastle epidemic (when he was a 19-year-old medical apprentice) and the 1848–1849 London epidemic, during which he had set out his views in his pamphlet: 'On the Mode of Communication of Cholera' (Snow, 1849). Although his views were considered at the time somewhat eccentric, spread of cholera by this route had certainly been suspected during the Exeter outbreak in 1832 (Shapter, 1849) (see above).

Soho had a superfluidity of water supplies; every house had had a mains supply since 1850. The New River Company supplied Berwick Street and the eastern area, water being derived from Hertfordshire (Howell and Ford, 1985). The Grand Junction Water Company supplied Golden Square and the western area, the origin of this water being the Thames, at Kew. Both companies turned on their mains for only 2 hours each day, with the exception of Sundays; therefore, each house had to store its water daily in a water-butt (cistern), situated in the basement or yard. Beneath the parish of St James, however, lay a stratum of impervious clay; this was covered by 25–30 feet of subsoil, consisting of sand and light gravel. These shallow wells produced abundant water, and throughout the district large cast-iron parish pumps yielded a *constant* supply of water; at least a dozen pumps lay within a quarter-mile radius of Broad Street. Water obtained from the Broad Street pump was invariably cool and clear; it was considered by many as the most palatable water then available in London.

Snow obtained a list of cholera deaths in the vicinity from the General Register Office for the week ending 2 September; almost all fell in a cluster within an area approximately 500 yards in diameter around the Broad Street pump. He had noted that the contents of the district's three main sewage systems were discharged directly into the Thames – at the end of Northumberland Avenue – close to Charing Cross; at high tide, sewage came flooding back into the drains. By 7 September, Snow was convinced that the pump was the source of infection. However, the epidemic was by this time already subsiding (see above) – the yellow flags were removed from the streets – and removal of the handle (recommended by Snow) was regarded by the Medical Committee of the General Board of Health (Pelling, 1978) as being irrelevant to the course of the outbreak; Snow's theory was therefore rejected (Howell and Ford, 1985)! Furthermore, after the pump had been examined, the Parish Paving Committee issued a report indicating that there was no evidence that its water played any part in the spread of the disease.

On 2 November 1854, Dr Edwin Lankester (father of E.R. Lankester), a member of the St James Vestry, demanded a full investigation into the outbreak (Chave, 1958; Howell and Ford, 1985). A committee was set up, but it received strong opposition from Sir Benjamin Hall, the President of the General Board of Health (which had 'cleared' the pump's water); all information gathered by the department was withheld. The St James Vestry was undeterred however, and expanded its 'Committee of Enquiry' – to include Snow and the Reverend Henry Whitehead (Figure 2.8) – the 29-year-old curate of the local St Luke's church. During the 3 months since the outbreak, Snow (1854) had now expanded his pamphlet into a 162-page book, *On the Mode of Communication of Cholera.*

Whitehead proceeded to interview 497 of the 896 original inhabitants of Broad Street, gathering facts about each one of the 90 residents and 28 non-residents who had died of cholera (Whitehead, 1865; Chave, 1958). Some 80 out of 137 individuals in Broad Street who drank from the pump had developed cholera; however, only 20 out of 299 of those who did not use this source of drinking-water suffered. While initially opposed to Snow's theory, Whitehead had now in effect confirmed it (Whitehead, 1865; Chave, 1958)! The Poland Street Workhouse had escaped the outbreak; it had its own pump in addition to a mains supply from the Grand Junction Company. Also, no one at the Broad Street brewery was affected; it had its own deep well and received a mains-supply from the New River Water Company. An isolated case,

Fig. 2.8 The Reverend Henry Whitehead (1825–1896).

Susannah Eley, in Hampstead was shown to have resulted from the fact that the 'victim' preferred water from the Broad Street Pump (Chave, 1958), and her sons sent her a large bottle daily by carrier's cart! Nevertheless, the Scientific Committee of the General Board of Health continued to reject Snow's theory (Pelling, 1978).

But, how had the pump-water become dangerous (Howell and Ford, 1985)? On 3 April 1855, Whitehead – after perusal of the Registrar's returns (of deaths) – was able to associate the outbreak with the 'index case' on 2 September at 40 Broad Street (a 5-month-old girl), a house next to the Broad Street pump. The house's 'privy', and cesspool, stood a mere 3 feet away from the pump. Examination showed the drain to leak like a sieve; furthermore, the cesspool was very badly constructed. Another case of cholera – a policeman, who also died – was also traced to number 40; on 8 September, the father of the 'index case' contracted the disease – *after* Snow had been successful in getting the pump handle removed. Two patients who did not contract the disease, had been nursed in an upstairs room (at the back of the house), and their excreta had been flung from the back window, thus avoiding the pump (Howell and Ford, 1985)!

On 9 August 1855, the Cholera Inquiry Committee presented its 175-page report to the Vestry, Lankester being the chairman (Anonymous, 1855):

> 'Anxious to give due weight to every fact and consideration that have offered themselves in the inquiry, the Committee is unanimously of the conclusion that the striking disproportionate mortality in the "cholera area" as compared with the immediately surrounding districts, which constitutes the sudden, severe and concentrated outbreaks, beginning on August 31st, and lasting for a few early days of September, was in some measure attributable to the use of the impure water of the well in Broad Street.'

The report was adopted, but by the narrowest of margins (the Chairman gave the casting vote). As a consequence, the committee recommended that all surface wells be abolished, all cisterns removed, and the water companies provide a continuous water supply; also, stand-pipes connected to the mains should be erected in the streets to provide free water. It was not until 1866, that Lankester succeeded in getting the Broad Street pump abolished.

By 1871, Snow's theory had become generally accepted; however, Whitehead's contribution had been crucial. Snow's classical 1854 monograph – an expansion of the pamphlet published in 1849 – is regarded to this day as a pioneering achievement in epidemiology, and remains a salutary reminder that contaminated water is a constant source of gastrointestinal pathogens. Furthermore, by preventing further use of the Broad Street pump, evidence had been clearly provided that a 'germ-borne disease' was preventable – some three decades before the germ theory was formulated. Snow also suggested that contaminated water could be purified by filtration through sand and/or gravel, or by allowing it to stand in reservoirs.

However, Whitehead's contribution to the saga should certainly not be underestimated; referring later to the investigation (carried out with Snow), he wrote (Whitehead, 1865; Chave, 1958):

> '...from St. Luke's pulpit on September 18th I congratulated the poor old women who formed a considerable proportion of the congregation upon their remarkable immunity from the pestilence. At that time I had been too busy to meddle with hypotheses and had not even heard of Dr. Snow's bill of indictment against the pump. The escape of these women, many of whom, living alone, had no-one to send to the well, was one of those 'eccentricities' which found their best explanation in the pump theory.
>
> One of the strongest facts in connection with this inquiry is that the impurity of the well-water was, in point of time, the very last discovery made by the investigation. We collected the evidence already described, not only in ignorance of the fact of the well having been contaminated, but in the face of positive and seemingly reliable evidence to the contrary. The sides of the well had been examined and declared in a report made by order of the Paving Board on November 24th, 1854, "to be free from any fissures or other communication with drains or sewers by which such matters could possibly be conveyed into the waters". Both chemical and microscopical analyses had "failed to detect anything which could be pronounced peculiar to the cholera period or capable of acting as a predisposing co-operating or specific agent in the production of that disease". We stand exonerated from the imputation of seeking to impugn the well-water on the ground of any previous knowledge of its impurity. Indeed for my own part I had a leaning the other way.'

And in his memoirs, published in 1874, Whitehead paid this tribute to Snow (Whitehead, 1874; Chave, 1958):

> 'Yet for wholly exceptional reasons, I may say a few words about Dr. John Snow – as great a benefactor in my opinion to the human race as has appeared in the present century. Dr. Snow had long believed that he had discovered the mode in which cholera is propagated and fortunately he was at hand to direct an inquiry into the cause of the Broad Street outbreak, which inquiry resulted in a remarkable confirmation of his hypothesis. The story of his researches and of this investigation in particular I have elsewhere related at some length. ...

What I chiefly wish to dwell upon is the calm prophetic way in which he would talk of the ultimate results of the doctrine which he laid down.

"You and I", he would say to me, "may not live to see the day, and my name may be forgotten when it comes, but the time will arrive when great outbreaks of cholera will be things of the past; and it is the knowledge of the way in which the disease is propagated which will cause them to disappear."

He died in 1858 and since his death we have seen a complete revolution in the mode of investigating the causes of cholera and typhoid, a revolution already fruitful in beneficial consequences and destined hereafter to achieve all the important results that he anticipated. He did not in his lifetime receive all the recognition which was due to his genius, though unstinted respect was paid to his character.

"Dr. Snow's views on cholera", said a medical friend to me in 1855, "are generally regarded in the profession as very unsound." "If that be the case", I replied, "heresy may be as good a thing in your profession as some of you are apt to suppose in mine."

A portrait of Dr. Snow hangs on my study wall and ever serves to remind me that in any profession the highest order of work is achieved not by fussy demand for "something to be done" but by patient study of the eternal laws.'

2.8 OTHER COUNTRIES FOLLOW THE ENGLISH EXAMPLE

In the USA it was not until 1866 that a Board of Health, comparable with that in England (see above), was established in New York (McNeill, 1976); this was both modelled on the British prototype, and inspired by apprehensions of a 'new' cholera epidemic.

Hamburg postponed costly improvements in its water supply until 1892 (Scott, 1939; McNeill, 1976); an old free city, Hamburg remained self-governing within the new German Reich and drew its water supply from the Elbe (without undergoing special treatment). Altona (part of the Prussian state) lay nearby; here the government installed a water filtration plant. When cholera broke out in 1892, the outbreak ran down one side of the street dividing the two cities, sparing the other completely (Van Heyningen and Seal, 1983). This too was clear proof of contamination of the water supply (and not air or earth – still favoured by miasmatists – which were similar on both sides). Cholera has not returned to a European city since!

Britain thus pioneered the new era of sanitation and water management (McNeill, 1976); this influenced the decline, not only of cholera, but of other water-borne infections, typhoid included. How relevant this 'message' pertains in the present-day developing world! Also, international medical cooperation had achieved a new momentum as a result of Europe's encounters with cholera.

2.9 A BACTERIUM AS THE CAUSATIVE AGENT

Recognition of a specific causative organism to account for the Asiatic cholera had to wait until the time was ripe! A mid-19th century supposition that 'cholera animalcules' were carried by the wind from India – the endemic centre – giving rise to epidemics and pandemics was attractive. However, it could not account for the different routes taken by the disease outside India, i.e. across Central Asia, Europe, and to America (Scott, 1939)! Evidence grew that the disease followed the routes (and rate) of travel; this did not advance the demonstration of a specific causative organism.

It was not until 1883 that Robert Koch (1843–1910) discovered a 'comma-shaped bacillus' (in the intestinal lumen, and not in blood or any tissue fluid) at Alexandria, Egypt; he confirmed this work in Calcutta – where he detected the organism in the dejecta of all cases he examined and termed it the *Kommabazillen* (Barua and Greenough, 1992). However, Klein and Lankester rejected Koch's conclusions as to the *specific* nature of this vibrio. The former considered that Koch was wrong because the organism depicted was not a bacillus, but a spirillum (Scott, 1939). He later withdrew most of his assertions and subsequently admitted that Koch '...has found the true cholera bacilli in all cases of cholera'. [It is of interest that Macnamara (1892) considered that he had demonstrated in 1866 that the disease was caused by 'specific organic matter' (destroyed by gastric acid); however, there is no evidence of this conclusion in any of his extant writing (see Scott, 1939).]

Little progress was made for 20–25 years; bacteriologists became increasingly involved in differentiation of true cholera from cholera-like vibrios (Scott, 1939). In 1894, Pfeiffer (1858–1945) discovered a test which differentiated *Vibrio cholerae* from organisms which resembled it (but did not produce symptoms) (Singer and Underwood, 1962); when a guinea-pig was immunized against cholera, vibrios could be injected into the peritoneal cavity without ill-effect. Peritoneal fluid then withdrawn a few minutes after injection confirmed motionless vibrios – which swelled and disintegrated, i.e. they had been lysed by a substance in the peritoneal fluid. Pfeiffer later recorded this phenomenon *in vitro*; it was the first demonstration of bacteriolysis.

A variant bacillus was shown to be distinct from true *V. cholerae* in 1905 (Greenough, 1995); these vibrios were haemolytic (did not fix complement), and could thus be separated from *V. cholerae*. They were isolated by Gotschlich from dead bodies of returned Mecca pilgrims at the quarantine camp of El Tor (Greenough, 1995). In 1937, a mild, cholera-like disease caused by this vibrio – later designated *V. cholerae* O1, biotype El Tor – was described in Indonesia. It was not until 1961 (arguably the date signifying the Seventh Pandemic) however, that this biotype produced an epidemic of major proportions in the Philippines (Greenough, 1995). In 1969, the El Tor variant appeared in the Ganges Delta, and by 1974 it had replaced the Classical biotype. In 1982, the Classical biotype resurfaced – in Bangladesh – in a new capacity (Greenough, 1990); it both produced a more severe illness, and rapidly replaced the El Tor variant. Recently, a non-O1 strain has emerged, in India, and is currently sweeping westwards across Asia (Cook, 1993; Swerdlow and Ries, 1993; Wachsmuth *et al.*, 1994).

Despite the major advances in understanding the infection which have been outlined in this chapter, cholera remains a major world scourge and continues to affect massive areas of the globe. In fact the global situation, with endemic areas scattered throughout Asia, Africa and many countries of South and Central America, is far worse than it was at the end of the first quarter of the 20th century, when the Sixth Pandemic started (Barua and Greenough, 1992). Snow and Whitehead would be dismayed to see what little progress has been made since their major discovery nearly a century-and-a-half ago.

2.9 REFERENCES

Anonymous. (1849a) History of the Origin, Progress and Mortality of the Cholera Morbus. *London Medical Gazette*, **a**, 507–11, 556–9, 600–2.

Anonymous. (1849b) The fungoid theory of cholera. *The Times, London*, 5 October, p. 3.

Anonymous. (1853) Seamen's Hospital Society – Cholera on the River. *Lancet*, **ii**, 406.

Anonymous. (1854) The cholera in the United Kingdom in 1854. *Lancet*, **i**, 459–61.

Anonymous. (1855) *Report on the Cholera Outbreak in the Parish of St James, during the autumn of 1854, Presented to the Vestry by the Cholera Inquiry Committee, July 1855*, Churchill, London.

Anonymous. (1859) 'Dreadnought' Hospital Ship. Cholera and choleroid diarrhoea. *Lancet*, **ii**, 137–8.

Anonymous. (1868) Dreadnought Hospital Ship. Two cases of cholera; recovery. *Lancet*, **ii**, 77.

Anonymous. (1871) The cholera. The Port of London. *Lancet*, **ii**, 413.

Barua, D. and Greenough, W.B. III, (eds) (1992) *Cholera*, Plenum Medical Book Co., New York, London.

Brittan, F. (1849) Report of a series of microscopical investigations on the pathology of cholera. *London Medical Gazette*, **9**: 530–42.

Chave, S.P.W. (1958) Henry Whitehead and Cholera in Broad Street. *Medical History*, **2**, 92–108.

Cook, G.C. (1992) *From the Greenwich Hulks to Old St Pancras: a History of Tropical Disease in London*, Athlone Press, London.

Cook, G.C. (1993) Cholera in Africa. *Africa Health*, **16**(1), 20–1.

Cook, G.C. (1994) Evolution: the art of survival. *Transactions of the Royal Society of Tropical Medicine and Hygiene*, **88**, 4–18.

De, S.N. (1961) *Cholera: its Pathology and Pathogenesis*, Oliver and Boyd, Edinburgh, London, pp. 41–61.

Editorial. (1854) *The Times, London*, 15 September, p 9.

Evans, R.J. (1987) *Death in Hamburg: Society and Politics in the Cholera Years 1830–1910*, Clarendon Press, Oxford.

Greenough, W.B. III. (1995) Vibrio cholerae and cholera, in *Principles and Practice of Infectious Diseases*, 4th edn, (eds G.L. Mandell, J.E. Bennett and R. Dolin), Churchill Livingstone, New York, Edinburgh, pp. 1934–45.

Howell, M. and Ford, P. (1985) Death in the parish, in *The Ghost Disease and Twelve other Stories of Detective Work in the Medical Field*, Penguin Books, London, pp. 138–64.

Hunt, R. (1849) The probable causes in operation to produce pestilential cholera. *London Medical Gazette*, **9**, 473–5.

Letter. (1849) The British Microscopical Society, *versus* the President of the Microscopical Society of London. *Lancet*, **ii**, 460.

Longmate, N. (1966) *King Cholera: the Biography of a Disease*, Hamish Hamilton, London.

McGrew, R.E. (1965) *Russia and the Cholera, 1823–1832*, University of Wisconsin Press, Kingsport, Tennessee.

McNeill, W.H. (1976) *Plagues and Peoples*, Penguin Books, London.

Pelling, M. (1978) *Cholera, Fever and English Medicine 1825–1865*, Oxford University Press, Oxford.

Pollitzer, R. (1959) History of the disease, in *Cholera*, World Health Organization, Geneva, pp. 11–50.

Rogers, L. (1926) The conditions influencing the incidence and spread of cholera in India. *Proceedings of the Royal Society of Medicine* (section on epidemiology and state medicine) **19**, 59–93.

Rogers, L. (1933) The methods and results of forecasting the incidence of cholera, smallpox and plague in India. *Transactions of the Royal Society of Tropical Medicine and Hygiene*, **27**, 217–38.

Rogers, L. (1950) *Happy Toil: Fifty-Five Years of Tropical Medicine*, Frederick Muller Ltd, London.

Rosenberg, C. (1960) The cause of cholera: aspects of etiological thought in nineteenth century America. *Bulletin of the History of Medicine*, **34**, 331–54.

Russell, A.J.H. (1926) The epidemiology of cholera. *Indian Journal of Medical Research*, **14**, 1–26.

Sachs, J. von. (1890) *History of Botany 1530–1860*, Oxford University Press, Oxford.

Sandwith, F. (1960) *Surgeon Compassionate: the story of Dr. William Marsden M.D., M.R.C.S. Founder of the Royal Free and Royal Marsden Hospitals*, Peter Davies, London.

Scott, H.H. (1939) *A history of tropical medicine based on the Fitzpatrick lectures delivered before the Royal College of Physicians of London 1937–38.* Arnold, London, vol. 2, pp. 649–701.

Shapter, T. (1849) *The History of the Cholera in Exeter in 1832,* John Churchill, London.

Sheppard, D.A.E. (1995) *John Snow; Anaesthetist to a Queen and Epidemiologist to a Nation: a Biography.* York Point Publishing, Cornwall, Edward Island, Canada.

Singer, C. and Underwood, E.A. (1962) *A Short History of Medicine,* 2nd edn, Clarendon Press, Oxford.

Snow, J. (1849) On the pathology and mode of communication of cholera. *London Medical Gazette, or Journal of Practical Medicine,* Longman, Brown, Green, and Longmans, Paternoster Row, London, pp. 730–2.

Snow, J. (1849) *On the Mode of Communication of Cholera,* John Churchill, London.

Snow, J. (1854) *On the Mode of Communication of Cholera,* John Churchill, London.

Swerdlow, D.L. and Ries, A.A. (1993) *Vibrio cholera* non-O1 – the eighth pandemic? *Lancet,* **342**, 382–3.

Tramont, E.C. Gangarosa E.J. (1995) Cholera, dysentery and diarrhea: lessons of history, in *Infections of the Gastrointestinal Tract* (eds M.J. Blaser, P.D. Smith, J.I. Ravdin, H.B. Greenberg and R.L. Guerrant) Raven Press, New York, pp. 1–11.

Tunstall, J. (1848) Petroleum in Asiatic cholera. *Provincial Medical and Surgical Journal,* **12**, 390–1, 471–2.

Van Heyningen, W.E. and Seal, J.R. (1983) *Cholera: the American Scientific Experience, 1947–1980,* Westview Press, Boulder, Colorado.

Wachsmuth, I.K., Blake, P.A. and Olsvik, Ø. (1994) Vibrio cholerae *and Cholera: Molecular to Global Perspectives,* ASM Press, Washington DC.

Whitehead, H. (1865) The Broad Street Pump: an Episode in the Cholera epidemic of 1854. *MacMillan's Magazine,* 113–22.

Whitehead, H. (1874) *Experience of a London Curate,* Clapham.

Wilkinson, J.S. (1849) Some remarks upon the development of epiphytes; with the description of a new vegetable formation found in connexion with the human uterus. *Lancet,* **ii**, 448–51.

Management of cholera: the vital role of rehydration

3

G.C. Cook

3.1 INTRODUCTION

Cholera is the archetypal disease with respect to small-intestinal watery (secretory) diarrhoea, which may be torrential. Management objectives are therefore: (i) rapid replacement of water and salt(s) (already lost); (ii) maintenance of normal hydration until cessation of diarrhoea; further losses should be replaced as they occur; (iii) reduction of both magnitude and duration of diarrhoea (the role of antibiotics remains controversial – see below); and (iv) prompt introduction of a normal diet to minimize the adverse nutritional effects resultant upon the illness. The logical (and correct) form of management therefore lies in rapid rehydration (Nalin and Morris, 1991; Mahalanabis *et al.*, 1992; Bennish, 1994; Greenough, 1995). But, a very long time was required historically for this 'message' to register (Cosnett, 1989; Carpenter, 1990; da-Cunha-Ferreira and Cash, 1990; Bonnici, 1993); thus, Cosnett (1989) has written:

> 'Because of the constant observation of "black, thick, cold blood" in collapsed cases, it was believed [in the early nineteenth century] that the cure must lie in the removal of this grumous blood from the patient. Of the three predominant modes of treatment the most universal was blood-letting [see Chapter 2]. An authority wrote: "In commencing the treatment of cholera, no time is to be lost in endeavouring to bleed the patient ... the effect of blood-letting would appear to be almost miraculous...". Of the other treatments, emesis was aimed at ridding the body of poisons, and vomiting was consid-

Cholera and the Ecology of Vibrio cholerae. Edited by B.S. Drasar and B.D. Forrest.
Published in 1996 by Chapman & Hall, London. ISBN 0 412 61220 8.

> ered "the best of restoratives for torpidity of the blood". Calomel was almost universally used, as a means of "unlocking the secretions". Other treatments included a large fraction of the pharmacopoeia of the time, mostly based on empiricism and superstition. *The Lancet* viewed this therapeutic miscellany with some scepticism when it declared: "All have their seasons of celebrity and subsequent neglect".
>
> A few doctors showed an inkling of scientific understanding. One thought that loss of water from the blood rendered it "difficult of circulation" and proposed that "distilled water be liberally poured into the stomach". Dr Clanny of Sunderland [where the first English outbreak of the 'Asiatic Cholera' began with deaths in October 1831 – Chapter 2] analysed the blood of cholera patients and began his report with: "This blood, on applying the tongue to it, had no taste nor any particular smell...". He found that the amount of water was decreased and the "colouring matter" increased, but he gave no specific advice about a remedy.'

Dr George Taylor, who had considerable experience of the 'blue cholera' in India during the early 19th century – writing to *The Lancet* in November 1831 – also felt that bleeding (via the radial artery if the veins failed to yield blood) was a valuable, and effective, method of management (Taylor, 1831):

> '...Theory would point out that free bleeding, the application of heat and stimulants to the surface of the body, and the exhibition of aromatic and exciting remedies internally, are the means best calculated to meet such a case; and practical theory amply sustains the correctness of the theory.
>
> All the practitioners who have seen much of, and have written upon, this disease, agree upon one point; viz., the immense advantage of bleeding at the commencement of the disease. Indeed, the recovery or death of the patient seemed, in the opinion of some writers upon this disease, to depend upon the success or failure of the attempt to bleed. The oppressed condition of the heart and lungs (in most cases of this disease in its worse form), seemed to render the ordinary operation of bleeding impracticable; and for this reason, that the veins upon the surface contained no blood, while the heart and lungs (as the examination of the dead verifies) were gorged, and the former consequently incapable of exerting sufficient contractile energy to propel it into the veins. Bleeding, then, being considered in the early stage of the severe form of the disease, the remedy and the common mode frequently failing, I beg to suggest a method by which I have succeeded in cases of oppression, and wherein the veins would afford no blood. I beg to recommend that the radial artery

> should be resorted to upon failure in the veins. The communication between that vessel and the heart being direct, it offers a facility for unloading the chest, which no other mode of bleeding possesses; and the operation is one which may be safely undertaken by any surgeon who has an average share of anatomical knowledge, and its efficacy I have proved in my own practice.'

However, Dr J.H. Stein, writing in the same issue of *The Lancet*, was of the opinion that inhalation of oxygen was likely to be accompanied by a far more beneficial result (Stein, 1831):

> 'Conversing with a friend upon that all-engrossing subject the cholera morbus, he suggested that the inhalation of nitrous oxide or other combinations of oxygen (from their actively-stimulating properties) might prove very efficient remedies in this dreadful malady.
>
> After having attentively considered the subject and consulted one or two eminent medical friends, we coincided entirely with his opinions; and I have his permission to submit the subject to the consideration of my medical brethren, through the medium of your valuable paper. ... I will merely premise, as some claim to consideration for my opinion, that when a surgeon in his Majesty's navy, I witnessed the cholera raging with its utmost virulence in India itself; and when in practice in London, I had considerable experience in the beneficial effects of inhalation in cases of chlorosis, paralysis and other such diseases; therefore I am no stranger to pneumatic medicines. It is self-evident, that the most active stimulants ought to be, on the instant, employed in an attack of cholera. In order to effect this, brandy, opium, peppermint, &c., have been administered. Now, in the then-defective state of the functions, it is not surprising that the largest doses of the above have proved ineffectual, while it is, most frequently, impossible even to keep them upon the stomach. After duly weighing all these things, it must strike every one, that inhalation is by far the quickest and surest means of introducing a stimulant, because from its exhibition into the lungs, it is instantly conveyed throughout the whole system. I have classed nitrous oxide first, because it is so active a stimulant, and so very rarely followed by the depression consequent upon the exhibition of other stimulants. In plethoric habits, I am aware that this gas has been found to produce mischievous effects; but this state of things is scarcely to be apprehended in the complete prostration which attends cholera. Other combinations of oxygen might be instanced. As for external applications, ineffective as they may be in such extreme cases, yet if baths were used of hot oxygen, direct from the retort, they might also be found effective.

As a preventive, the best effects might be reasonably expected from inhalation, by preventing that depression which predisposes for contagion.'

A report in *The Lancet* the following year (Anonymous, 1832) outlined methods used by Dr M'Duffal for treating cholera at Constantinople, Turkey:

'The writer considered the great indication in the treatment of this formidable disease was to restore the lost balance of the circulation; and this was to be done, in their way, by the exhibition of internal stimuli, by free venesection, and by the use of such external applications, as mustard cataplasms, hot-bricks, and various rubefacients. This mode of treatment he had found successful, in the great majority of instances. When a patient was brought to him with the hollow eye, cold surface, leaden hue, and feeble pulse, characteristic of this malady, he immediately administered brandy, in large quantities, qualified with opium. He then would bleed the patient to sixteen or twenty ounces. As the blood flowed, the surface grew gradually warm, the pulse rose, and the eye and countenance regained their tranquillity. The patient after this was put into a warm bed, and treated with rubefacients, and recovered in a few days. In the village of Terapia, in the vicinity of Constantinople, out of 204 patients attacked with cholera, and treated in this manner, 200 recovered; and the proportion in other places was nearly the same. A native merchant of great wealth affected with the incipient approaches of cholera, would not suffer his barber to bleed him, being buoyed up by the hope that the disease would not run its course. Dr. M'Duffal was sent for, and found him with leaden extremities, cold skin and breath, imperceptible pulse, and the peculiar facial expression of this epidemic. He was bled immediately to sixteen ounces, and the symptoms gradually subsided; stimulants internal as well as external were freely resorted to. About an hour after this a tendency to relapse was observed, when sixteen ounces more blood were abstracted, and the other treatment continued. He ultimately recovered.

An Armenian man, who had long been subject to frequent attacks of gastro-enteritis, was attacked with cholera to a severe degree; stimulants were freely given, and copious venesection resorted to. The collapse was subdued, but it was followed by a sharp gastro-enteric affection, which was overcome by leeches, mucilaginous potions, and enemata. Dr M'Duffal opines, that there is no hope if more than four hours have elapsed before the administration of remedies; and he recommends that the public in general should be made acquainted with the importance of immediate measures being had recourse to. The writer used himself to counsel the people within his sphere of observation, to give freely to the affected person any spirits he might have in their dwelling; to place him in a warm situation, and immediately to send to a barber to bleed him. When these prompt measures were resorted to, the patient, on the arrival of the physician, was in a much more favourable condition for treatment; for the stage of collapse being by those means subdued, the stage of reaction alone remained to be grappled with. When marks of reaction were observed, it became expedient to intermit the stimulants, for in all cases a great tendency to inflammatory affections of the head, chest, and abdomen, existed during the stage of reaction. These affections were managed much in the same manner as in England, by leeches, antiphlogistics alternatives, such as calomel, opium, and others, and epispastics...'

3.2 EARLY SCIENTIFIC APPROACHES TO MANAGEMENT

On 3 December 1831, Dr William Brooke O'Shaughnessy, a young Edinburgh graduate who had been born in Limerick, read a paper before the Westminster Medical Society. His percipient observations heralded the introduction of *intravenous* infusions to restore the 'arterial properties' of the blood (O'Shaughnessy, 1831a):

'...It seems to me...that, if we could bring certain salts of a highly oxygenated constitution fairly into contact with the black blood of cholera, we would certainly restore its arterial properties, and most probably terminate the bad symptoms of the case. It is however obvious, that in a disease so electric-like in its rapidity, and one moreover in which circulation is almost suspended, and the intestines generally covered with a thick pultaceous paste, it would be a vain expectation were we to imagine that, as in the more protracted yellow fever of *Santa Cruz* ... time would permit of the absorption of the saline oxygenating materials from the alimentary canal. I therefore conceived the idea of injecting into the veins such substances as an examination of the blood in cases of cholera would show to be most capable of restoring it to the arterial qualities. Before so novel, and apparently startling, a practice as the injection of such remedies into the veins of the human subject could be prudently put into execution, it is necessary to have distinct notions of the effects of the individual salts on the blood without the body, and also while still circulating in the living frame....

The medical journal [it is unclear as to which one he alludes] recommends the injection of common salt, the *chloride* of *sodium*. I, on the other hand, advise that of the nitrate or chlorate of potassa. I would venture to remind the editor that *chloride* of *sodium*, though it does redden venous blood, cannot possibly oxygenate it, as *it contains no oxygen itself*. There are, indeed, many substances capable of reddening blood, which cannot arterialise or oxygenate it. Hydrogen gas, for example, is stated by Dr. Davy to possess this quality. I believe it will, therefore, be conceded, that as far as theory is concerned, the salts I mention are likely to ensure more satisfactory therapeutic results than that which the writer in question has proposed.'

O'Shaughnessy (1832) undertook a series of detailed investigations in order to determine the precise composition of the blood in the cholera sufferer. He had published his preliminary observations in a letter to *The Lancet* in December 1831 (O'Shaughnessy, 1831b); the main findings were summarized:

'1. The blood drawn in the worst cases of *the* cholera, is unchanged in its anatomical or globular structure.
2. It has *lost a large proportion of its water*, 1000 parts of cholera serum having but the average of 860 *parts of water*.
3. *It has lost also a great proportion of its* NEUTRAL *saline ingredients*.
4. *Of the free alkali contained in healthy serum, not a particle is present in some cholera cases, and barely a trace in others*. [The blood and dejected substances were obtained in one analysis, from the same patient, and the blood was drawn half an hour after the evacuation occurred.]
5. Urea exists in the cases where suppression of urine has been a marked symptom.
6. *All the salts deficient in the blood, especially the carbonate of soda, are present in large quantities in the peculiar white dejected matters*. [See comment following 4.]

O'Shaughnessy's (1832) definitive account (this also utilized analytical data provided by a Dr Thomson of Glasgow) can hardly be bettered today and appeared in *The Lancet* in May of that year:

'...It is thus seen that my experiments here agree with Dr. Thomson's in again denoting a great deficiency of water, and a *corresponding increase of the albumen in the serum* compared with the normal standard.'

And the first and foremost of his 'deductions' was:

'...1. That the universality of the diminution of the quantity of water in cholera blood is extensively and satisfactorily confirmed.'

O'Shaughnessy followed this with a more detailed report, addressed to the Central Board of Health; he both confirmed his earlier findings (see above) and also recorded results obtained on excreta of affected individuals (Cosnett, 1989): 'The ingredients deficient in the blood were detected in the dejections or in other words, *the addition of the dejection to the blood, in due proportion, would have restored the latter to its normal constitution*'. He stressed that the changes 'should not be regarded as primary *causes*' but rather as the '*result* of an external impression ... we still remain in darkness as to the mode in which that impression is communicated...'. Pursuing his theme logically to the 'therapeutic conclusions', he wrote:

> '...the indications of cure ... are two in number – viz. 1st to restore the blood to its natural specific gravity; 2nd to restore its deficient saline matters ... The first of these can only be effected by absorption, by imbibition, or by the injection of aqueous fluid into the veins. The same remarks, with sufficiently obvious modifications, apply to the second ... When absorption is entirely suspended ... in those desperate cases ... the author recommends the injection into the veins of tepid water holding a solution of the normal salts of the blood.'

O'Shaughnessy ended:

> 'I am therefore entitled to conclude that the exudation of the colourless part of the blood constitutes one of the chief diagnostic characteristics of the malignant cholera ... in the fluidity, alkalesence and albuminous nature of the dejections we have the means of forming a certain and chemical diagnosis between this disease and others with which ... it may be confounded.'

3.3 THE FIRST APPLICATION OF INTRAVENOUS REHYDRATION

Hermann (1831) (who was aware of haemoconcentration in patients affected by the cholera) attempted intravenous rehydration during the Moscow cholera epidemic of 1830 (Pollitzer, 1959). However, credit for the first successful use of intravenous rehydration is invariably accorded to Dr Thomas Latta (Pollitzer, 1959; Cosnett, 1989). His experience is first recorded in a letter to the Secretary of the Central Board of Health, London, and published in *The Lancet* in June 1832 (Latta, 1832):

> 'Leith, May 23, 1832.
>
> SIR, – My friend Dr. Lewins has communicated to me your wish for a detailed account of my method of treating cholera by saline injection into the veins, with which I now most willingly comply. My scope for observation, since I commenced this treatment, has been too

limited to allow me to be very copious on the subject, but I think I can adduce sufficient proof to the unprejudiced, not only of its safety, but of its unquestionable utility. I have never yet seen one bad symptom attributable to it, and I have no doubt that it will be found, when judiciously applied, to be one of the most powerful, and one of the safest remedies yet used in the second stage of cholera, or that hopeless state of collapse to which the system is reduced.

Before entering into particulars, I beg leave to premise that the plan which I have put in practice was suggested to me on reading in THE LANCET, the review of Dr. O'Shaughnessy's report on the chemical pathology of malignant cholera, by which it appears that in that disease there is a very great deficiency both of the water and saline matter of the blood. On which deficiency, the thick, black, cold state of the vital fluid depends, which evidently produces most of the distressing symptoms of that very fearful complaint, and is, doubtless, often the cause of death. In this opinion I am abundantly borne out by the phenomena produced on repletion by venous injection.'

Rehydration by the rectal route?:

'So soon as I learnt the result of Dr. O'Shaughnessy's analysis, I attempted to restore the blood to its natural state, by injecting copiously into the larger intestines warm water, holding in solution the requisite salts, and also administered quantities from time to time by the mouth, trusting that the power of absorption might not be altogether lost, but by these means I produced, in no case, any permanent benefit, but, on the contrary, I thought the tormina, vomiting, and purging, were much aggravated thereby, to the further reduction of the little remaining strength of the patient; finding thus, that such, in common with all the ordinary means in use, was either useless or hurtful, I at length resolved to throw the fluid immediately into the circulation. In this, having no precedent to direct me, I proceeded with much caution.'

Latta then described his first dramatic, but transient, success:

'The first subject of experiment was an aged female, on whom all the usual remedies had been fully tried, without producing one good symptom; the disease, uninterrupted, holding steadily on its course. She had apparently reached the last moments of her earthly existence, and now nothing could injure her – indeed, so entirely was she reduced, that I feared I should be unable to get my apparatus ready ere she expired. Having inserted a tube into the basilic vein, cautiously – anxiously, I watched the effects; ounce after ounce was injected, but no visible change was produced. Still persevering, I thought she began to breathe less laboriously, soon the sharpened features, and sunken eye, and fallen jaw, pale and cold, bearing the manifest impress of death's signet, began to glow with returning animation; the pulse, which had long ceased, returned to the wrist; at first small and quick, by degrees it became more and more distinct, fuller, slower, and firmer, and in the short space of half an hour, when six pints had been injected, she expressed in a firm voice that she was free from all uneasiness, actually became jocular, and fancied all she needed was a little sleep; her extremities were warm, and every feature bore the aspect of comfort and health. This being my first case, I fancied my patient secure, and from my great need of a little repose, left her in charge of the hospital surgeon; but I had not been long gone, ere the vomiting and purging recurring, soon reduced her to her former state of debility. I was not apprised of the event, and she sunk in five and a half hours after I left her. As she had previously been of a sound constitution, I have no doubt that the case would have issued in complete reaction, had the remedy, which already had produced such effect, been repeated.'

He continued in more detailed terms:

'Not having by me the Number of THE LANCET containing Dr. O'Shaughnessy's analyses, I adopted that of Dr. Marcet, only allowing a smaller proportion of saline ingredients. This I now find to be considerably less than natural, according to the more recent analyses. I dissolved from two to three drachms of muriate of soda and two scruples of the subcarbonate of soda in six pints of water, and injected it at temperature 112° Fah. If the temperature is so low as a hundred, it produces an extreme sense of cold, with rigors; and if it reaches 115°, it suddenly excites the heart, the countenance becomes flushed, and the patient complains of great weakness. At first there is but little felt by the patient, and symptoms continue unaltered, until the blood, mingled with the injected liquid, becomes warm and fluid; the improvement in the pulse and countenance is almost simultaneous, the cadaverous expression gradually gives place to appearances of returning animation, the horrid oppression at the praecordia goes off, the sunken turned-up eye, half covered by the palpebrae, becomes gradually fuller, till it sparkles with the brilliancy of health, the livid hue disappears, the warmth of the body returns, and it regains its natural colour, – words are no more uttered in whispers, the voice first acquires its true cholera tone, and ultimately its wonted energy, and the poor patient, who but a few minutes before was oppressed with sickness, vomiting, and burning thirst, is suddenly relieved from every distressing symptom; blood now drawn exhibits on exposure to air its natural florid hue.

Such symptoms, so gratifying both to the sick and the physician, must never allow the latter to relax in his care – the utmost vigilance is still necessary. At first the change is so great that he may fancy all is accomplished, and leave his post for a while. The diarrhoea recurring, he may find his patient, after the lapse of two or three hours, as low as ever. As soon as reaction by the first injection is produced, mild warm stimulants, such as weak gin-toddy, mixed with some astringent, should be freely and assiduously administered. An attempt should be made to fill the colon with some astringent fluid. That such is requisite, is evident from the watery diarrhoea returning with violence, and if not restrained, death will ultimately make sure of his victim, therefore so soon as the pulse fails, and the features again shrink, the venous injection must be repeated, taking care that the fluid in use retains its proper temperature. The injection should be carried on very slowly, unless the patient is much exhausted, when it may be used more rapidly at first, until a little excitement is produced, after which it should not exceed two or three ounces per minute, and now is the time for the exhibition of astringents by the mouth, which will be retained, for in general the sickness entirely leaves during the operation.

Such remedies must be persisted in, and repeated as symptoms demand, or until reaction is permanently established. I have witnessed no violent symptoms accompanying the rapid injection of the fluid, but I have thought that the hasty repletion of the system was followed by great increase of the evacuations, and consequently a more sudden depression of the powers of life. The quantity to be injected depends on the effect produced, and the repetition on the demands of the system, which generally vary according to the violence of the diarrhoea; the greater the degree of collapse, the greater will be the quantity needed, though not uniformly, for a very slight loss produces much depression in some systems; hence there is often great collapse, without much vomiting, purging, or cutaneous discharge.'

But Latta also experienced failures using the intravenous rehydration technique, in addition to his successes:

'Although in every case, even the most desperate, the cholera symptoms were removed, some of my cases failed, which I attributed to one or other of the following causes; either the quantity injected was too small, or its effects were rendered abortive by extensive organic disease, or its application was too late.

I have already given an instance where deficiency in quantity was the cause of failure, which I will now contrast with one in which it was used freely. A female, aged 50, very destitute, but previously in good health, was on the 13th instant, at four a.m., seized with cholera in its most violent form, and by half-past nine was reduced to a most hopeless state. The pulse was quite gone, even in the axilla, and strength so much exhausted, that I had resolved not to try the effects of the injection, conceiving the poor woman's case to be hopeless, and that the failure of the experiment might afford the prejudiced and the illiberal an opportunity to stigmatise the practice; however, I at length thought I would give her a chance, and in the presence of Drs. Lewins and Craigie, and Messrs. Sibson and Paterson, I injected one hundred and twenty ounces, when like the effects of magic, instead of the pallid aspect of one whom death had sealed as his own, the vital tide was restored, and life and vivacity returned; but diarrhoea recurred, and in three hours she again sunk. One hundred and twenty ounces more were injected with the same good effect. In this case 330 ounces were so used in twelve hours, when reaction was completely reestablished; and in forty-eight hours she smoked her pipe free from distemper. She was then, for better accommodation, carried to the hospital, where probably, from contagion, slight typhoid symptoms were produced. She is now, however, convalescent.'

Underlying organic disease was, Latta recognized, a factor which adversely influenced outcome:

'The second cause of want of success is the presence of organic disease; this probably, renders the possessor very liable to attacks of cholera; and the latest evil which previously gave but little uneasiness, suffers aggravation in all its symptoms, more especially after reaction has been produced, and has evidently in many cases been the cause of death. A delicate young female, of strumous habit, who had been for some years subject to pectoral complaints, was rescued from a state of collapse by the injection of sixty ounces of the saline fluid, administered in separate portions, within the space of twelve hours. After lingering for ten days she died; the heart was found in a state of atrophy, covered with strong evidence of the existence of ancient disease, and floating in eight ounces of pus. In another case every internal organ was diseased; some of them so much so, that it was astonishing the individual lived so long.'

The fact that rehydration had to commence *before* the disease had progressed to a late (and irreversible) stage was also well recognized by Latta:

'The third cause of the occasional want of success, is the late application of the remedy. Hitherto I have had opportunity of injecting only in extreme cases, after every other means had entirely failed, cases which apparently would have soon proved fatal. Here the obstacles to be overcome have been of no ordinary kind, notwithstanding the result of the practice is of the most encouraging nature, and the number of cases now convalescent or doing well highly gratifying. In every fatal case we have had an opportunity of examining, independent of organic disease, I have found a large quantity of fibrin in the cavities of the heart, especially on the right side, where it had extended from the auricle through the ventricle into the pulmonary artery. Such deposition must have formed a certain obstacle to recovery, and is, no doubt, from the interruption it gives to the pulmonary circulation, the cause of the heavings of the chest, and the inordinate action perceptible in the centre of circulation many hours before death. Now surely it is reasonable to suppose, that if this, the most simple of all remedies, were applied early, before the blood drained of its water has collected in the larger vessels, in fact before such fibrinous depositions have taken place in the cavities of the heart, is it not reasonable to suppose that such would be entirely prevented?'

Latta concluded his letter with the following remarks:

'But not only is early injection advisable on this account, not only is stagnation of the blood prevented by it, and the laborious breathing, and the praecordial oppression, the intense sickness, the burning thirst, the extreme depression of the vital powers, and the chances of aggravating chronic disease, or of producing new organic lesion, in a great measure avoided; but it is rational to suppose that the consecutive fever will be rendered much milder, and that this is the case, is supported by my own experience, even though the remedy has not been applied earlier, indeed the fact is very evident. In an ordinary attack of cholera, much fluid is lost; and if the individual is so fortunate as to get out of the stage of collapse, if consecutive fever of typhoid type comes on, the system, left to its own resources to replace the lost serum, must be but ill fitted for the task, for the debility is extreme, absorption goes on slowly, the fever will be much aggravated by the irritation of internal congestion; local inflammation will thereby be produced, and the chance of recovery will be but small. Much of this evil is to be mitigated or entirely avoided by injection into the veins, of which circumstance I can adduce living instances; and where the patient, who had been injected, has sunk under organic disease, the usual marks of congestion are not perceptible.

The apparatus I have used, is Read's patent syringe, having a small silver tube attached to the extremity of the flexible injecting tube. The syringe must be quite perfect, so as to avoid the risk of injecting air;

the saline fluid should never be injected oftener than *once* into the same orifice, and the vein should be treated with much delicacy to avoid phlebitis. The wound should be poulticed and carefully watched, if it does not heal by the first intention.

I am Sir,
Your most obedient servant,
THOMAS LATTA, M.D.'

This letter was followed by several case reports in which intravenous rehydration had yielded a successful result; the last, and in many respects most dramatic of all was written by Dr Robert Lewins MD FRCP (whose first successful result was also recorded in *The Lancet*, 1832) (Cosnett, 1989); this letter was also addressed to the Secretary of the Central Board of Health:

'Sir,–The urgency of my present private and public duties prevents me from communicating more to you to-night than the following brief particulars of a case that occurred at the Leith Cholera Hospital yesterday:–

A woman of about forty years of age was admitted on Sunday evening at seven o'clock. She was *pulseless* even at the axilla, *sightless*, *cold*, and *blue* over almost the whole body. *Respiration* was very slow and irregular – in a word, she was all but lifeless. It was feared she would be dead before the operation of injecting could be commenced. Between seven at night and two o'clock next morning, there were thrown in 284 ounces, upwards of twenty three pounds. The report of her situation at two o'clock on Monday morning, in the hospital book, is as follows:– "A change for the better that appears almost miraculous has taken place. The action of the heart is greatly improved; respiration not in the last laborious, but quicker than natural; pulse 120, small, but distinct. She can articulate distinctly; countenance natural; lips red; tongue moist and warm; she perspires freely; heat over the whole body natural."

A full report of this wonderful case shall be forwarded soon. I remain, Sir, &c.,

ROB. LEWINS

P.S. – In one the pulse had ceased at the wrist eight hours before the injection. Dr. Alison had seen the patient eight hours before the operation, and the pulse was then imperceptible.

Quality Street, Leith, May 29, 1932,
1 o'clock a.m.'

O'Shaughnessy was obviously delighted with the early results of intravenous rehydration, but remained cautious (Cosnett, 1989), emphasizing that:

> 'Although by the injection of water and salts ... we may restore the deficient fluids of the body, and bring back the blood to its normal state, ... we must still remember that the unknown *remote* cause, and other agents, ... still in operation, and require to be remedied before a perfect cure be performed.'

The Lancet of June 2, 1832, contained a leading article in which the history of intravenous injections was reviewed. The intravenous saline treatment of a cholera patient was described as 'more like the workings of a miraculous and supernatural agent...'. It seems there was no previous record of water and salts being given deliberately to restore constituents lacking in the blood, although there is a record of blood transfusion, and of the injection of 6 ounces of water into a Russian cholera patient, with death in 2 hours. Over the ensuing months there was a flurry of reports of cases treated with intravenous saline in Britain. Almost all reported dramatic, but often temporary, improvement, but relapse was frequent as the purging continued. Various modifications in composition of the intravenous fluid were tried, even 'milk, boiled once, skimmed and strained'. Some patients were reported to have rigors [probably caused by pyrogens – see below] or laboured respiration after the saline injections. Of the first 25 reported cases so treated, eight recovered. There was much severe criticism, often outside medical journals. This provoked Latta to defend his treatment against 'members of the medical profession guilty of scribbling on medical matters in the news-papers of the day'.

Despite these generally favourable reports, scepticism of the new technique persisted; thus Griesinger (1857) maintained that results 'have been so unfavourable thus far that it cannot at all be recommended...'. Rehydration by other routes was also attempted, e.g. inhalation of water vapours, or introduction of saline solution into the urinary bladder (Pollitzer, 1959)! Use of the subcutaneous route (Cantani, 1892) also received a good deal of attention.

Gaertner and Beck (1893), in extrapolating from results obtained in dog experiments, recommended 'over-salting' blood of cholera patients – via the intravenous route. Following up this report, Rosner (1895) used a 10% sodium chloride solution (300 ml); however, four out of seven patients died. Sir Leonard Rogers (1908, 1909) first drew attention to the use of *hypertonic* saline rather than the normal or hypotonic fluids which were introduced by the pioneers.

Although advocated by some of the earlier workers, the use of alkalis – by injection – was pioneered by Sellards (1910); this was later recommended by Rogers (1915, 1916).

The use of glucose solution(s) in intravenous rehydration was first recommended by Kausch (1911, 1916), and later by Whyte (1913, 1915), Gaertner (1915) and Strauss (1915).

3.4 REHYDRATION FLUIDS

3.4.1 CONTAMINATION AND COMPOSITION OF EARLY FLUIDS

Many early intravenous infusion fluids were contaminated by pyrogens; work designed to determine their origin (and nature) was initiated by Hort and Penfold (1911) and followed up by Seibert (1923). Subsequently, various methods for the production of pyrogen-free solutions were published (Pollitzer, 1959); the first to use such solutions in cholera patients were Thomas and Ting (1938), who distilled surface (canal) water in a special apparatus and sterilized the resultant saline solutions within 3 hours of manufacture. A more refined method was later outlined by Napier (1946).

Table 3.1 summarizes the composition of several intravenous solutions used in the first half of the 20th century. Rogers (1921), Napier (1946) and Shattuck (1951) gave details of the preparation of an alkaline solution, sodium bicarbonate being unstable when subjected to heat.

3.4.2 ROUTE OF ADMINISTRATION

Clearly, since Latta's work (see above), the intravenous route was favoured by the majority of physicians. However, other routes were also explored. Oral fluids combined with alkalis were used by Rogers and Shorten (1915), but these writers emphasized that the technique did not combat alkalosis reliably.

Rogers (1921) also considered that rectal fluid was valuable in *mild* cholera attacks, i.e. when the blood pressure had not fallen below 70 mmHg, because:

> 'the large bowel retains its powers of absorption as long as there is a fair pulse, and the patient may often be tided over the danger of collapse by frequently repeated copious saline enemata.'

He recommended that ½–1 pint (270–540 ml) be given every 2 hours *per rectum* during the 'evacuation stage', but that after the urine output had become satisfactory in the 'stage of reaction', the same amount(s) should be given at 4-hourly intervals. The solutions had to be injected slowly and as high up as possible (with the aid of a long, soft tube), and the patients were urged to retain the enemata as long as they could. In the case of patients with a blood pressure only marginally above 70 mmHg – especially in the case of children or elderly people –

Table 3.1 Composition of several solutions used historically for intravenous rehydration in cholera (Pollitzer, 1959)

Rogers (1921, 1952)		
(i)	Hypertonic saline	
	Sodium chloride	8 g
	Calcium chloride	0.25 g
	Water	568 ml
(ii)	Alkaline saline	
	Sodium bicarbonate	10.7 g
	Sodium chloride	6.0 g
	Water	568 ml
Napier (1946)		
(i)	Hypertonic saline	
	Sodium chloride	16 g
	Pyrogen-free distilled water	1 l
(ii)	Alkaline saline	
	Sodium chloride	9 g
	Sodium bicarbonate	20.5 g
	Pyrogen-free distilled water	1 l
(iii)	Alkaline hypotonic saline	
	Sodium chloride	6.8 g
	Sodium bicarbonate	20.5 g
	Pyrogen-free distilled water	1 l
(iv)	Bicarbonate	
	Sodium bicarbonate	50 g
	Pyrogen-free distilled water	1 l
Shattuck (1951)		
(i)	Isotonic saline	
	Sodium chloride	9 g (8.5 g usually recommended)
	Water	1 l
(ii)	Hypertonic saline	
	Sodium chloride	14 g
	Freshly distilled water	1.1 l
	(Reduce volume by boiling to 1 l)	
(iii)	Alkaline saline	
	Sodium bicarbonate	18 g
	Sodium chloride	6 g
	Freshly distilled water	1.1 l
	(Before addition of sodium bicarbonate, reduce volume to 1 l by boiling)	
(iv)	Dextrose (glucose) saline	
	50 g of the sugar are dissolved in 1 l of isotonic or hypertonic saline. Desirable but not essential to add 2 mg of thiamine chloride	

the rectal injections were best administered by a continuous method, the rate of flow being regulated at between ½ and 1 ounce (15–30 g) per minute. Isotonic saline had to be used – to which, as long as the urine was acid, '100 grains of sodium bicarbonate per pint' (ca. 150 g in 540 ml) were added. The addition of calcium chloride 4 grains per pint (ca. 16 g in 540 ml) was also considered advantageous. These patients had to be closely monitored so that, if necessary, prompt additional use could be made of intravenous saline administration.

Rectal fluid administration was again recommended for the treatment of cholera by Rogers (1952), and also by Strong (1944) and Napier (1946). Strong stressed that intravenous infusions should be used only when strictly indicated and that, even when they became necessary, they should be combined with 'intrarectal' fluid administration.

Although he did not favour the use of the peritoneal route, Rogers (1921) was aware that as long as the radial pulse could be felt, fluid injected into the peritoneal cavity is rapidly absorbed; he concluded however, that mortality was higher when this route was used. Later, Napier (1946) considered that this route had a use in children (and adults) in whom a suitable vein (for infusion) could not be found. Napier felt that subcutaneous (see below), intramuscular, or bone marrow (sternum or tibia) routes could all be utilized when intravenous infusion was considered impossible. Subcutaneous administration of fluid(s) was certainly favoured by several investigators, although the technique had its critics, of whom Rogers was one. An answer to this dilemma had been provided in 1908 during an epidemic at Manila (Philippines) when comparative results of the subcutaneous and intravenous were published by Nichols and Andrews (1909) in 450 cholera patients (Table 3.2).

The optimal temperature for fluid used in intravenous infusion was also the subject of considerable debate. While Nichols and Andrews (1909) heated 'flasks' to 43°C, Rogers (1921) considered that the temperature should be adjusted to that of the patients' rectal temperature; in fact, administration of fluid at body temperature was he felt

Table 3.2 Comparison of the subcutaneous and intravenous routes for rehydration. These regimens were used in a cholera epidemic at Manila, Philippines in 1908 (Nichols and Andrews, 1909)

Method of treatment	No. of patients	No of deaths	Mortality rate (%)
'Stimulants'	145	47	32.4*
Subcutaneous rehydration	175	117	66.8
Subcutaneous and intravenous rehydration	36	19	52.7
Intravenous rehydration	94	41	43.6

* Mortality in the group treated wtih 'stimulants' (e.g. strychnine and digitalin) was lowest because it included only mildly affected patients. Isotonic saline solution was used for the hydration in the three other groups.

ideal, but when the rectal temperature was reduced (<36°C) the solution(s) should be warmed to 39–40°C, and alternatively, when elevated it should be comparatively cool.

The ideal volume of intravenous fluid to be infused also received a great deal of attention (Pollitzer, 1959). While Nichols and Andrews (1909) based requirement(s) on the clinical state of the patient (0.1 l was adequate if it restored a 'strong pulse', although 2–3 l were sometimes necessary), Rogers (1921) considered that the specific gravity of the blood was the essential guide to the required volume (Pollitzer, 1959). Other means of measuring haemoconcentration were also applied; thus, onset of urine production (in the catheterized patient) was used with success, as was restoration of a detectable blood pressure (Pollitzer, 1959). However, the potential dangers of over-hydration with saline – especially in patients with compromised cardiac function – were stressed (Napier, 1946; Shattuck, 1951). The serious complication, pulmonary oedema, resulting from over-enthusiastic management (see below), was stressed by Chakravarta and Chaudhuri (1954).

3.4.3 COMPOSITION – LATER DEVELOPMENTS

The importance of an alkaline solution (to counteract acidosis) was first established by Sellards (1910); a comparison of early treatment using this approach with a sodium chloride infusion demonstrated not only a lower mortality rate, but also the virtual elimination of uraemic deaths. Rogers (1921), among others, subsequently confirmed this work, and alkaline solutions became widely established in rehydration therapy.

Concentration of the saline solution used for intravenous rehydration had for long been the subject of debate; Rogers (1921) was again at the forefront of research and advocated a hypertonic solution, later combined with potassium permanganate and alkalis. Table 3.3 summarizes results from his Calcutta experience, in which this strategy proved superior. Numerous subsequent workers, including Napier (1951), have confirmed the advantage rendered by hypertonic solutions (Pollitzer, 1959).

Various investigators recorded their experience(s) using the addition of glucose to the infusion fluid (Pollitzer, 1959; Nalin and Morris, 1991). Whyte (1913, 1915) was the first to use this technique; Napier (1946) also advocated the addition of glucose. A note of caution was issued, however, by Shattuck in 1951: '...no more than 50 g of glucose should be given in 1 h or 400 g in 24 h. It is desirable, but not essential, to add thiamin chloride 1 mg for every 25 g of glucose'.

Although oral rehydration therapy (ORT) has recently come to the fore in management of cholera (see below), rapid intravenous rehydration remains to this day, life-saving in severe disease (Glass *et al.*, 1991;

Table 3.3 Statistics gathered at Calcutta by Rogers (1921) emphasizing the superior value of hypertonic saline (Pollitzer, 1959). Later results demonstrated the value of adding potassium (permanganate) and alkalis

Date	Method of treatment	Total number		Percentage	
		Patients	Deaths	Recoveries	Deaths
1895–1905	Saline intrarectally and subcutaneously	1243	783	37.0	63.0
1906	Normal saline, intravenously	112	57	49.1	50.9
1907	Saline intrarectally and subcutaneously	158	94	40.5	59.5
1908–1909	Hypertonic saline, intravenously	294	96	67.4	32.6
1910–1914	Hypertonic saline + potassium permanganate	858	222	74.1	25.9
1915–1919	Hypertonic saline + potassium permanganate + alkalis	1429	298	79.2	20.8

Nalin and Morris, 1991; Mahalanabis *et al.*, 1992; van Loon, 1993); fluid composition should be based on an underlying concept that it directly replaces the fluid/electrolyte loss(es) in cholera stool: sodium (100–140 mmol/l), bicarbonate (30–50 mmol/l), and potassium (15–30 mmol/l). Table 3.4 compares the composition of cholera stool with that of six intravenous fluids used in management; Ringer's lactate, DTS, and 'Dhaka solution' correspond most closely to this ideal.

Use of intravenous rehydration should be confined to patients with severe disease, associated with signs of hypovolaemia; complete rehydration should be accomplished in 2–4 hours, after which time ORT (see below) is usually adequate. It should not be used in mild cases. Some 10–15% of (severe) hospitalized patients require an additional short course(s) of intravenous rehydration fluid.

3.4.4 WHY DID INTRAVENOUS REHYDRATION TAKE SO LONG TO BECOME ESTABLISHED?

Two major pandemics swept through the cities of England in 1849 and 1854 (Chapter 2); however, the use of intravenous saline infusion(s) was simply not accepted by the medical establishment, despite the fact that

Table 3.4 Comparison of faecal composition in cholera, with that of fluids used for intravenous rehydration (Mahalanabis *et al.*, 1992)

	mmol/litre				
	Na^+	K^+	Cl^-	HCO_3^-	Glucose
Cholera stool					
Adults	130	15	100	45	–
Children	105	25	90	30	–
Intravenous fluids					
Ringer's lactate (Hartmann's fluid)*	131	4	111	29	–
Diarrhoea treatment solution (DTS)	118	13	83	48	50
'Dhaka solution'	113	13	98	48	–
2 Saline: 1 lactate	158	–	103	56	–
Normal saline	154	–	154	–	–
Half-Darrow's solution with 2.5% glucose[+]	61	17.5	52	26	150

* Commercially available.
[+] Full-strength Darrow's solution contains excessive potassium; concentration of sodium in half-strength Darrow's is too low for correction of dehydration in cholera.

there were many reports of its successful use (see above). The reasons for this have been summarized (Cosnett, 1989):

1. The technique had only been applied to patients who were considered moribund; therefore, while its proponents were satisfied that lives were being saved, others (the public included) considered that deaths were in fact being hastened by this form of management. It was also felt by many, that rehydration provoked further purging.
2. The technique was usually not repeated often enough to maintain fluid balance.
3. The fluid used was invariably unsterile, chemically impure, and (very) hypotonic. Therefore, the more fluid administered, the greater the likelihood of bacteraemia, pyrogen reaction(s) and haemolysis.

Perhaps the principles underlying the technique were so far ahead of contemporary knowledge in physiological chemistry and microbiology, that it was inevitably doomed to failure! Also, when these two epidemics struck Britain, the major protagonists were no longer on the scene. Latta had already died (in 1833), and in the same year, O'Shaughnessy had joined the East India Company and sailed for India, where he immediately became fascinated by developments in chemistry, electricity and telegraphy (Cosnett, 1989); O'Shaughnessy received a knighthood in 1856 as a reward for establishing a telegraphic service between the major cities of India (which was considered to have influenced the outcome of the mutiny). His obituary notices failed to mention his pioneering achievement(s) in the management of cholera which were, of course, catalytic to the introduction of intravenous rehydration by Latta and Lewins (see above).

3.5 ORAL REHYDRATION THERAPY (ORT)

3.5.1 GLUCOSE-FACILITATED INTESTINAL ABSORPTION OF SALT(S) AND WATER

Oral rehydration fluids (a form of management used by Hippocrates) have been used for many centuries; solutions have ranged from apple juice to yoghurt. However, an understanding of the physiological principles underlying ORT is a modern development (Mahalanabis *et al.*, 1992). Physiologists have studied the relationship(s) between glucose and sodium absorption from the small-intestinal lumen for many years; for example, in 1902, the Scottish physiologist Reid (1902) used mammalian small-intestinal loops to demonstrate enhanced sodium absorption in the presence of glucose. These findings were later confirmed by Barany and Sperber (1939) and Fisher and Parsons (1953) using rabbit small-intestine. Linkage of the sodium ion to glucose absorption was first demonstrated by Riklis and Quastel in 1958, and the mechanism underlying it was described by Schultz and Zalusky (1964). At that time, Schedl and Clifton, by using a small-intestinal intubation technique in human volunteers, demonstrated a significant increase in sodium chloride and water absorption from Ringer's solution (see above) in both jejunum and ileum when glucose (1 g per 100 ml) was added to the perfusing fluid (Schedl and Clifton, 1963). Subsequent *in vivo* studies delineated quantitative relationships between glucose, and enhanced sodium and water absorption (Malawar *et al.*, 1965; Levison and Schedl, 1966; Fordtran *et al.*, 1968).

The potential importance of the use of ORT in relation to cholera patients originated when Phillips – working on adult patients suffering from severe cholera in Bangladesh – confirmed that intestinal mucosa remained intact in this infection, and furthermore that glucose-facilitated absorption of sodium and water remained unimpaired (Phillips, 1964; Phillips *et al.*, 1965). Phillips however, remained sceptical throughout his life, regarding the potential value of the technique in management of severe cholera because several deaths occurred during his early trials with ORT (Van Heyningen and Seal, 1983). Hirschhorn *et al.* (1968) (also working at Dhaka), and Pierce *et al.* (1968) (working at

Calcutta), demonstrated that when an isotonic solution containing approximately equal concentrations of glucose and sodium was given orally, significant intestinal absorption resulted, and the necessity for intravenous fluid(s) in rehydration was significantly lessened.

Subsequent work at Bangladesh, in refugee camps in India, and at Calcutta (involving large numbers of patients suffering from both cholera, and non-cholera diarrhoea), confirmed the efficacy of ORT (Mahalanabis *et al.*, 1992). An extremely valuable addition to intravenous rehydration had therefore been achieved in all but severe cholera sufferers (Casteel and Fiedorek, 1990; Sircar *et al.*, 1990; Carpenter, 1992). In practice, however, the composition of the oral rehydration solution (ORS) must be precise, and under field conditions this is often extremely difficult to achieve (see below). Clinical application and success did not therefore immediately match up with early theoretical enthusiasm for the new technique!

3.5.2 COMPOSITION OF ORAL REHYDRATION SOLUTIONS

Early trials of ORT in cholera utilized a solution containing: sodium 90–120 mmol/l; potassium 10–25 mmol/l; sodium bicarbonate 30–48 mmol/l; glucose 110–120 mmol/l (Nalin and Morris, 1991; Mahalanabis *et al.*, 1992). This formula was adapted by WHO in 1971 to contain: sodium 90 mmol/l; potassium 20 mmol/l; chloride 80 mmol/l; bicarbonate 30 mmol/l; glucose 111 mmol/l. Table 3.5 summarizes the composition of this fluid, which has now been modified with the addition of citrate (see below). In the initial formulation, bicarbonate was utilized as the base, in order to correct the metabolic acidosis resulting from cholera. However, ORT sachets containing this electrolyte possess a relatively short 'shelf-life' in a hot/humid climate (a few months only) (Nalin and Morris, 1991). Therefore, bicarbonate has more recently been replaced with citrate (as base) (Table 3.5). Bicarbonate and potassium, in addition to sodium and water, are lost in the isosmotic choleric stool. After 'coupling' (at the surface of the enterocyte), sodium and glucose are absorbed on an equimolar basis; therefore, it is essential that sodium and glucose concentrations are similar. It is also important that the solution is not excessively hypertonic; an excessive osmotic load both reduces absorption and exacerbates diarrhoea. ORT has been extensively tested under both controlled hospital and field conditions (Sircar *et al.*, 1990); it is furthermore of value in all acute diarrhoeal infections (not merely cholera) (Casteel and Fiedorek, 1990), and in all age-groups. Optimal formulation of ORS for use both in cholera and non-choleric diarrhoea, remains the subject of debate; a recent report indicates that a relatively hypotonic solution might be preferable (Hunt *et al.*, 1994).

Table 3.5 Composition of WHO-recommended oral rehydration solutions (ORS); first used in 1971 (Mahalanabis *et al.*, 1992)

Oral rehydration constituent	Weight (g) /l ORS
Sodium chloride	3.5
Trisodium citrate, dihydrate	2.9
or	
Sodium hydrogen carbonate (sodium bicarbonate)	2.5
Potassium chloride	1.5
Glucose, anhydrous*+	20.0

Component	Bicarbonate-containing solution (mmol /l ORS)	Citrate-containing solution (mmol /l ORS)
Sodium	90	90
Potassium	20	20
Chloride	80	80
Citrate	–	10
Bicarbonate	30	–
Glucose	111	111

* Or glucose, monohydrate, 22.0 g; or sucrose, 40.0 g.
+ 50 g rice powder can replace 20 g glucose. To prepare a rice powder solution: place 50 g rice powder (i.e. rice flour) in 1100 ml of water and bring it to the boil. Continue boiling for approximately 7 min, when the mixture becomes opalescent. Allow it to cool; then add and mix the three salts. Administer solution warm. After 8 h discard and prepare a fresh batch.

Clinical trials were subsequently undertaken in order to compare the citrate-containing with the original formulation (Islam *et al.*, 1984; Hoffman *et al.*, 1985; Salazar-Lindo *et al.*, 1986). Acidosis was corrected at a similar rate, and furthermore, in one study – in adults suffering from cholera – faecal output was significantly decreased in the group receiving citrate; this observation has not, however, been confirmed in children. All centrally produced and distributed ORT products produced by UNICEF, WHO and other major agencies are now citrate-based.

A further modification which has received a great deal of attention is the substitution of sucrose (readily available in most homes in a developing country) for the glucose component, the latter often being difficult to obtain in developing countries. The major theoretical objection was that sucrase (required to hydrolyse sucrose to its constituent monosaccharides) concentration is significantly depressed in secondary disaccharidase deficiency. Although this frequently accompanies small-intestinal mucosal damage in many other acute small-intestinal infections, it is an insignificant problem in cholera. Also, the possibility that fructose (which is absorbed far less rapidly than glucose) might exacerbate diarrhoea caused consternation. Despite these theoretical limitations, clinical studies have clearly shown that sucrose is virtually (although not entirely) as satisfactory as glucose (Palmer *et al.*, 1977; Sack *et al.*, 1980). When incorporated, sucrose should be present at 40 g/l in order to obtain a glucose yield (after hydrolysis) of 20 g/l.

Cereals (Khin-Maung and Greenough, 1991) – especially rice (Gore *et al.*, 1992) – have also been utilized as substitutes for glucose, owing to their local availability. Amino acids (Khin-Maung, 1993) and short-chain peptides (Khin-Maung *et al.*, 1991) [which result from hydrolytic (digestive) processes] can themselves potentiate sodium and water absorption. Early evidence (Mahalanabis *et al.*, 1992) suggested that rice was a suitable substitute, and a subsequent controlled clinical study in children suffering from cholera and non-choleric diarrhoea indicated that rice-powder-containing ORS (50 g pre-cooked rice powder per litre) is effective in reducing both volume and duration of diarrhoea (Patra *et al.*, 1982); this finding has been confirmed in adults and older children, using 80 g rice-powder/l (Molla *et al.*, 1985). A combination of cooked rice-powder (50 g/l packet) and glycine (111 mmol/l) in ORS failed, however, to potentiate the improvement in efficacy (Patra *et al.*, 1986). Wheat-based ORS [115 g protein, 870 g carbohydrate (60% maltose, 15% maltotriose, and 10% maltodextrins; Khin-Maung *et al.*, 1991) per l] – following thorough hydrolysis – is also effective in reducing diarrhoea in childhood cholera (Alam *et al.*, 1987). The major drawback with cereal-based ORS is that thorough cooking is essential, and the 'shelf-life' of the final product is of necessity short (6–12 h), especially in a hot/humid climate.

3.5.3 IMPROVED (SUPER) ORS FORMULATIONS

Although currently used ORS preparations are extremely successful in achieving rehydration, they do not (with the possible exception of the cereal-based formulations), reduce faecal volume, frequency or duration; therefore, the practical result is far from dramatic in the eyes of the bystander (Mahalanabis *et al.*, 1992)! The following modifications have been suggested in order to create a 'super ORS': (i) organic nutrient-linked absorption; (ii) use of glucose-polymers rather than glucose; (iii) weak organic acid-linked absorption; and (iv) colonic salvage of salt(s) and water by short-chain fatty acids (Mahalanabis *et al.*, 1992). ORS containing an amino acid in addition to glucose, has undergone clinical trials; initial studies indicated that it was associated with substantial reduction both in faecal volume and duration of diarrhoea (Mahalanabis *et al.*, 1992). A double-blind, controlled study using glucose (111 mmol/l) + glycine (111 g/l)-based ORS in children with cholera and non-cholera diarrhoea demonstrated a reduction in faecal volume, duration of diarrhoea, and thence volume of ORS required for rehydration, when compared with standard ORS (Patra *et al.*, 1984). Also, a trial utilizing L-alanine in adults and older children suffering from cholera demonstrated a reduction in faecal output and duration of diarrhoea compared with standard ORS (Patra *et al.*, 1989). If these additional substrates are to be added to the standard ORT sachet, satisfactory formulation will be dependent on a clinically significant improvement in efficacy without an appreciable increase in cost (Mahalanabis *et al.*, 1992); at present, cereal-based preparations (Khin-Maung and Greenough, 1991; Gore *et al.*, 1992) come closest to meeting these requirements.

3.5.4 'INCOMPLETE' ORS

A 'sugar–salt' rehydration solution, prepared from ingredients found in most homes, lacks potassium and added base; there is therefore an increased associated risk of persisting acidosis and hypokalaemia (Islam *et al.*, 1980). While such a solution is effective in non-cholera diarrhoea, it can only result in a temporary improvement in severe 'toxigenic' diarrhoea resulting from cholera; it should therefore only be used until an appropriate 'complete' ORS (see above) can be administered. When used, an 'incomplete' solution should be supplemented with a diet rich in potassium-containing foodstuffs, in order to compensate for associated potassium loss.

3.6 THE ROLE OF ANTIBIOTICS IN CHOLERA

The raison d'etre for antibiotic usage is to decrease the volume and duration of diarrhoea (Bennish, 1994); these agents are not in themselves life-saving! By virtue of this effect(s) they also increase efficacy and reduce the cost of ORT by reducing the length of administration. By reducing the period of treatment therefore, more patients can be resuscitated, assuming that supplies of ORS are readily available! Antibiotics should be given orally, after completing the initial rehydration exercise, i.e. approximately 4–6 hours after starting therapy in a severely dehydrated patient.

Table 3.6 summarizes various antibiotic regimens which have been used. Tetracycline is the agent of choice; its introduction into the management strategy reduces:

(a) the duration of diarrhoea by approximately 50%, to an average of around 2 days;

Table 3.6 Antibiotic and antimicrobial regimens which have been used in *Vibrio cholerae* serogroup O1* infection(s) (Bennish, 1994)

Drug	Dose	
	Adult	Paediatric
Tetracycline	500 mg four times daily – 3 days, or 1 g single dose	50 mg/kg body weight/day; four divided doses for 3 days
Doxycycline	300 mg – single dose	(see text)
Furazolidone	100 mg four times daily – 3 days	5 mg/kg body weight/day; four divided doses for 3 days, or 7 mg/kg of body weight single dose
Trimethoprim–sulfamethoxazole	320 mg trimethoprim–1600 mg sulfamethoxazole twice daily – 3 days	8 mg of trimethoprim–40 mg of sulfamethoxazole/kg body weight/day; two divided doses for 3 days
Norfloxacin+	400 mg twice daily – 3 days	Not recommended for children

* When possible, therapy should be based on a knowledge of the susceptibility of isolates of *V cholerae* O1 in the community. *V cholerae* O1 isolates resistant to all listed antimicrobial agents (with the exception or norfloxacin) have been identified.

+ Other second-generation quinolones are likely to be as effective as norfloxacin, but have not been adequately evaluated.

(b) the volume of diarrhoea by approximately 60% after initiation of treatment; and

(c) the duration of *V. cholerae* excretion to an average of 24 (maximum 48) hours.

Single-dose doxycycline (300 mg) has recently been shown to be as effective as the accepted tetracycline regimen(s) in reduction of faecal output, duration of diarrhoea, vomiting and ORS requirement(s) (Alam *et al.*, 1990). Furazolidone [also administered as a single dose in both adults (Rabbani *et al.*, 1989a) and children (Rabbani *et al.*, 1991)] is also effective in the management of cholera. Norfloxacin has proved superior to trimethoprim–sulfamethoxazole (TMP–SMX) (which proved to be no better than a placebo) in reducing faecal output, duration of diarrhoea, fluid requirement(s) and vibrio excretion (Bhattacharya *et al.*, 1990). In addition to the antibiotics listed in Table 3.6, erythromycin and chloramphenicol are also effective; the former agent was shown to be of value in a locality where TMP–SMX-resistant *V. cholerae* was emerging (Burans *et al.*, 1989).

The role of antibiotic prophylaxis remains controversial. When administered to family contacts, both tetracycline and doxycycline prevent the occurrence of secondary cases (Gupta *et al.*, 1978). However, in developing countries, it is logistically impossible to provide a blanket cover to whole communities; therefore, high-risk individuals only should be selected for antibiotic chemoprophylaxis (Mahalanabis *et al.*, 1992). Family members are certainly at increased risk in an epidemic; however, they represent a very small fraction of the population who will ultimately become infected. Risk of cholera in an endemic area also extends over a prolonged period; therefore, in this situation also, antibiotic chemoprophylaxis is impracticable. Also, widespread use of antibiotic chemoprophylaxis results in the emergence of 'resistant' strains of *V. cholerae* (see below). Overall, therefore, there are several good reasons why antibiotic chemoprophylaxis is not widely recommended.

3.6.1 ANTIBIOTIC RESISTANCE

V. cholerae resistance to tetracycline and other antibiotics has been widely reported in East Africa (Mhalu *et al.*, 1979) and Bangladesh (Glass *et al.*, 1980); in these two locations, resistance patterns were different, but in both of them mediation was via conjugative plasmids. Resistance has also been recorded at other geographical locations (Tabtieng *et al.*, 1989; Rossi *et al.*, 1993; Threlfall *et al.*, 1993). Resistant strains invariably disappear with time, however, and are replaced with sensitive ones. When antibiotics are to be administered, it is always essential to determine the local pattern(s) of resistance – at that particular time – in order to select the most appropriate agent (see above; Table 3.6). Multi-resistance in *V. cholerae* is common. In practice, suspicion of resistance is usually heralded by observations indicating an impaired clinical response.

3.7 ANTIDIARRHOEAL/ANTISECRETORY AGENTS

In practice, no antisecretory compound has proved efficacious in the clinical management of cholera (Mahalanabis *et al.*, 1992); this is disappointing, because any agent which impairs small-intestinal secretory function would be expected *a priori* to exert a beneficial effect(s). Furthermore, kaolin, pectin, activated charcoal and bismuth subsalicylate are also ineffective; charcoal has in fact been shown to interfere with the beneficial effect(s) of tetracycline (see above) (Sack *et al.*, 1970). Opiates and opiate-like compounds have been clearly shown to be useless, and may actually render harm, especially in children (Mahalanabis *et al.*, 1992). Other agents which have been tested include:

nicotinic acid, aspirin, indomethacin, chloroquine and somatostatin. The first of these had a possible, albeit marginal, effect (Rabbani *et al.*, 1989b) but the others have not been shown to impart a detectable benefit.

3.8 NUTRIENT INTAKE DURING *V. CHOLERAE* INFECTION

Food does not interfere with recovery from *V. cholerae* infection (it can even be absorbed during the acute phase of disease), and may also provide a useful substrate for ORT, i.e. water and electrolyte absorption. Pancreatic trypsin and amylase activities are well preserved in adults suffering from cholera (Molla *et al.*, 1986). Many studies have addressed the issue of early feeding (Alam *et al.*, 1992). Mean calorie intake in children during the acute phase has been estimated at 75 cal/kg/24 h; this increased to 111 cal/kg/24h 2 weeks later (Molla *et al.*, 1983). In young children suffering from the infection (including those with severe disease), a nitrogen-balance study during the acute phase demonstrated that positive result could be achieved promptly with liberal milk-feeding, even before control of the diarrhoea (Mahalanabis, 1981). In a controlled study carried out in Myanmar, rice meals (together with appropriate ORT) were given to a group of children with severe cholera (Benyajati *et al.*, 1960); when compared with controls they showed a significant increase in faecal output, but demonstrated superior weight-gain. Therefore, early and liberal feeding is entirely compatible with a rapid and uneventful recovery (Mahalanabis *et al.*, 1992).

3.9 OVERALL MANAGEMENT OF THE CHOLERA PATIENT

It is essential to ascertain at the outset, the level of dehydration and degree of fluid deficit (Table 3.7) (Nalin and Morris, 1991; Mahalanabis *et al.*, 1992; Greenough, 1995); under field conditions this is entirely dependent upon clinical observation and assessment. However, when available, estimation of haematocrit, plasma specific gravity and/or serum protein concentration are valuable adjuncts. When acidosis is profound, Kussmaul's reaction is present (Nalin and Morris, 1991; Mahalanabis *et al.*, 1992). Fluid replenishment is a matter of the greatest possible urgency; a severely dehydrated patient may be literally only minutes from death – unless hydration is initiated immediately (Greenough, 1995). When treatment is commenced promptly, fatality rate should not exceed 1%; most deaths occur in very young or aged individuals, often those with a major complicating disease (Nalin and Morris, 1991; Siddique *et al.*, 1992).

At least 5% of body weight has been lost before signs of dehydration

Table 3.7 Clinical assessment of dehydration and fluid deficit (Bennish, 1994)

Clinical index	Sign/symptom/laboratory finding in:		
	Mild/no dehydration	Moderate dehydration	Severe dehydration
Cerebral status	Alert	Restless/lethargic	Apprehensive/lethargic/stuporous/comatose
Thirst	Present	Present	Marked
Radial pulse rate and character	Normal	Rapid	Rapid and feeble/impalpable
Respiratory rate and character	Normal	Tachypnoeic	Tachypnoeic, deep, laboured
Elasticity of subcutaneous tissues (determined by skin pinch)	Immediate retraction	Slow retraction	Very slow retraction
Eyes	Normal	Sunken	Dramatically sunken
Urine flow	Normal	Scant/dark	Scant/absent
Serum specific gravity*	<1.027	1.028–1.034	>1.034
Approximate fluid deficit (mg/kg body weight)	$\leqslant 50$	$\geqslant 51$–90	>90
Preferred method of rehydration	ORT	ORT or IV, depending on presence of vomiting and rate of continued faecal loss	IV

* Patients with malnutrition often have a lower baseline specific gravity; therefore, the listed values may not be applicable.
ORT, oral rehydration therapy; IV, intravenous.

become clinically apparent (Table 3.7); those with moderate dehydration have a deficit of about 7.5%, whereas those severely affected have a reduction of the order of $\leqslant$10%. As a vital baseline observation, therefore, admission weight should always be obtained, and carefully recorded.

When dehydration is either mild or moderate, ORT (Table 3.5) is invariably effective (Nalin and Morris, 1991; Mahalanabis *et al.*, 1992); however, those severely affected require urgent (intravenous) (Table 3.4) management. A severely dehydrated adult weighing 40 kg requires 4 l of rehydration fluid to correct his/her state of hydration (i.e. 10% of 40 kg = 4 kg = 4 l), whereas a mildly dehydrated child needs 500 ml rehydration fluid for initial resuscitation (Mahalanabis *et al.*, 1992). In a severely dehydrated patient, 50% of the estimated fluid requirement should be given within the first hour – initially as rapidly as possible (until the radial pulse is palpable). Such a patient should be fully rehydrated in 2–4 hours, at which time the intravenous line can be removed. It may be necessary to commence an intravenous line at several sites as an initial life-saving approach. ORT can be commenced immediately the oral route can be tolerated; this should be continued (in order to replace on-going faecal losses) until the diarrhoea ceases.

Vomiting may be a significant feature during the early stage(s) of management, thus rendering ORT impossible. However, persistence with frequent small volumes usually succeeds. In this exercise, nurses should provide supervision, encouragement, and guidance to any family members present, and assist generally in the management strategy.

Adequacy of fluid replacement can be readily assessed clinically (Nalin and Morris, 1991; Mahalanabis *et al.*, 1992 Greenough; 1995):

(a) return of the rapid pulse to normal strength and rate; in adults < 90 min;
(b) return of skin turgor to normal; if this normalizes while a rapid pulse rate continues, other causes of shock, e.g. sepsis and cardiac infarction, should be sought;
(c) the patient feels comfortable and cyanosis, muscular cramps, nausea and vomiting have resolved; children who are initially drowsy or stuporous may not become fully alert for 12–18 hours, despite adequate hydration;
(d) return of normal 'fullness' of the jugular vein(s);
(e) satisfactory weight gain; a severely dehydrated patient should gain 8–10% body weight following rehydration; and
(f) return of urinary output; in a severe case, this usually occurs within 12–20 hours after initiating rehydration.

3.9.1 MAINTENANCE THERAPY

After the initial fluid and electrolyte deficits have been corrected, all on-going loss(es) must be replaced. In a severely affected patient, ORT should replace intravenous therapy immediately the oral route can be used. The basic underlying principle is to provide ORT at a rate adequate to maintain fluid lost in diarrhoeic stool and vomit, and in addition to provide hypotonic fluid(s) for normal daily fluid/electrolyte requirement(s). In older children and adults, normally consumed fluids can usually be allowed as desired in addition to ORS; in infants and young children, breast-feeding should be continued, and in non-breast-fed babies other fluids offered at regular intervals.

In an unsophisticated scenario, the classical 'cholera cot' is valuable (Figure 3.1) (Greenough, 1995) and faecal loss(es) can be easily measured in a calibrated bucket placed beneath the central hole; this provides an extremely valuable guide to subsequent fluid requirement(s). Faecal volume should be measured every 4 hours, or thereabouts, and matched by fluid intake. Urine should be collected separately, and the patient must be instructed to keep faecal output and urine apart – a difficult task, especially in children. A simple input–output chart (at the bedside) is of great value in management.

Occasionally, intravenous rehydration must be reinstated during the 'maintenance period', but ORT should be continued and used whenever possible.

Antibiotics should be commenced within 3–6 hours of starting rehydration (see above) (Nalin and Morris, 1991; Mahalanabis *et al.*, 1992; Greenough 1995); their use conserves valuable intravenous fluids (see above) and in an epidemic, maximizes use of available hospital beds, with a substantial reduction in numbers of trained staff.

3.9.2 DIET DURING THE REHYDRATION PERIOD

Food appropriate to age should be offered/allowed during the maintenance period; the dietary item(s) depend entirely on the patient's choice/appetite. Nothing can be gained from withholding food, thereby 'resting' the bowel (Alam *et al.*, 1992; Mahalanabis *et al.*, 1992).

3.9.3 PROBLEMS AND COMPLICATIONS DURING REHYDRATION THERAPY

Most complications result either from a delay in initiating rehydration, or in the provision of appropriate fluid(s). Although such factors most commonly account for death (which should now be a rare event;

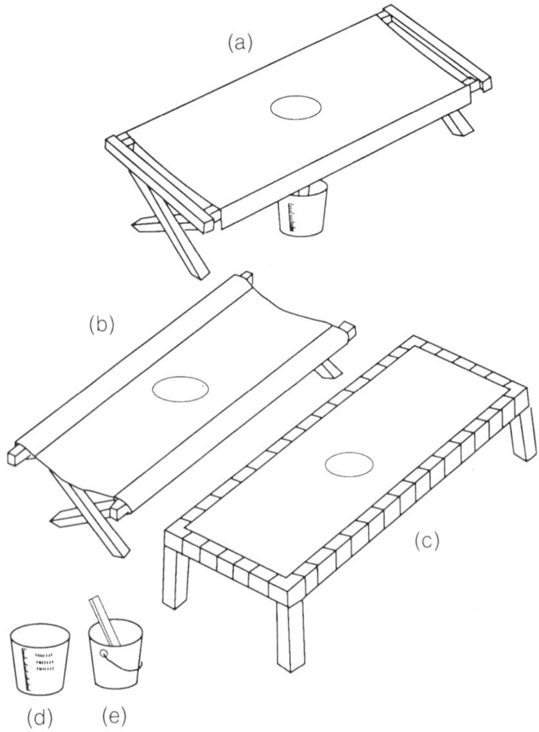

Fig. 3.1 Several simple cholera cots for patient comfort and ease in measuring faecal output during cholera (Greenough, 1995). (a) Camp cot with hole on which plastic sheet can be placed; the sleeve channels stool to calibrated bucket. (b) Camp cot with hole on which plastic sheet can be placed to channel stool to bucket. (c) Jute on wood frame bed. (d) Calibrated bucket for stool to be placed under cholera cots. (e) Bucket with calibrated stick (alternative method). The plastic sheet is desirable but not essential for effective use of these cots. The size of the cots can be adjusted for children or adults.

Siddique *et al.*, 1992), another underlying condition, e.g. malnutrition or pneumonia, may herald significant problems (Rabbani and Greenough, 1992). Complications directly or indirectly related to the rehydration regimen include (Mahalanabis *et al.*, 1992):

1. Persistence/recurrence of signs of dehydration and hypovolaemic shock, usually resulting from inadequate fluid replacement; other causes of 'shock' (see above) should be kept in mind! Occasionally, cholera patients accumulate watery secretions (which are not evacuated) within the intestinal lumen ('cholera sicca'); symptoms may mimic acute peritonitis; intravenous therapy should be continued and a rectal tube passed.

2. Persisting vomiting (a component of the 'cholera syndrome') may continue and give rise to acidosis and/or hypovolaemia. Whereas ORS administered in small volumes is usually feasible, intravenous therapy may be indicated. Gastric emptying is not compromised in acute cholera (Collins *et al.*, 1989).
3. Mildly elevated urea concentration is a common feature in the 'cholera syndrome'; this results from dehydration (pre-renal azotaemia) (Rabbani and Greenough, 1992). Occasionally, acute renal failure supervenes, usually following a period of prolonged hypotension; this results from inadequate fluid replacement consequent upon several underlying causes, e.g. shortage of intravenous infusion fluid, or a serious delay in initiating rehydration. Acute renal failure should be managed as for acute tubular necrosis; careful monitoring of fluid and electrolytes is of paramount importance, and dialysis occasionally indicated (Greenough, 1995). When the diarrhoeic phase has supervened, it is essential to match fluid intake with output; this can usually be accomplished with ORT, although the electrolyte content of ORS should be adjusted to balance urinary loss(es); 2% glucose added to the solution facilitates absorption.
4. When potassium loss is excessive, and the electrolyte concentration in rehydration fluid is present at an inadequate concentration, hypokalaemia may result. This may precipitate severe symptoms in childhood, e.g. abdominal distension, paralytic ileus (Rabbani and Greenough, 1992), urinary retention, muscular weakness, cardiac arrhythmias and even death.
5. Overhydration is an occasional sequel when excessive intravenous (very rarely ORS) fluid is administered. Swelling of the eyelids, distended jugular veins, and a slow 'full' pulse are present. Excessive administration of intravenous fluid can also precipitate cardiac failure; this does not, however, result from ORT.
6. Pulmonary oedema is a further complication of excessive use of intravenous fluid (Rabbani and Greenough, 1992; Greenough, 1995); it has been recorded especially when metabolic acidosis has been left uncorrected, i.e. an excessive usage of normal saline, rather than intravenous fluid incorporating base.
7. A mild fall in blood glucose concentration (a non-specific response in the correction of 'stress') is usual in acute cholera. Severe hypoglycaemia occasionally occurs (unrelated to malnutrition), nearly always in children (Rabbani and Greenough, 1992); this usually manifests itself with sudden onset of seizures. Ideally, reduced blood glucose concentration should be demonstrated directly, but in any case an intravenous bolus of glucose should be administered immediately (Greenough, 1995). If confirmed by

appropriate investigations, hypoglycaemia should be managed using continuous administration of intravenous glucose-containing fluid(s); subsequent blood glucose concentration(s) should be monitored (Greenough, 1995).

8. Severe dehydration in the third trimester of pregnancy may precipitate abortion and still-birth (Rabbani and Greenough, 1992); principles of management are however, similar to those in the non-pregnant woman.
9. Administration of incorrectly prepared fluid(s), or repeated use of intravenous administration sets can give rise to rigors and fever; pyrogen-free fluids and disposable sets should always be used.
10. During a cholera epidemic, there is a tendency to diagnose this disease in all severely ill patients; it is, however, important to appreciate that a substantial list of alternative differential diagnoses, e.g. meningitis, septic shock, myocardial infarction (Greenough, 1995), and *Salmonella typhi* infection, also exists.

3.9.4 MANAGEMENT OF CHOLERA IN A RURAL SETTING

Before the introduction of ORT, management of cholera was difficult or impossible in an unsophisticated setting; this was revolutionized by the 'new technique' and all but severely affected cases can be managed in a treatment centre rather than hospital (Quick *et al.*, 1993).

Cholera epidemics usually occur in remote areas, and hence often go unrecognized. An outbreak is often suspected when a cluster of adults dies as a result of watery diarrhoea (Rogers, 1952); infant deaths following diarrhoeal diseases are generally common in developing countries and do not raise suspicion of this infection. When an outbreak has been identified, relatively inexpensive management can now be implemented and numerous unnecessary deaths prevented. Table 3.8 summarizes essential items of apparatus which have been recommended for cholera management in a rural health centre; the most essential prerequisite lies in a liberal stock of ORS (Quick *et al.*, 1993). Use of cereal-based ORS in this setting is surrounded by substantial logistical problems (Nalin and Morris, 1991). A small organizing team should be set up to achieve optimal results, utilizing the available staff/apparatus, its functions are (Collins *et al.*, 1989; Mahalanabis *et al.*, 1992):

(a) rapid assessment of the nature and severity of the problem;
(b) to obtain necessary resources to establish a treatment centre, if appropriate. Any available building (e.g. school premises, a community centre, huts or houses) will suffice; if the climate is favourable, a building may not be strictly necessary;

Table 3.8 Suggested fluid, antibiotics, and apparatus for management of cholera under field conditions (Mahalanabis *et al.*, 1992)*

Rehydration supplies
240 l intravenous polyelectrolyte solution, e.g. Ringer's solution (480 bags of 0.5 l each) with giving sets
10 disposable syringes (5 or 10 ml)
10 needles (18 G) for adults
20 needles (21–23 G) for children
10 scalp vein sets
10 nasogastric tubes (adult size 16 Fr, 50 cm long; child size 8 Fr, 40 cm long)
1300 packets ORS (for 1 l each)
Antibiotics
Adults and older children: 3200 tetracycline capsules, 250 mg (16 capsules per case) or 600 doxycycline capsules, 100 mg (3 capsules per case) and furazolidone tablets (for pregnant women)
Younger children: doxycycline syrup or furazolidone syrup
Other supplies
Two large water dispensers with tap (marked at 5-, 8-, 10-l intervals) for making ORS solution in bulk
20 bottles (1 l) for ORS solution (e.g. IV bottles)
20 bottles (0.5 l)
40 tumblers 200 ml
3 kidney dishes
5 spirit lamps
Flashlights (2) (or hurricane lantern)
Cotton (5 kg)
Alcohol (250 ml)
Adhesive tape (3 reels)
Soap
Disinfectant
Bleaching powder
If possible
Cholera cots (adult and infant)
Plastic drain sheets
Plastic buckets
Patient scales

* Recommendations based on approximately 200 cases (a typical figure for an outbreak in a total population of 100 000) with the assumption that 40 cases require initial intravenous fluid followed by ORT, and that the remaining 160 cases are managed exclusively with ORT.
ORS, oral rehydration solution; ORT, oral rehydration therapy; IV, intravenous.

(c) to work with community leaders in order to assist in cholera control;

(d) to train health-care workers – especially in rehydration techniques (Collins *et al.*, 1989). Although not always practicable, all drinking water should ideally be boiled. [Widespread immunization and prophylactic antibiotic usage are not indicated; Nalin and Morris, 1991; Greenough, 1995];

(e) to ensure that the local population is aware of the available treatment facilities; and
(f) to continue assisting the community – by 'trouble shooting'.

Records of all recognized cholera outbreaks should be recorded centrally in order to assist future management teams, and help in overall surveillance.

3.10 REFERENCES

Alam, A.N., Sarkar, S.A., Molla, A.M. *et al.* (1987) Hydrolysed wheat based oral rehydration solution for acute diarrhea. *Archives of Disease in Childhood*, **62**, 440–4.

Alam, A.N., Alam, N.H., Ahmed, T. and Sack, D.A. (1990) Randomised double blind trial of single dose doxycycline for treating cholera in adults. *British Medical Journal*, **300**, 1619–21.

Alam, N.H., Ahmed, T., Khatun, M. and Molla, A.M. (1992) Effects of food with two oral rehydration therapies: a randomised controlled clinical trial. *Gut*, **33**, 560–2.

Anonymous. (1832) Royal College of Physicians. May 28th, 1832: treatment of cholera at Constantinople. *Lancet*, **ii**, 281–2.

Barany, E.H. and Sperber, E. (1939) Absorption of glucose against a concentration by the small intestine of the rabbit. *Scandinavian Archives of Physiology*, **81**, 290–7.

Bennish, M.L. (1994) Cholera: pathophysiology, clinical features, and treatment, in Vibrio cholerae *and Cholera: Molecular to Global Perspectives*, (eds I.K. Wachsmuth, P.A. Blake and Ø. Olsvik), American Society for Microbiology, Washington, DC, pp. 229–55.

Benyajati, C., Keoplug, M., Beisel, W.R. *et al.* (1960) Acute renal failure in Asiatic cholera: clinicopathologic correlations with acute tubular necrosis and hypokalaemic nephropathy. *Annals of Internal Medicine*, **52**, 960–75.

Bhattacharya, S.K., Bhattacharya, M.K., Dutta, P. *et al.* (1990) Double-blind, randomized, controlled clinical trial of norfloxacin for cholera. *Antimicrobial Agents and Chemotherapy*, **34**, 939–40.

Bonnici, W. (1993) The blue epidemic cholera – some aspects of treatment in the mid 19th century. *Journal of the Royal Army Medical Corps*, **139**, 76–8.

Burans, J.P., Podgore, J., Mansour, M.M. *et al.* (1989) Comparative trial of erythromycin and sulphatrimethoprim in the treatment of tetracycline-resistant *Vibrio cholerae* O1. *Transactions of the Royal Society of Tropical Medicine and Hygiene*, **83**, 836–8.

Cantani, A. (1892) Cholerabehandlung. *Berl Klinische Wochenschrift*, **29**, 913.

Carpenter, C.C. (1990) The erratic evolution of cholera therapy: from folklore to science. *Clinical Therapeutics*, **12** (Suppl. A), 22–7.

Carpenter, C.C.J. (1992) The treatment of cholera: clinical science at the bedside. *Journal of Infectious Diseases*, **166**, 2–14.

Casteel, R.B. and Fiedorek, S.C. (1990) Oral rehydration therapy. *Pediatric Clinics of North America*, **37**, 295–311.

Chakravat, H.S. and Chaudhuri, R.N. (1954) Plasma sodium, potassium and chloride changes in cholera and their significance in prognosis and treatment. *Journal of the Indian Medical Association*, **23**, 488.

Collins, B.J., Van-Loon, F.P.L. Molla, A., Molla, A.M. and Alam, N.H. (1989) Gastric emptying of oral rehydration solutions in acute cholera. *Journal of Tropical Medicine and Hygiene*, **92**, 290–4.

Cosnett, J.E. (1989) The origins of intravenous fluid therapy. *Lancet*, **i**, 768–71.

da-Cunha-Ferreira, R.M. and Cash, R.A. (1990) History of the development of oral rehydration therapy. *Clinical Therapeutics*, **12** (Suppl. A), 2–11, 11–13.

Fisher, R.B. and Parsons, D.S. (1953) Glucose movements across the wall of the rat small intestine. *Journal of Physiology (London)*, **119**, 210–23.

Fordtran, J.S., Rector, F.C. and Carter, N.W. (1968) The mechanism of sodium absorption in the human small intestine. *Journal of Clinical Investigation*, **47**, 884–900.

Gaertner, G. and Beck, A. (1893) Über den Einfluss der intravenösen Kochsalzeinspritzung auf die Resorption von Flüssigkeiten. *Wien Klinische Wochenschrift*, **6**, 563–6.

Gaertner, G. (1915) Bemerkungen zur Pathologie und Therapie der Cholera asiatica. *Wien Medische Wochenschrift*, **65**, 182–5.

Glass, R.I., Claeson, M., Blake, P.A., Waldman, R.J. and Pierce, N.F. (1991) Cholera in Africa: lessons on transmission and control for Latin America. *Lancet*, **338**, 791–5.

Glass, R.I., Huq, I., Alim, A.R.M.A. and Yunus M. (1980) Emergence of multiply antibiotic resistant *Vibrio cholerae* in Bangladesh. *Journal of Infectious Diseases*, **142**, 939–42.

Gore, S.M., Fontaine, O. and Pierce, N.F. (1992) Impact of rice based oral rehydration solution on stool output and duration of diarrhoea: meta-analysis of 13 clinical trials. *British Medical Journal*, **304**, 287–91.

Greenough, W.B. III (1995) *Vibrio cholerae and cholera*, in *Principles and Practice of Infectious Diseases*, 4th edn, (eds G.L. Mandell, J.E. Bennett and R. Dolin), Churchill Livingstone, New York, Edinburgh, pp. 1934–45.

Griesinger, W. (1857) Infektionskrankheiten, Malariakrankheiten, gelbes Fieber, Typhus, Pest, Cholera, in *Handbuch der speciellen Pathologie und Therapie*, (ed. R. Virchow), Erlangen 2 (part 2), p. 242.

Gupta, P.G.S., Sircar, B.K., Mondal, S. *et al.* (1978) Effect of doxycycline on transmission of *Vibrio cholerae* infection among family contacts of cholera patients in Calcutta. *Bulletin of the World Health Organization*, **56**, 323–6.

Hermann, R. (1831) Über die Veraenderungen, die Sekretionen des menschlichen Organizmus durch die Cholera erleiden. *Poggendorff's Ann*, **22**, 161.

Hirschhorn, N., Kinzie, J., Sachar, D.B. *et al.* (1968) Decrease in net stool output in cholera during intestinal perfusion with glucose-containing solutions. *New England Journal of Medicine*, **279**, 176–81.

Hoffman, S.L., Moechtar, M.A., Simanjuntak, C.H. *et al.* (1985) Rehydration and maintenance therapy of cholera patients in Jakarta: citrate-based versus bicarbonate-based oral rehydration solution. *Journal of Infectious Diseases*, **152**, 1159–65.

Holmberg, S.D. (1992) Cholera and related illnesses caused by vibrio species and aeromonas, in *Infectious Diseases*, (eds S.L. Gorbach, J.G. Bartlett and N.R. Blacklow), W.B. Saunders Co., Philadelphia, London, pp. 605–12.

Hort, E.C. and Penfold, W.J. (1911) The dangers of saline injections. *British Medical Journal*, **ii**, 1589.

Hunt, J.B., Thillainayagam, A.V., Carnaby, S., Fairclough, P.D., Clark, M.L. and Farthing, M.J.G. (1994) Absorption of hypotonic oral rehydration solution in a human model of cholera. *Gut*, **35**, 211–14.

Islam, M.R., Greenough, W.B., Rahaman, M.M. *et al.* (1980) Labon-gur (common salt and brown sugar) oral rehydration solution in the treatment of diarrhoea in adults. *Journal of Tropical Medicine and Hygiene*, **83**, 41–5.

Islam, M.R., Samadi, A.R., Ahmed, S.M. *et al.* (1984) Oral rehydration therapy: efficacy of sodium citrate equals sodium bicarbonate for correction of acidosis in diarrhoea. *Gut*, **25**, 900–4.

Kausch, W. (1911) Über intravenöse und subkutane Ernährung mit Traubenzucher. *Deutsch Medische Wochenschrift*, **37**, 8.

Kausch, W. (1916) Traubenzuckerinfusionen bei Cholera. *München Medische Wochenschrift*, **63**, 544.

Khin-Maung, U. and Greenough, W.B. (1991) Cereal-based oral rehydration therapy. 1. Clinical studies. *Journal of Pediatrics*, **118**, S72–9.

Khin-Maung, U., Myo-Khin, Nyunt-Nyunt-Wai and Tin, U. (1991) Comparison of glucose–electrolyte and maltodextrin/glycine/glycyl-glycine/electrolyte oral rehydration solutions in cholera and watery diarrhoea in adults. *Annals of Tropical Medicine and Parasitology*, **85**, 645–50.

Khin-Maung, U. (1993) *In vitro* uptake of amino acids in the jejunal mucosa of patients with cholera. *Journal of Diarrhoeal Diseases Research*, **11**, 67–74.

Latta, T. (1832) Letter from Dr Latta to the Secretary of the Central Board of Health, London, affording a view of the rationale and results of his practice in the treatment of cholera by aqueous and saline injections. *Lancet*, **ii**, 274–7.

Levinson, R.A. and Schedl, H.P. (1966) Absorption of sodium, chloride, water and simple sugars in rat small intestine. *American Journal of Physiology*, **211**, 939–42.

Lewins, R. (1832) Latest communication from Dr Lewins to the Secretary of the Central Board of Health. *Lancet*, **ii**, 280.

Mahalanabis, D. (1981) Nitrogen balance during recovery from secretory diarrhea of cholera in children. *American Journal of Clinical Nutrition*, **34**, 1548–51.

Mahalanabis, D., Molla, A.M. and Sack, D.A. (1992) Clinical management of cholera, in *Cholera*, (eds D. Barua and W.B. Greenough), Plenum Medical Book Co., New York, London, pp. 253–83.

Malawer, S.J., Ewton, M., Fordtran, J.S. *et al.* (1965) Interrelation between jejunal absorption of sodium, glucose and water in man. *Journal of Clinical Investigation*, **44**, 1072–3.

Mhalu, F.S., Mmari, P.W. and Ijumba, J. (1979) Rapid emergence of El Tor *Vibrio cholerae* resistant to antimicrobial agents during first six months of fourth cholera epidemic in Tanzania. *Lancet*, **i**, 345–7.

Molla, A., Molla, A.M., Sarkar, S.A. *et al.* (1983) Whole-gut transit time and its relationship to absorption of macronutrients during diarrhoea and after recovery. *Scandinavian Journal of Gastroenterology*, **18**, 537–43.

Molla, A.M., Ahmed, S.M. and Greenough, W.B. (1985) Rice based oral rehydration solution decreases the stool volume in acute diarrhoea. *Bulletin of the World Health Organization*, **63**, 751–6.

Molla, A., Gyr, K., Majid, A.M. *et al.* (1986) Preserved exocrine function in patients with acute cholera and acute non-cholera diarrheas. *International Journal of Pancreatology*, **1**, 259–264.

Nalin, D.R. and Morris, J.G. (1991) Cholera and other vibrioses, in *Hunter's Tropical Medicine*, 7th edn, (ed. G.T. Strickland), W.B. Saunders Co., Philadelphia, London, pp. 366–77.

Napier, L.E. (1946) Cholera, in *The Principles and Practice of Tropical Medicine*, New York.

Napier, L.E. (1951) Cholera, in *Modern Practice in Infectious Fevers*, (ed. H.S. Banks), New York.

Nichols, H.J. and Andrews, V.L. (1909) The treatment of Asiatic cholera during the recent epidemic. *Philippines Journal of Science, Section B*, **4**, 81.

O'Shaughnessy, W.B. (1831a) Proposal of a new method of treating the blue epidemic cholera by the injection of highly-oxygenised salts into the venous system. *Lancet*, **i**, 366–71.

O'Shaughnessy, W.B. (1831b) Experiments on the blood in cholera. *Lancet*, **i**, 490.

O'Shaughnessy, W.B. (1832) Chemical pathology of cholera. Remarks on 'Dr Thomson's analysis of the blood of individuals affected with malignant cholera'. *Lancet*, **ii**, 225–32.

Palmer, D.L., Koster, F.T., Rafiqul Islam, A.F.M. *et al.* (1977) Comparison of sucrose and glucose in the oral electrolyte therapy of cholera and other severe diarrheas. *New England Journal of Medicine*, **297**, 1107–10.

Patra, F.C., Mahalanabis, D., Jalan, K.N. *et al.* (1982) Is oral rice electrolyte solution superior to glucose electrolyte solution in infantile diarrhoea? *Archives of Disease in Childhood*, **57**, 910–12.

Patra, F.C., Mahalanabis, D. and Jalan, K.N. (1984) In search of a super solution: controlled clinical trial of glycine–glucose oral rehydration solution in infantile diarrhoea. *Acta Paediatrica Scandinavica*, **73**, 18–21.

Patra, F.C., Mahalanabis, D., Jalan, K.N. *et al.* (1986) A controlled clinical trial of rice and glycine containing oral rehydration solution in acute diarrhoea in children. *Journal of Diarrhoea Diseases Research*, **4**, 16–19.

Patra, F.C., Sacks, D.A., Islam, A., Alam, A.N. and Mazumder, R.N. (1989) Oral rehydration formula containing alanine and glucose for treatment of diarrhoea: a controlled trial. *British Medical Journal*, **298**, 1353–6.

Phillips, R.A. (1964) Water and electrolyte losses in cholera. *Federal Proceedings*, **23**, 705–12.

Phillips, R.A., Wallace, C.K. and Blackwell, R.Q. (1963) Failure of an oral solution comparable to stool in volume and electrolyte composition to replace stool losses in cholera. Research Report MR005.09.1040.1.14. US Naval Medical Research Unit no. 2, September 24.

Phillips, R.A., Wallace, C.K. and Blackwell, R.Q. (1965) Water and electrolyte absorption by the intestine in cholera, in *Proceedings of the Cholera Research Symposium, Honolulu, Hawaii, January 24–29*, US Government Printing Office, Washington, DC, pp. 299–311.

Pierce, N.F., Banwell, J.G., Mitra, R.C. *et al.* (1968) Effect of intragastric glucose–electrolyte infusion upon water and electrolyte balance in Asiatic cholera. *Gastroenterology*, **55**, 333–43.

Pollitzer, R. (1959) Treatment, in *Cholera*, World Health Organization, Geneva, pp. 750–819.

Quick, R.E., Vargas, R., Moreno, D. *et al.* (1993) Epidemic cholera in the Amazon: the challenge of preventing death. *American Journal of Tropical Medicine and Hygiene*, **48**, 597–602.

Rabbani, G.H., Islam, M.R., Butler, T., Shahrier, M. and Alam, K. (1989a) Single-dose treatment of cholera with furazolidone or tetracycline in a double-blind randomized trial. *Antimicrobial Agents and Chemotherapy*, **33**, 1447–50.

Rabbani, G.H., Butler, T., Bardhan, P.K. *et al.* (1989b) Reduction of fluid loss in cholera by nicotinic acid: a randomized controlled clinical trial. *Lancet*, **ii**, 1439–42.

Rabbani, G.H., Butler, T., Shahrier, M., Mazumdar, R. and Islam, M.R. (1991) Efficacy of a single dose of furazolidone for treatment of cholera in children. *Antimicrobial Agents and Chemotherapy*, **35**, 1864–7.

Rabbani, G.H. and Greenough, W.B. (1992) Pathophysiology and clinical aspects of cholera, in *Cholera*, (eds D. Barua and W.B. Greenough), Plenum Medical Book Co., New York, London, pp. 209–28.

Reid, E.W. (1902) Intestinal absorption of solutions. *Journal of Physiology (London)*, **18**, 241–56.

Riklis, E. and Quastel, J.H. (1958) Effects of cations on sugar absorption by isolated surviving guinea pig intestine. *Canadian Journal of Biochemistry and Physiology*, **35**, 347–62.

Rogers, L. and Mackelvie, M. (1908) Note on the value of large quantities of hypertonic salt solutions in transfusion for cholera. *Indian Medical Gazette*, **43**, 165–7.

Rogers, L. (1909) The treatment of cholera by injections of hypertonic saline solutions with a simple and rapid method of intraabdominal administration. *Philippines Journal of Science, Section B*, **4**, 99.

Rogers, L. (1915) The results of the hypertonic and permanganate treatment in 1000 cases of cholera: with remarks on the value of alkalies in the prevention of uraemia and the role of atropine. *Lancet*, **ii**, 219.

Rogers, L. and Shorten, A.J. (1915) The alkalinity of the blood in kala-azar and cholera and the technique of its estimation. *Indian Journal of Medical Research*, **2**, 867–81.

Rogers, L. (1916) Further work on the reduction of the alkalinity of the blood in cholera; and sodium bicarbonate injections in the prevention of uraemia. *Annals of Tropical Medicine and Parasitology*, **10**, 139–49.

Rogers, L. (1921) *Bowel Diseases in the Tropics – Cholera, Dysenteries, Liver Abscess and Sprue*, London.

Rogers, L. (1952) Cholera, in *Tropical Medicine*, 6th edn, (eds L. Rogers and J.W.D. Megaw), London.

Rosner, M. (1895) Das Übersalzen des Blutes bei Cholerakranken. *Wien Medische Wochenschrift*, **45**, 376–8, 431–3.

Rossi, A., Galas, M., Binztein, N. *et al.* (1993) Unusual multiresistant *Vibrio cholerae* O1 El Tor in Argentina. *Lancet*, **342**, 1172–3.

Sack, R.B., Cassells, J., Mitra, R. *et al.* (1970) The use of oral replacement solutions in the treatment of cholera and other severe diarrhoeal disorders. *Bulletin of the World Health Organization*, **43**, 351–60.

Sack, D.A., Islam, S., Brown, K.H. *et al.* (1980) Oral therapy in children with cholera: a comparison of sucrose and glucose electrolyte solutions. *Journal of Pediatrics*, **96**, 20–5.

Salazar-Lindo, E., Sack, R.B., Chea-Woo, E. *et al.* (1986) Bicarbonate versus citrate in oral rehydration therapy in infants with watery diarrhea: a controlled clinical trial. *Journal of Pediatrics*, **108**, 55–60.

Schedl, H.P. and Clifton, J.A. (1963) Solute and water absorption by the human small intestine. *Nature*, **199**, 1264–7.

Schultz, S.G. and Zalusky, R. (1964) Ion transport in isolated rabbit ileum. II. The interaction between active sodium and active sugar transport. *Journal of General Physiology*, **47**, 1043–59.

Seibert, F.B. (1923) Fever-producing substance found in some distilled waters. *American Journal of Physiology*, **67**, 90–104.

Sellards, A.W. (1910) Tolerance for alkalies in Asiatic cholera. *Philippines Journal of Science, Section B*, **5**, 363.

Shattuck, G.C. (1951) Cholera, in *Diseases of the Tropics*, New York.

Siddique, A.K., Zaman, K., Baqui, A.H. *et al.* (1992) Cholera epidemics in Bangladesh: 1985–1991. *Journal of Diarrhoeal Diseases Research*, **10**, 79–86.

Sircar, B.K., Saha, M.R., Deb, B.C., Singh, P.K. and Pal, S.C. (1990) Effectiveness of oral rehydration salt solution (ORS) in reduction of death during cholera epidemic. *Indian Journal of Public Health*, **34**, 68–70.

Stein, J.H. (1831) Treatment of cholera by inhalation of oxygen. *Lancet*, **i**, 364–5.

Strauss, H. (1915) Zuckerinfusionen bei Cholera. *Ther d Gegenw*, **56**, 370.

Strong, R.P. (1944) Cholera, in *Stitt's Diagnosis, Prevention and Treatment of Tropical Diseases*, 7th edn, Philadelphia.

Tabtieng, R., Wattansri, S., Echeverria, P. *et al.* (1989) An epidemic of *Vibrio cholerae El Tor Inaba* resistant to several antibiotics with a conjugative group C plasmid coding for type II dihydrofolate reductase in Thailand. *American Journal of Tropical Medicine and Hygiene*, **41**, 680–6.

Taylor, G. (1831) Treatment of the blue cholera in India. *Lancet*, **i**, 365.

Thomas, H. and Ting, L.C. (1938) Innocuous intravenous infusion with an appeal for the establishment of central solution laboratories for cholera relief. *Chinese Medical Journal*, **54**, 358.

Threlfall, E.J., Said, B., Rowe, B. and Dávalos–Pérez, A. (1993) Emergence of multiple drug resistance in *Vibrio cholerae* O1 El Tor from Ecuador. *Lancet*, **342**, 1173.

Van Heyningen, W.E. and Seal, J.R. (1983) *Cholera: the American Scientific Experience, 1947–1980*, Westview Press, Boulder, Colorado, pp. 215–48.

van-Loon, F.P. (1993) Cholera: developments in prevention and cure. *Tropical and Geographical Medicine*, **45**, 269–73.

Whyte, G.D. (1913) The treatment of an epidemic of cholera by Rogers' method: based on a study of 215 cases which required intravenous infusions of saline. *Chinese Medical Journal*, **27**, 107.

Whyte, G.D. (1915) The treatment of cholera by hypertonic saline solutions during an epidemic at Swatow, South China. *British Medical Journal*, **ii**, 425.

Intestinal immunity to *Vibrio cholerae* 4

B.D. Forrest

For most infectious diseases, elucidation of the mechanisms involved in either protection against subsequent clinical disease, or in the modification of the clinical outcome following infection, requires an understanding of the specific immune responses that follows infection. The application of this understanding is seen in the development of effective vaccines for prophylaxis, or in effective immunotherapeutic intervention. Historically, an understanding of 'natural' immunity has played, at best, only a limited role in the development of suitable vaccines. As a result, vaccines developed in the past have covered the range from being either highly effective, such as the inactivated or live attenuated poliovirus vaccine, or have been of practically no value at all, such as the inactivated parenterally administered cholera vaccine.

Much of the failure to better define the immune response has been attributable to the lack of the appropriate facilities and methods needed to obtain the necessary knowledge. The modern era has seen a proliferation in the techniques available to dissect the immune response, and as a result has seen an accompanying surge in our understanding of the immune mechanisms responsible in mediating protection from disease, and the modification of disease outcome, for a variety of diseases, including those caused by *Vibrio cholerae*.

4.1 PROTECTIVE IMMUNITY FOLLOWING NATURAL INFECTION

In 1884, Robert Koch published his observations that convalescent anti-cholera protective immunity was only of short duration. He observed

Cholera and the Ecology of Vibrio cholerae. Edited by B.S. Drasar and B.D. Forrest.
Published in 1996 by Chapman & Hall, London. ISBN 0 412 61220 8.

that a person suffering an attack of cholera during one epidemic could have another attack during another outbreak, but that this person was not attacked twice during the same epidemic, despite the fact that recovery meant that the patient returned home and was again exposed to the disease (Koch, 1884). Today, it is recognized that clinical cholera invokes a significant degree of protection upon convalescent patients against subsequent disease following reinfection, as determined by the observation that in endemic areas, the incidence of clinical disease is maximal in children aged 2–4 years, decreasing in older age groups despite continuing exposure to the organism (Mosley *et al.*, 1968; Merson *et al.*, 1980). In North American volunteer subjects, clinical infection has been demonstrated to evoke solid protection against clinical disease that lasted for at least 3 years against re-challenge with the homologous or a heterologous serotype (Cash *et al.*, 1974; Levine *et al.*, 1979, 1981; Levine, 1980).

However, although in endemic areas it has been demonstrated that primary infection confers solid protection against second attacks for several years (Glass *et al.*, 1982), more recent evidence suggests that the efficacy of this protection may vary according to biotype. In the long-term follow-up of 2214 convalescent patients in Bangladesh recovering from a primary episode of clinical cholera, seven had a second episode (Clemens *et al.*, 1991). The incidence of cholera was found to be 61% lower in subjects who had an earlier episode of cholera than in those who did not. This variation in long-term protection appears to be attributable, in part, to the biotype involved in the initial disease, as initial episodes of classical cholera were associated with complete protection against subsequent clinical cholera, whereas only negligible protection was seen following an initial episode of El Tor cholera.

4.2 SECRETORY IgA AND THE MECHANISM OF INDUCTION OF LOCAL INTESTINAL IMMUNITY

It was Jaime Ferran y Clua in Spain, who in 1885 first espoused the use of a live attenuated *V. cholerae* vaccine for widespread protection against cholera (Bornside, 1981). Although Klemperer, in 1892, claimed to be able to protect guinea pigs against oral infection with *V. cholerae* by the administration of small doses of viable organisms into their stomachs (Klemperer, 1892), it was Sawtschenko and Sabolotny, in 1893, who orally vaccinated themselves and one other human subject with numerous doses of agar-grown heat-inactivated *V. cholerae*, and were able to demonstrate that their serum, obtained 25 days after vaccination was completed, was able to protect guinea pigs against virulent challenge (Pollitzer, 1959). Furthermore, these latter two investigators, following a further oral dose of the vaccine preparation, ingested 0.1 ml of a 24-hour culture of *V. cholerae* O1. Neither of them showed any clinical signs of cholera, despite both excreting viable *V. cholerae* up to 3 days post-challenge. This study represents the first study in humans showing the protective effect of large multiple orally administered doses of killed whole cells.

Since those pioneering studies in humans, it has been repeatedly demonstrated that after infection or oral vaccination with attenuated organisms, the immune response generated at the intestinal mucosal surface is mainly of the IgA antibody class, and always of the secretory type (Mestecky *et al.*, 1986; Mestecky, 1987). It is now generally considered that the intestinal secretory IgA (SIgA) system is quantitatively the most important humoral immune system in the body (Plaut and Keonil, 1969; Brandtzaeg *et al.*, 1985; Czinn and Lamm, 1986). SIgA, the predominant immunoglobulin in the human gastrointestinal tract, exists in a polymeric, frequently dimeric, form, and not in the monomeric form of bone marrow-derived serum IgA (Tomasi and Bienenstock, 1968; Walker and Isselbacher, 1977). Serum IgA, irrespective of its form – be it monomeric or polymeric – does not appear to contribute significantly to the total IgA in secretions (Mestecky *et al.*, 1986; Bartholomeusz *et al.*, 1990). SIgA has been found to have several roles in the maintenance of mucosal defences, through its ability to neutralize toxins and viruses, and its ability to block bacterial adherence to surface epithelium (Targan, 1987). In fact, SIgA *in vitro*, is able specifically to inhibit bacterial adherence and colonization of cultured respiratory epithelial cells (Williams and Gibbons, 1972). One possible reason for the effectiveness of SIgA is its retention within the mucous layer over the epithelial cells after its secretion into the intestinal lumen, achieved by interacting with cysteine residues contained in the mucins within the glycocalyx (Bienenstock, 1974). It is this relative juxtaposition to the intestinal epithelial cell which is responsible for its efficient interaction with intestinal antigens.

An important step in the pathogenesis of many microbial diseases is the successful penetration of the protective mucosal epithelium of the respiratory or gastrointestinal tracts. Many organisms are able to breach the epithelial barrier through their selective uptake by the epithelial M cell (Wolf and Bye, 1984), including *V. cholerae* (Owen *et al.*, 1986). In addition to particles and organisms, some 'soluble' antigens, such as cholera toxin (CT), require uptake into the Peyer's patch (PP) in order to induce a local intestinal immune response (Owen, 1977; Enders *et al.*, 1987). The successful uptake of an antigen into the underlying PP or lymphoid aggregate, has been shown to result in the stimulation of IgA-secreting precursor plasma cells (Craig and Cebra, 1971; Rudzik *et al.*, 1975), which upon stimulation, migrate into the blood circulation via the mesenteric lymph nodes and the lymphatics (McWilliams *et al.*,

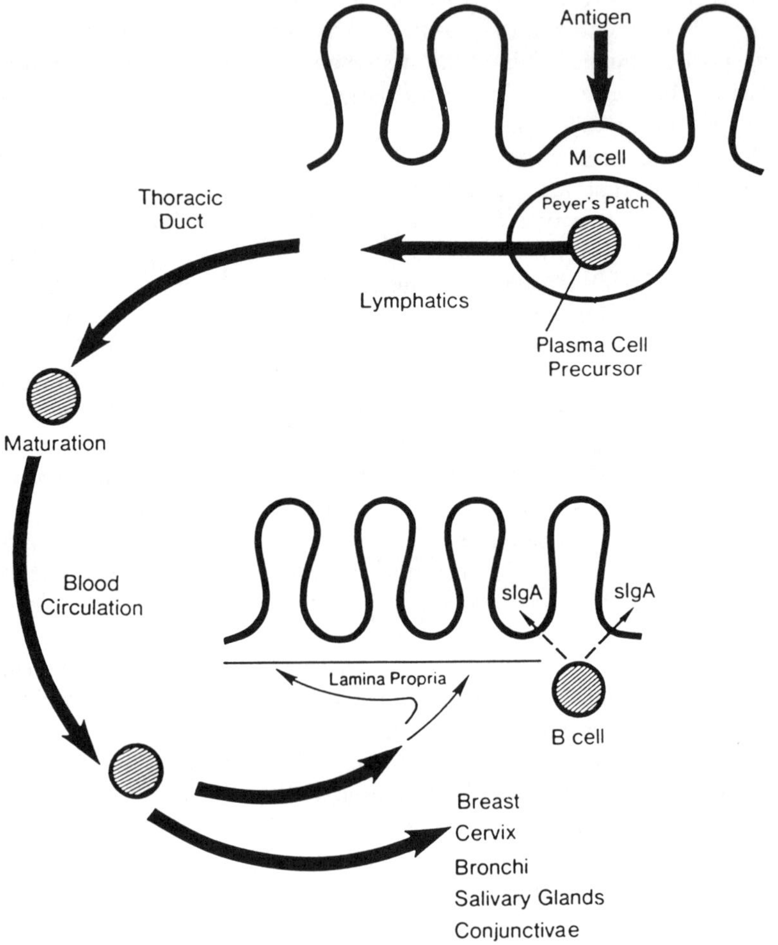

Fig. 4.1 Maturation cycle of Peyer's patch precursor plasma cells following local mucosal antigenic stimulation. This figure depicts schematically the process of induction of a local intestinal immune response following antigenic uptake by the epithelial M cells that overly Peyer's patch lymphoid follicles. This stimulation induces plasma cell precursors to migrate via the lymphatics into the systemic circulation. During this migration these precursors maturate and seed to the lamina propria of mucosal surfaces where they produce polymeric antigen-specific IgA. Mucosal T-lymphocytes have been demonstrated to undergo a similar process of circulation following antigen exposure.

1977) (Figure 4.1). During this migration these cells undergo maturation and eventually seed to the lamina propria of a variety of mucosal surfaces where they secrete specific IgA (Mestecky, 1987; Forrest, 1988a, 1992). However, it has been determined that not all the migrating lymphocytes that derived from the gut-associated lymphoid tissue (GALT) necessarily return to the intestinal mucosa. Certainly the bulk of these cells (70%) have been observed to return, but many in fact are found to seed to other mucosal surfaces, such as respiratory tract and genital tracts (McDermott and Bienenstock, 1979) (Figure 4.1).

In rats, the effect of deprivation of PP significantly decreased the number of IgA containing cells in the intestinal mucosa of PP-excised rats, while IgM- and IgG-containing cells showed no statistically significant changes in number (Serizawa *et al.*, 1993). Examination of subsequent CT-specific immune rats using intraduodenal priming and challenge with CT, showed a significant reduction of CT-specific antibody production in PP-excised rats compared with controls. In addition, there was a significant reduction in the number of cells in the intestinal mucosal lamina proprias producing CT-specific antibody, demonstrating the dependence on PP uptake for the successful induction of an intestinal immune response to CT.

4.3 INTESTINAL IMMUNITY FOLLOWING NATURAL INFECTION

Serum antibodies directed against cholera antigens can be found following cholera infection. However, since *V. cholerae* O1 is non-invasive – remaining confined to the lumen and mucosal epithelium of the small intestine – these circulating antibodies probably play a minor role in protective immunity, and it is most likely SIgA antibodies produced locally in the small intestine make the main contribution to protective antibody responses (Svennerholm, 1980). Local intestinal immunity to cholera is probably now generally accepted as being important in the prevention of clinical disease. However, this may not necessarily be so for *V. cholerae* O139, where invasive disease has been reported (Johnson *et al.*, 1994), probably mediated through its possession of a polysaccharide capsule (Weintraub *et al.*, 1994; Comstock *et al.*, 1995). This capsule appears to afford *V. cholerae* O139 with a predilection for binding to epithelial M cells (Yamamoto *et al.*, 1994). Therefore it is possible that anti-capsular antibody may provide protection against systemic disease caused by *V. cholerae* O139.

Early studies in cholera, using guinea pigs, suggested that coproantibody (faecal antibody) induced by previous infection, conferred protection against subsequent challenge with pathogenic *V. cholerae* O1 (Burrows *et al.*, 1947). This led to the discovery that acutely ill cholera patients also produced coproantibody against cholera (Freter *et al.*, 1965). Furthermore, early estimations of the IgA antibody content in the stools of 25 acute cholera patients supported the view that the coproantibody being measured was produced locally in the intestine and was not a transudate from serum; the serum levels of IgA being no lower than convalescent levels (Northrup *et al.*, 1970; Waldman *et al.*, 1971). Subsequently, it was determined that much of the coproantibody IgA was in fact the polymeric secretory form, and that there was a significant proportion of monomeric IgA and smaller IgA fragments present, possibly representing the enzymatic breakdown products of SIgA (Northrup *et al.*, 1970).

When specific anti-cholera antibody levels were examined in the jejunal fluid obtained by the intestinal intubation of cholera patients, it was observed that during the early stage of infection the predominant intestinal immunoglobulin was initially IgG, but by 7 days after recovery the predominant immunoglobulin had become IgA (Northrup *et al.*, 1970), peaking 2–4 weeks post-convalescence (Waldman *et al.*, 1972). Furthermore, the increase in specific anti-cholera IgA was demonstrated to be a mean 30-fold in the small intestinal secretions after clinical cholera, with no concomitant rise in the total IgA; this in marked contrast to a mean rise in specific serum IgA of only two-fold (Waldman *et al.*, 1972). This further supported the importance of specific local antibody production in clinical cholera, and also determined that the intestinal IgG and IgM responses appeared to be a transudate from serum.

More recently, animal studies have supported the importance of the role of the generation of local immunity, for example that the elimination of *V. cholerae* from the intestine was promoted in mice fed live *V. cholerae* but not following feeding with killed organisms (Fujisawa *et al.*, 1987). This was attributed to the finding that the total IgA in the intestinal contents of mice fed live *V. cholerae* were significantly higher than those of mice fed with killed antigens.

4.4 LOCAL IMMUNITY TO CHOLERA TOXIN

For infections caused by *V. cholerae*, despite the identification of a variety of other toxins, CT has been held largely responsible for the major contribution to clinical disease and protective immunity (Levine *et al.*, 1983). Since De (1959) demonstrated convincingly that pathogenic strains of *V. cholerae* O1 produced an enterotoxin, by showing that a cell-free culture filtrate could stimulate fluid secretion in the rabbit intestinal-loop model, CT has been purified and its mechanism of action elucidated at the molecular level. CT is a protein with a molecular mass (M_r) of 84 kDa, comprising two immunologically different regions designated 'A' and 'B'. The B region (previously known as 'choleragenoid') has an approximate M_r 56 kDa, and is composed of five non-covalently associated subunits each of M_r 11.5 kDa arranged in a circular fashion (WHO Scientific Working Group, 1980). The B subunit (CTB) encircles the A region, which comprises two peptides, A_1 (M_r 20.5 kDa) and A_2 (M_r 7.5 kDa) linked by a disulphide bond (Kurosky *et al.*, 1975). The CTB is responsible for the binding of the CT holotoxin to host-cell membrane receptors that contain the glycolipid, the GM_1-ganglioside that leads to the enzymatic cascade responsible for the induction of clinical disease manifested in the secretion of water and electrolytes from the intestinal crypt cells and inhibits the absorption by the villous cells – the cholera stool (Gill and Meren, 1978; Fishman, 1980; WHO, 1980).

CT is produced by both biotypes and serotypes of *V. cholerae* O1, and by *V. cholerae* O139 (Hall *et al.*, 1994; Nair *et al.*, 1994). In experimental animal studies, it has been demonstrated that anti-CT antibodies induced by active immunization were able to protect against symptomatic infection (Curlin *et al.*, 1970; Fujita and Finkelstein, 1972; Pierce and Reynolds, 1974; Holmgren *et al.*, 1975; Svennerholm and Holmgren, 1976; Pierce *et al.*, 1977). In mice, oral administration of CT has been demonstrated to induce a brisk IgA anti-CT antibody response in faecal extracts, which has been associated with elevated IgG anti-CT antibody in serum (Xu-Amano *et al.*, 1994), and repeated oral administration of CT gave rise to high numbers of anti-CT IgA antibody-secreting cells located in the lamina propria, as well as conferring production against challenge with CT in ligated intestinal loops (Lycke *et al.*, 1989). Similar observations were made in rats immunized three times intra-intestinally with CT (Pierre *et al.*, 1988). It was also observed that, on a molar basis, rat biliary SIgA (SIgA) antibodies were roughly seven-fold more efficient than serum IgG antibodies in neutralizing CT in the ligated intestinal loop assay. However, intravenous immunization of CT to mice has also not been demonstrated to induce intestinal anti-CT antibody responses (Xu-Amano *et al.*, 1994), further supporting the importance of oral delivery in the induction of local mucosal immunity.

Supporting evidence for the role of mucosal immunoglobulin in conferring protection against toxin-mediated disease has been provided from studies using anti-CT immunoglobulins derived from lactating animals fed either live *V. cholerae* or purified CT. Milk collected from lactating rabbits immunized enterally with toxinogenic, live *V. cholerae* O1, was able to inhibit water secretion induced by *V. cholerae* in rat ileal loops. The active component in the milk was believed to be anti-CT antibodies, as specific protective activity could be removed by immunoabsorption with CT, but only partially removed by incubation with whole *V. cholerae* (Yoshiyama and Brown, 1987). Similarly, bovine derived colostrum obtained from lactating cows immunized with CT, contained anti-CT immunoglobulins that elicited statistically significant protection against diarrhoea in infant rabbits challenged intra-intestinally with virulent *V. cholerae* (Boesman-Finkelstein *et al.*, 1989).

Mouse hybridoma clones that produced polymeric or dimeric monoclonal IgA and IgG antibodies directed against CTB were generated through the fusion of PP lymphocytes with mouse myeloma cells, following the oral-systemic immunization of mice with CT and CTB. These hybridoma cells have been used to further demonstrate the ability of mucosally derived anti-CT IgA to neutralize CT (Apter *et al.*, 1993a). Despite none of three selected polymeric or dimeric anti-CTB IgA monoclonal antibodies being directed against the binding site of the GM_1 ganglioside toxin receptor, all blocked CT-induced chloride secretion *in vitro* in a dose-dependent manner and completely inhibited binding of rhodamine-labelled CT to apical cell membranes (Apter *et al.*, 1993b). Therefore, CTB presented itself as a possible contributor to an oral vaccine, and this eventually led to the development of a more efficacious purified CTB vaccine.

4.4.1 LOCAL IMMUNITY TO CHOLERA TOXIN B SUBUNIT

In humans, the use of 'native' CT as a possible oral vaccine candidate is unacceptable due to the toxicity associated with its oral administration in human subjects (Levine *et al.*, 1983). In recognition of this, a variety of toxoid-based vaccine candidates were developed and assessed with limited success (Rappaport *et al.*, 1974; Curlin *et al.*, 1975). Although less immunogenic than the CT holotoxin (Pierce, 1978), anti-CTB antibodies have been demonstrated to have greater cholera toxin-neutralizing capacity than anti-CT A subunit antibodies (Holmgren and Svennerholm, 1977).

An oral vaccine comprising the CTB administered with killed whole *V. cholerae* (B-WCV), has been extensively evaluated in clinical studies (Czerkinsky *et al.*, 1991). In healthy Swedish adults, and post-cholera convalescent Bangladeshi adults, both oral and parenteral immunization with the B-WCV induced anti-CT antibodies of all IgG subclasses, while the IgA antibodies were restricted to the IgA_1 subclass. A single oral dose of B-WCV induced higher levels of IgG_4 anti-CT in previously immunized volunteers than in unvaccinated control subjects. This suggests that repeated immunization or exposure preferentially stimulates the formation of IgG_4 antibodies. There was little difference between the IgG and IgA subclass distribution of antitoxin antibodies in orally vaccinated Swedes or post-cholera convalescent Bangladeshis (Jertborn *et al.*, 1988). Further studies have indicated that the source of specific anti-CT antibody-secreting cells following oral immunization is not the bone marrow, further lending support to the mucosal origin of this specific anti-CTB immunity (Lewis *et al.*, 1991).

The protective efficacy of one particular preparation of this vaccine was assessed in North American volunteer subjects in a small efficacy study (Black *et al.*, 1987). The B-WCV was administered orally to 19 subjects while another 14 ingested the WCV without CTB. Of the B-WCV vaccine group, four-fold or greater rises in anti-cholera lipopolysaccharide (LPS) IgA antibody titres were observed in the jejunal fluid of 53% of vaccinated subjects, and in the serum of 68%. Anti-CT antibody responses of the same magnitude were observed in serum of 100% of B-WCV subjects but in only 74% were jejunal SIgA anti-CT antibody responses detectable. In the group receiving the whole-cell vaccine without CTB, specific anti-cholera LPS jejunal SIgA responses were not detected in any subject, yet serum IgA anti-LPS responses were elevated in 57% of subjects. No anti-CT antibody response was detected in either the serum or jejunal fluid of any volunteer. Both groups failed to produce a measurable antibody response to the outer membrane protein (OMP) in jejunal fluid, yet such a response was detected in the serum of 58% of B-WCV subjects and in 57% of WCV recipients only.

Following oral challenge with 2×10^6 viable pathogenic *V. cholerae* El Tor Inaba organisms the protective efficacy of the combined B-WCV vaccine was 64% against a dose that induced clinical illness in all control subjects ($P = 0.01$), and the WCV without CTB was 56% against a dose that induced clinical illness in 75% of control subjects ($P = 0.11$). There was no significant difference between the protective efficacy of the two vaccine preparations, with or without CTB (Black *et al.*, 1987).

Subsequently, a large-scale, randomized, double-blind, placebo-controlled field trial of efficacy was undertaken in the Matlab area in Bangladesh (Clemens *et al.*, 1986), in a study population comprising 63 498 children aged 2–15 years, and women aged over 15 years. The B-WCV, WCV and an inactivated *E. coli* K12 placebo vaccine, were administered as single doses once weekly for 3 weeks, although there were some variations on this schedule in a small number of participants. The incidence of cholera and the protective efficacy (PE) of the respective vaccine groups versus the placebo group during the following 6 months post-vaccination surveillance, are presented in Table 4.1. This study demonstrated that the combined B-WCV was able to induce significantly greater protective immunity in the residents of an endemic area than the orally administered WCV alone, and that this protective immunity remained significantly elevated above that of the WCV alone for at least 6 months. During the subsequent 3 years of observation vaccine efficacy gradually declined, so that by the end of the third year of post-vaccination surveillance the protective efficacy of the WCV alone was 43% and on the B-WCV only 17% (Table 4.1) (WHO, 1989).

More recently, a B-WCV based on recombinantly produced CTB has been developed, and in studies performed in Swedish volunteers,

Table 4.1 Efficacy of the oral killed whole-cell/B subunit combination cholera vaccine and the oral killed whole-cell vaccine alone (Dhaka, Bangladesh). (Adapted from WHO Diarrhoeal Disease Control Programme, 1989.)

Vaccinated subjects and follow-up period	Protective efficacy (%)	
	WCV	B-WCV
All vaccinated subjects		
Year 1	53	62
Year 2	57	57
Year 3	43	17
All three years	52	50
Protection by biotype*		
Classical	60	58
El Tor	40	39
Protection by age*		
<6 years old	23	26
>6 years old	68	63

* Protection over 3-year period. Three doses of the vaccines or placebos were given at 1-month intervals. None of the levels of protection for the WCV or B-WCV was significantly different ($P > 50.5$).
WCV, killed whole-cell vaccine only; B-WCV, killed whole-cell vaccine with CT B subunit.

appears to retain the characteristics of the previously extensively evaluated B-WCV (Jertborn *et al.*, 1992).

4.5 INDUCTION OF INTESTINAL MEMORY IMMUNE RESPONSES BY CHOLERA TOXIN AND CTB SUBUNIT

Local mucosal immunological memory may be responsible for the long duration of protection against reinfection and disease seen in convalescent cholera patients. In mice that had previously received a series of oral immunizations with CT 6 months to 2 years previously, a further single oral dose of CT in sub-microgram amounts, induced a three-fold increase in anti-CT IgA lamina proprial antibody-secreting cell (ASC), evident within 16 hours of exposure, and with a steady increase in these numbers observed over the subsequent several days (Lycke and Holmgren, 1987). This anamnestic anti-CT IgA ASC response was associated with a substantial increase in protection against challenge with CT, and is consistent with the existence of the presence of intestinal mucosal memory cells.

CT naive mice challenged orally with CT, following transfer of mesenteric lymph node (MLN) cells from mice that had been orally primed with CT 1 year earlier, demonstrated anti-CT ASC detected by ELISPOT in the spleen, MLN and lamina propria, but not in PP (Lycke and Holmgren, 1989). The magnitude of the anti-CT ASC response in the recipient mice following oral CT immunization equalled that observed in the original donor mice following primary exposure to CT, and correlated directly with the number of MLN cells transferred. Demonstration of the memory anti-CT ASC response in MLN and in the lamina propria required oral challenge with CT. However, the splenic anti-CT ASC response could be triggered by oral or intravenous challenge with CT. The isotype distribution of the anti-CT ASC response in recipient mice was similar to that observed in donor mice, with predominantly IgA (96%) in lamina proprial ASC and IgG (66%) in MLN and spleen. The investigators were able to completely abrogate the anamnestic ASC response to CT in the recipient mice by incubating the isolated MLN cells with J11d monoclonal antibody and complement before transfer. Further, the anamnestic ASC response was unaffected by treatment with anti-Thy-1.2 antibody and complement. These findings were suggestive that in mice, intestinal mucosal memory is probably B-lymphocyte-mediated.

Human subjects vaccinated orally with the previously described B-WCV, have also been found to possess anti-CT memory cells that could be isolated from peripheral blood for at least 1 year after immunization (Lycke *et al.*, 1987b; Lebens and Holmgren, 1994). Anti-CT ASC were identifiable in the peripheral blood of four or five previously vaccinated subjects, with IgA and IgM isotypes predominating. These peripheral blood cells could be induced *in vitro* to produce anti-CT antibodies in the presence of autologous T cells and monocytes by CTB. In contrast to the observations in mice described above, where the evidence suggested that the anamnestic mucosal immune responses were independent of T-lymphocyte control, in humans the dose dependency and requirements for a specific ratio of T- to B-cells for activation of the memory cells *in vitro* implied T-cell control of anti-CT ASC responses. It is likely that immunological memory is induced in convalescent patients following infection or a clinical episode of cholera, and that this is a major contributor to the long-term protection previously observed.

4.6 T-LYMPHOCYTE CONTROL OF INTESTINAL IMMUNITY TO CT

In contrast to immunologically intact BALB/c mice, nude, athymic mice do not generate an antibody response following oral immunization with CT and are not protected against subsequent challenge with CT (Lycke *et al.*, 1987a). However, following the reconstitution of nude mice through grafting of syngeneic thymus glands, IgA anti-CT ASC responses in the lamina propria can be stimulated with orally administered CT. These responses are of a similar magnitude to those of immunologically intact mice after oral immunizations. Protection against CT challenge is also restored. These observations imply that intestinal anti-CT antibody formation and protection against CT challenge after oral immunization with CT are both thymus-dependent and under T-lymphocyte control.

The treatment of normal mice with specific anti-CD4 monoclonal antibodies has produced mice that demonstrate effective elimination of CD4+ T-lymphocytes in the spleen, MLN and PP, as determined by flow cytometry analysis. Lamina propria lymphocyte numbers have been found to be only slightly decreased following such treatment, although CD4 expression on individual lamina propria lymphocytes has been shown to be strongly down-regulated (Hornqvist *et al.*, 1991). The depletion of CD4+ T-lymphocytes before oral immunization with CT has been demonstrated to completely inhibit anti-CT antibody responses, with no evidence of anti-CT antibody production in spleen, MLN, PP, nor serum, compared with untreated control mice. Treatment with the anti-CD4 monoclonal antibodies was also found to effectively inhibit the anti-CT IgA antibody response in the lamina propria, and inhibited the ability to develop intestinal protection against CT challenge after oral CT immunization.

In 50-day-old rats malnourished during lactation, a substantial decrease in the CD4+ and CD8+ T-cell subsets was found to be associated with a marked reduction in the induction of intestinal antibody response to CT (Flo *et al.*, 1993). This suppression of intestinal antibody formation, which persisted after feeding commenced, and remained even after a CT booster was given to animals 113 days after feeding recommenced, is strongly supportive of the role of CD4 cells in the control of intestinal immunity to CT.

In humans, oral immunization with CTB has been shown to induce a lymphoproliferative response in peripheral blood in all subjects receiving CTB, with a peak stimulation index of 20, and that was detectable for up to 1 year. The proliferative response kinetics suggested the appearance in the blood of primed T-cells from the gut coinciding with the disappearance of primed plasmablasts from the circulation (Lewis *et al.*, 1993). Collectively, both human and animal studies, support a critical role for functional CD4+ T-cells in the development of a protective primary mucosal immune response to CT-induced disease.

The role of cytokines in the generation of intestinal antibody responses to CT is less understood. Interleukin-4-targeted mice (IL-4-/-), defective in T-helper 2 (Th2) cytokine production, have been used to further explore the cellular basis of immunity to CT. While Th2 cells appear to be selectively induced by oral immunization, as demonstrated by IL-4-mediated mucosal antibody responses to orally administered KLH and ovalbumin, in contrast, CT stimulated strong mucosal and systemic antibody responses following oral immunization of IL-4-/- mice (Vajdy *et al.*, 1995). Both systemic responses, and mucosal responses including total mucosal IgA levels, were indistinguishable in magnitude to those observed in wild-type mice (Haq and Szewczuk, 1993). However, anti-CT IgA antibody production by PP B-lymphocytes has been shown to be augmented by both LPS and IL-5 stimulation irrespective of the age of the animal. In addition, both anti-IL-4 and anti-IL-5 monoclonal antibodies blocked the respective IL-4 and IL-5-induced enhancement of antigen-specific IgG or IgA antibody production. Although IL-4 production is important in the generation of mucosal immune responses following oral immunization, it is apparent that both IL-4 and IL-5 independently are able to augment specific intestinal anti-CT antibody responses following oral immunization with CT.

4.7 ROLE OF ANTI-CT LOCAL IMMUNITY IN LONG-TERM PROTECTION AGAINST CHOLERA

Despite the considerable quantity of animal and convalescent human experience accumulated to support the role of CT in the generation of durable, protective, intestinal specific antibody responses, anti-CTB immunity in humans appears to be of only a short duration and requires repeated doses/exposure to maintain a significant level of specific intestinal or serum antibodies. This is supported by one observation that each recurrent orally administered dose of CTB, failed to stimulate a local antibody response of greater magnitude and duration than that observed following the primary vaccination (Svennerholm *et al.*, 1984). Further, the combination B-WCV that was subsequently developed for use as an oral vaccine against cholera has met with variable degrees of success (Svennerholm and Holmgren, 1976; Jertborn *et al.*, 1984; Svennerholm *et al.*, 1984; Clemens *et al.*, 1986; Black *et al.*, 1987). Although the combined B-WCV was able to induce significantly greater protective immunity in the residents of an endemic area than the orally administered killed whole-cell cholera vaccine alone, and that this protective immunity remained significantly elevated above that of the WCV alone for at least 6 months, during the subsequent 3 years of observation vaccine efficacy gradually declined, so that by the end of the third year of post-vaccination surveillance the protective efficacy of the WCV alone was 43% and on the B-WCV only 17% (Table 4.1) (Clemens *et al.*, 1986).

The observation that subjects previously challenged with *V. cholerae* when re-challenged were protected against clinical disease and had a significant and substantial reduction in the number of pathogenic organisms that could be isolated from their jejunum, implied that local antibacterial immunity played a more significant role in providing protection against reinfection than did anti-CT immunity (Craig, 1981).

The development of a variety of attenuated *V. cholerae* vaccines through the identification and genetic deletion of toxin genes, has resulted in a variety of recombinant vaccine candidate strains (Michalski *et al.*, 1993; Tacket *et al.*, 1993), one of which – *V. cholerae* CVD-103HgR – has demonstrated the best risk–benefit, balancing protective efficacy with reduced adverse events, having been extensively evaluated for safety and immunogenicity (Levine *et al.*, 1988; Tacket *et al.*, 1992; Gotuzzo *et al.*, 1993; Levine and Kaper, 1993; Lagos *et al.*, 1995). It is from the development of these attenuated toxinless vaccine candidates, that our knowledge of the role in the induction of local intestinal immunity of various toxin gene-products has derived. The most important finding has been that protection against clinical cholera does not require antitoxin immunity. This has been demonstrated by the strong protective immunity shown by human subjects receiving the toxinless mutants of *V. cholerae* (e.g. *V. cholerae* CVD-103HgR) and subsequently orally challenged with pathogenic organisms (Tacket *et al.*, 1992). This protection was equal to that conferred by pathogenic organisms (90%) (Levine *et al.*, 1981) and significantly greater than that observed following the CTB whole-cell cholera vaccine (64%), and from ingested killed whole cells alone (56%) (Black *et al.*, 1987). Therefore, the role of non-toxin antigens in the conferring of protective immunity against clinical cholera needs to be further examined.

4.8 LOCAL IMMUNITY TO *V. CHOLERAE* NON-CT ANTIGENS

4.8.1 LIPOPOLYSACCHARIDE

The role of non-CT *V. cholerae* antigens in protection against infection and disease remains relatively underinvestigated in contrast to CT. However, there exists considerable supportive data suggesting that anti-CT or anti-CTB immunity may not necessarily provide the sole basis for local immunity against *V. cholerae* (Apter *et al.*, 1993b). For example, using the neonatal mouse model, it has been demonstrated that antibodies raised in rabbits against the O1 polysaccharide antigen of *V. cholerae* were able to confer protection on the animals against challenge with pathogenic *V. cholerae* O1 (Watanabe *et al.*, 1969; Manning *et al.*, 1986). An oral dose of 80 μg of a polymeric anti-CTB IgA monoclonal antibody, known to inhibit epithelial cell binding by CT *in vitro*, when given orally failed to protect neonatal mice against challenge with either virulent *V. cholerae* or a lethal (5 μg) oral dose of CT. In addition, secretion of the same anti-CTB IgA monoclonal antibodies into the intestines of mice bearing IgA hybridoma 'backpack' tumours, also failed to protect against lethal oral doses of either CT (5 μg) or *V. cholerae* (10^7 cells) (Apter *et al.*, 1993b).

Rabbits intestinally primed with live *V. cholerae* O1 can be boosted by intraduodenal administration of killed, smooth *V. cholerae* cells. Rabbits which had been immunized intraduodenally with strain R5 (a derivative of a smooth strain C5, but lacking the O-antigen of the LPS, and having a minimal lethal dose 100 000 times higher than that of the parent strain C5), were subsequently challenged with the smooth strain C5. All animals developed diarrhoea and two out of eight died, indicating the importance of the specific O antigen of the LPS for the induction of protective immunity in the rabbit (Jansen *et al.*, 1988).

Examination of specific anti-cholera antibody responses in the intestinal fluid obtained from bacteriologically confirmed cholera convalescent patients in Thailand, using a technique of collection of jejunal fluid by intestinal intubation, found that 100% of patients had anti-cholera LPS-specific IgA antibodies predominating in the intestinal fluid, with much lower levels of LPS-specific IgM and no detectable IgG antibody (Chongsa-Nguan *et al.*, 1987). The anti-cholera LPS-specific IgA antibodies were observed to increase during the convalescent period. Therefore, it became apparent that elevated anti-cholera LPS IgA antibodies occurred in the intestine following clinical cholera, and that these responses were similar to responses detected following the oral administration of a B-WCV (Svennerholm *et al.*, 1984).

Further evidence for the role of non-CT antigens in the mediation of protective immunity was demonstrated following oral administration of a monoclonal IgA antibody specifically directed against a lipopolysaccharide component of *V. cholerae* O1, that was demonstrated to protect neonatal mice against an oral challenge, as measured by reduced intestinal colonization (Lee *et al.*, 1994). A single dose of 0.1 μg of anti-LPS polymer IgA monoclonal antibody given 1 hour before challenge reduced the number of recoverable organisms by at least 100-fold. A 10-fold higher concentration of anti-LPS monomeric IgA was required to achieve the same level of protection as that conferred by polymeric IgA. This protective effect was retained by lower molecular weight fragments following digestion of the polymeric IgA with trypsin, or human intestinal aspirates.

Dimeric and polymeric IgA anti-CT monoclonal antibodies (Apter *et al.*, 1993b), directed against an Ogawa-specific lipopolysaccharide carbohydrate antigen exposed on the bacterial surface have been generated from hybridomas produced from PP lymphocytes following oral immunization with *V. cholerae* Ogawa O395 (Winner *et al.*, 1991; Apter *et al.*, 1993b). These hybridoma cells were used to produce subcutaneous 'backpack' tumours in syngeneic mice, that resulted in the secretion of anti-O antigen monoclonal SIgA onto mucosal surfaces. These were found to confer specific protection upon neonatal mice against oral challenge with 100 50% lethal doses of virulent *V. cholerae* Ogawa O395 organisms. Additionally, as mentioned previously, a monoclonal anti-CTB IgA, either delivered orally or secreted onto mucosal surfaces in mice bearing hybridoma tumours, did not significantly enhance protection over that provided by oral anti-LPS IgA alone (Apter *et al.*, 1993b), thus demonstrating that a single non-anti-CT monoclonal SIgA can be sufficient to protect against intestinal disease caused by toxigenic *V. cholerae* (Winner *et al.*, 1991). These results demonstrate that anti-LPS SIgA is much more effective than anti-CT IgA in prevention of *V. cholerae*-induced diarrhoeal disease (Apter *et al.*, 1993b).

The protective role of the O antigen of the *V. cholerae* LPS was further evaluated in humans using the live attenuated *Salmonella typhi* Ty21a oral typhoid vaccine strain, into which genes for the O antigen of *V. cholerae* LPS were inserted and expressed. Initially, the genes required for the biosynthesis of the O antigens of the Inaba serotypes of *V. cholerae* were initially transferred into *E. coli* K-12. The anti-Inaba O antigen antiserum raised in rabbits using the *E. coli*/Inaba hybrid was demonstrated to provide the same degree of protection against classical *V. cholerae* Inaba 569B challenge in the neonatal mouse protection test as the antiserum raised using the parent cholera strain (Manning *et al.*, 1986). The subsequent *S. typhi* Ty21a strains into which the *V. cholerae* Inaba O antigen genes had been inserted were all shown *in vitro* to produce *V. cholerae* O antigen when grown in the absence of exogenous galactose, and both *V. cholerae* and *S. typhi* O antigens when grown in the presence of exogenous galactose (LaBrooy *et al.*, 1987; Forrest, 1988b; Attridge *et al.*, 1990; Forrest and LaBrooy, 1991). Once such construction, termed EX645, was demonstrated to be both safe and immunogenic when administered to subjects in doses up to 6×10^{10} viable organisms (Forrest *et al.*, 1989), despite the validity of the vaccine strain and its plasmid retention being considerably adversely affected by the process of lyophilization.

However, EX645 was unable to confer a significant degree of protection upon recipients of three alternate-day oral doses of 6×10^{10} viable organisms, when challenged with 1.1×10^9 viable pathogenic *V. cholerae* Inaba strain N16961. The protective efficacy was determined to be 25% ($P = 0.13$, Fisher's exact test, 2-tailed) (Tacket *et al.*, 1990), although the vaccine did significantly reduce the mean stool volume of the vaccinated subjects with respect to the control subjects ($P = 0.0013$) (Table 4.2), with maximal protection being afforded against moderate (>2.01)

Table 4.2 Mean clinical and bacteriological responses of vaccinated and control subjects following challenge with *V. cholerae* El Tor Inaba N16961. (Adapted from Forrest, 1994.)

Subjects	Diarrhoeal stool volume (ml)	No. of diarrhoeal stools	Incubation period (h)	Peak *V. cholerae* excretion ($\times 10^7$)
Vaccinees	867*	6	25.6	3.5^+
Controls	2603*	11.6	20.5	6.2^+

*, P = 0.0013, Wilcoxon Rank Sum Test; $^+$, P = 0.053, Wilcoxon rank sum test.

to moderately-severe (>3.0 l) diarrhoea. Although the vaccine failed to solidly protect against clinical cholera, the results are suggestive of a protective effect attributable to the low levels of anti-*V. cholerae* O antigen intestinal antibody induced by this vaccine strain (Forrest, 1994).

4.8.2 OUTER MEMBRANE PROTEIN ANTIGENS

The specific outer membrane proteins (OMP) of *V. cholerae* have been implicated in contributing to the generation of a protective immune response in experimental animal infection (Attridge and Rowley, 1983; Kabir, 1983; Rijpkema *et al.*, 1987), and following clinical infection (Sears *et al.*, 1984). Antibodies to LPS, OMP antigens, CT and neuraminidase, have been detected in the rabbit model in intestinal extracts, sera and bile (Kabir, 1987). While both specific IgG and IgA antibodies can be detected, specific IgA has been shown to predominate, with anti-OMP and cell surface antigens appearing to be the major cross-reacting somatic antigens detectable.

V. cholerae O1 organisms belonging to different biotypes and serotypes have been shown to express other OMPs with molecular masses of 48–50, 40–43, 35–36, 27–28 and 20 kDa (Sengupta *et al.*, 1992a). Antisera raised against the individual OMPs of a single *V. cholerae* O1 strain were shown to recognize OMPs of corresponding molecular masses in other *V. cholerae* O1 and *V. cholerae* non-O1 strains, and inhibited *in vitro* the intestinal adhesion of the homologous and heterologous strains of *V. cholerae* irrespective of biotype, serotype and serovar, and induced passive protection against *V. cholerae* challenge in rabbit ileal loop experiments (Sengupta *et al.*, 1989). In other animal studies, bovine anti-OMP antibodies obtained from colostrum of newly parturient cows immunized with *V. cholerae* OMP extracts, were demonstrated to confer statistically significant protection against diarrhoea in infant rabbits challenged intra-intestinally with virulent *V. cholerae* O1 (Boesman-Finkelstein *et al.*, 1989). Of these five OMPs, two, the 40–43 kDa and 20 kDa cell surface proteins, were shown to induce significant protection against *V. cholerae* O1 El Tor Ogawa strain A17 challenge in the neonatal mouse model. Antisera raised against the 27–28 kDa OMP conferred only partial protection (Sengupta *et al.*, 1992b). Subsequent studies have demonstrated that these antisera, as well as IgG Fab fragments, induced significant inhibition of intestinal colonization of *V. cholerae* O1. The 40–43 kDa and 27–28 kDa proteins appeared to be porin-like, while the 20 kDa protein was found to be antigenically related to the toxin-co-regulated pilus (TCP). All these results demonstrate the involvement of more than one cell surface antigen of *V. cholerae* in the induction of protective immunity through inhibition of intestinal colonization of *V. cholerae*.

In previously unexposed adult volunteer subjects who were challenged with a variety of strains of *V. cholerae* and developed clinical illness, 50% of the 79 symptomatic subjects challenged with the *V. cholerae* El Tor biotype and 54% of those challenged with the Classical biotype had significant serum IgG rises against an OMP preparation. Significant rises were identified against the OMP preparation of both the homologous and heterologous strains (Sears *et al.*, 1984). A particular protein antigen of M_r 45–48 kDa found within both bio- and serotypes may represent the immunodominant protein antigen (Kabir, 1980). Local antibody responses to the OMP in post-challenge convalescent human subjects have indicated that the specific SIgA response is directed against components of the OMP of M_r 25 kDa or less (Richardson *et al.*, 1989). Additionally, serum IgG and jejunal fluid SIgA also reacted with antigens unique to *in vivo*-grown cells and several antigens in OMP preparations, suggesting the presence of unique *in vivo* expressed non-CT antigens may have a role in conferring protection against *V. cholerae* O1 (Richardson *et al.*, 1989).

In 1975, Dodin and Wiart described the isolation and purification of an antigenic component from *V. cholerae* lysate (Dodin and Wiart, 1975a). This antigen was characterized and found to be a 16-amino acid serotype-specific peptide of approximate M_r 57.5 kDa. Antibodies raised to this antigen were found to be protective in mice against pathogenic *V. cholerae* challenge; and capable of inducing an agglutinating and vibriocidal response in the serum of rabbits, donkeys and humans greater than that induced by the heat-inactivated subcutaneously administered cholera vaccine (Dodin and Wiart, 1975b).

In human volunteer subjects, 50 mg of this preparation administered orally induced serum agglutinins and vibriocidal antibodies that persisted for 4 months, as long as that induced by 100 μg of this preparation given subcutaneously (Dodin and Wiart, 1975b; Dodin, 1980). The efficacy of this orally administered preparation was evaluated in a controlled field trial performed in the Malemba-NKulu district of Shaba province in Zaire in 1983 (Dodin *et al.*, 1984; Masengo, 1984) and was found to induce a statistically significant level of protective efficacy against clinical disease; the PE was 98% compared with 22% for the commercial parenterally administered inactivated cholera vaccine, and further supported the role of *V. cholerae* OMP in protection against clinical cholera.

4.8.3 PILI/FIMBRIAL ANTIGENS

A fimbrial antigen of *V. cholerae* has been described that appears to play a major role in the ability of pathogenic strains of *V. cholerae* to adhere to the intestinal mucosae (Taylor *et al.*, 1988). This pilus, designated the toxin co-regulated pilus (TCP), and CT are regulated by the *toxR* gene complex. The importance of this pilus in the pathogenesis of clinical cholera has been evaluated in human subjects challenged orally with either a classical *V. cholerae* Ogawa strain O395-N1, or with a mutant of the O395 strain (designated JJM43) with a deletion of the *ctxA* gene rendering it unable to produce the CT A subunit, or with a third strain (designated TCP2) carrying defined deletions in the *toxR* and *tcpA* genes. Subjects receiving the O395-N1 strain reported abdominal cramps, malaise diarrhoea, anorexia and generated a good immune response indicating active colonization. Those receiving the strain TCP2 remained asymptomatic, with no evidence being found that the strain had colonized, nor had it generated an immune response. Recipients of strain JJM43 remained asymptomatic, although there was some evidence of poor colonization and the strain did stimulate an immune response.

Post-cholera challenge convalescent human subjects have been shown to generate a local intestinal SIgA antibody response to CT, and LPS, but not to TCP. However, three of six convalescent cholera patients in Indonesia were shown to seroconvert to TCP, suggesting that although seroconversion to TCP can occur in a population in an endemic area, it probably does not play a significant role in conferring protection against disease as protection has been demonstrated in the absence of a detectable anti-TCP immune response (Hall *et al.*, 1991).

4.9 CONCLUSION

It has been well accepted that clinical cholera caused by *V. cholerae* O1 strains confers a substantial degree of protection against subsequent attacks. However, there does not appear to be a clear direction in defining the mechanism of this protection, best represented by the fact that two quite divergent approaches to *V. cholerae* vaccine development currently exist: one being the identification and systematic removal of all toxin genes resulting in an attenuated live *V. cholerae* vaccine; the other being based on a purified or recombinant CTB co-administered with inactivated organisms. Both approaches have resulted in candidate vaccines that have demonstrated variable degrees of protection. The attenuated live toxin-deleted candidate vaccines represent a series of strains in which efficacy has had to be compromised to address concerns relating to the frequency and severity of adverse events.

However, with CT being demonstrated to be largely responsible for the diarrhoea so characteristic of clinical cholera, CT was an obvious initial target in vaccine development through either its removal, or through its attenuation or inactivation. While orally administered CT induces long-lasting solid protection in a variety of animal models, in its native form CT holotoxin cannot be used in human subjects as an oral immunogen. Although the CTB has performed well in a variety of animal models as both a potent immunogen inducing quite satisfactory protection against experimental *V. cholerae* infection, in humans it has only been demonstrated to confer partial, short term (<6 months) protection in the field.

However, only more recently – as attention has shifted away from CT and CTB as useful immunogens – has the role that other proteins play in conferring a degree of protective immunity against *V. cholerae* been identified. Although in limited studies to date, specific immunity against TCP has been considered to have a minor role in protection, this may not be the case for other surface antigens such as LPS and constituent proteins of the OMP. In a variety of experimental models anti-LPS specific IgA antibody has been demonstrated to confer solid protection against challenge with pathogenic strains, and in one small human study low-level anti-LPS antibodies did reduce the severity of disease induced by subsequent challenge. Similarly, local intestinal anti-OMP specific antibody has been repeatedly demonstrated to confer protection against challenge with toxigenic *V. cholerae*. The steady progression of *V. cholerae* O139 towards its first pandemic has resulted in an increase in interest in these alternatives to CT as specific protective antigens, especially since *V. cholerae* O139 is able to cause disease in individuals otherwise protected against disease caused by *V. cholerae* O1 strains. *V. cholerae* O139 also contributes a polysaccharide capsule to the range of non-CT antigens to be considered in further defining what is needed in order to confer solid protection against *V. cholerae*.

Perhaps *Bordetella pertussis* provides a model for vaccine development in this instance. The failure of the inactivated whole-cell pertussis vaccine to confer protection against a largely mucosal infection

(Marchant *et al.*, 1994; Wintermeyer *et al.*, 1994), led to the re-examination of the role of various antigens in the protective immune response. A variety of acellular pertussis vaccines, composed of a mixture of pertussis toxin (structurally similar to CT) and various surface antigens including fimbriae, and some other OMP component, have been developed in order to address the failure of the whole-cell vaccine (Halperin *et al.*, 1994; Podda *et al.*, 1994), many of which are still under evaluation for protective efficacy. Should they prove to confer a substantial degree of protection with minimal adverse events, they may represent a useful model for vaccine development for cholera. Although in cholera it is unlikely that any parenterally administered multicomponent vaccine will be able to confer protection against disease, applying the strategy used in the development of the acellular pertussis vaccines, combined with the novel technologies for the delivery of antigens to the intestinal mucosal immune system, may lead to the development of an orally administrable acellular cholera vaccine possibly including a genetically inactivated CT (Burnette and Kaslow, 1992), combined with detoxified LPS and other OMP antigens. The final make-up would be determined by further identification and definition of the role varying non-CT antigens have in protection. However, to move forward in vaccine development, additional efforts are required in elucidating the role of a variety of non-CT antigens in protective immunity.

4.10 REFERENCES

Apter, F.M., Lencer, W.I., Finkelstein, R.A., Mekalanos, J.J. and Neutra, M.R. (1993a) Monoclonal immunoglobulin A antibodies directed against cholera toxin prevent the toxin-induced chloride secretory response and block toxin binding to intestinal epithelial cells *in vitro*. *Infection and Immunity*, **61**, 5271–8.

Apter, F.M., Michetti, P., Winner, L.S. III, Mack, J.A., Mekalanos, J.J. and Neutra, M.R. (1993b) Analysis of the roles of anti-lipopolysaccharide and anti-cholera toxin immunoglobulin A (IgA) antibodies in protection against *Vibrio cholerae* and cholera toxin by use of monoclonal IgA antibodies *in vivo*. *Infection and Immunity*, **61**, 5279–85.

Attridge, S.R. and Rowley, D. (1983) Prophylactic significance of nonlipopolysaccharide antigens of *Vibrio cholerae*. *Journal of Infectious Diseases*, **148**, 931–9.

Attridge, S.R., Daniels, D., Morona, J.K. and Morona, R. (1990) Surface co-expression of *Vibrio cholerae* and *Salmonella typhi* O-antigens on Ty21a clone EX210. *Microbial Pathogenesis*, **8**, 177–88.

Bartholomeusz, R.C.A., Forrest, B.D., LaBrooy, J.T., Ey, P.L., Pyle, D., Shearman, D.J.C. and Rowley, D. (1990) The serum polymeric IgA antibody response to typhoid vaccination; its relationship to the intestinal IgA response. *Immunology*, **69**, 190–4.

Bienenstock, J. (1974) The physiology of the local immune response in the gastrointestinal tract. *Progress in Immunology*, **4**, 197–207.

Black, R.E., Levine, M.M., Clements, M.L., Young, C.R., Svennerholm, A.-M. and Holmgren, J. (1987) Protective efficacy in humans of killed whole-vibrio oral cholera vaccine with and without the B subunit of cholera toxin. *Infection and Immunity*, **55**, 1116–20.

Boesman-Finkelstein, M., Walton, N.E. and Finkelstein, R.A. (1989) Bovine lactogenic immunity against cholera toxin-related enterotoxins and *Vibrio cholerae* outer membranes. *Infection and Immunity*, **57**, 1227–34.

Bornside, G.H. (1981) Jaime Ferran and preventive inoculation against cholera. *Bulletin of the History of Medicine*, **55**, 516–32.

Brandtzaeg, P., Valnes, K., Scott, H., Rognum, T.O., Bjerke, K. and Baklien, K. (1985) The human gastrointestinal secretory immune system in health and disease. *Scandinavian Journal of Gastroenterology*, (Suppl. 114), 17–38.

Burnette, W.N. and Kaslow, H.R. (1992) Molecular engineering of cholera toxin. *Seventh International Congress of Mucosal Immunology, Prague, Czechoslovakia, August 16–20*, Abstracts, p. 37.

Burrows, W., Elliot, M. and Havens, I. (1947) Studies on immunity to Asiatic cholera. IV. The excretion of copro-antibody in experimental enteric cholera in the guinea pig. *Journal of Infectious Diseases*, **81**, 261–81.

Cash, R.A., Music, S.I., Libonati, J.P., Craig, J.P., Pierce, N.F. and Hornick, R.B. (1974) Response of man to infection with *Vibrio cholerae*. II. Protection from illness afforded by previous disease and vaccine. *Journal of Infectious Diseases*, **130**, 325–33.

Chongsa-Nguan, M., Chaicumpa, W., Alambaheti, T., Surachedchai, Y. and Thanungkul, B. (1987) Antibodies against *Vibrio cholerae* lipopolysaccharide, cell-bound haemagglutinin and toxin in the intestinal fluid during convalescence. *Southeast Asian Journal of Tropical Medicine and Public Health*, **18**, 33–8.

Clemens, J.D., Sack, D.A., Harris, J.R. *et al.* (1986) Field trial of oral cholera vaccines in Bangladesh. *Lancet*, **2**, 124–7.

Clemens, J.D., van Loon, F., Sack, D.A. *et al.* (1991) Biotype as determinant of natural immunising effect of cholera. *Lancet*, **337**, 883–4.

Comstock, L.E., Maneval, D., Jr, Panigrahi, P., Joseph, A., Levine, M.M., Kaper, J.B., Morris, J.G. Jr and Johnson, J.A. (1995) The capsule and O antigen in *Vibrio cholerae* O139 Bengal are associated with a genetic region not present in *Vibrio cholerae* O1. *Infection and Immunity*, **63**, 317–23.

Craig, J.P. (1981) Protective antigens of *Vibrio cholerae*, in Proceedings of the conference on experimental cholera vaccines, Dhaka, (ed. S. Rahman), *International Centre for Diarrhoeal Disease Research Special Publication No. 15*, pp. 14–22.

Craig, S.W. and Cebra, J.J. (1971) Peyer's patches: an enriched source of precursors for IgA-producing immunocytes in the rabbit. *Journal of Experimental Medicine*, **134**, 188–200.

Curlin, G.T., Craig, J.P., Subong and Carpenter, C.C. (1970) Antitoxic immunity in experimental canine cholera. *Journal of Infectious Diseases*, **121**, 463–70.

Curlin, G., Levine, R., Aziz, K.M.A., Mizanur Rahman, A.C.M. and Verwey, W.F. (1975) Field trial of cholera toxoid, in *Proceedings of the Eleventh Joint Conference on Cholera, US–Japan Co-operative Medical Science Program*, DHSS, Washington, DC, pp. 314–29.

Czerkinsky, C., Svennerholm, A.M., Quiding, M., Jonsson, R. and Holmgren, J. (1991) Antibody-producing cells in peripheral blood and salivary glands after oral cholera vaccination of humans. *Infection and Immunity*, **59**, 996–1001.

Czinn, S.J. and Lamm, M.E. (1986) Selective chemotaxis of subsets of B lymphocytes from gut-associated lymphoid tissue and its implications for the recruitment of mucosal plasma cells. *Journal of Immunology*, **136**, 3607–11.

De, S.N. (1959) Enterotoxicity of bacteria free culture filtrate of *Vibrio cholerae*. *Nature*, **183**, 1533.

Dodin, A. (1980) Un vaccin oral contre le cholera. *La Recherche*, **11**, 468–9.

Dodin, A. and Wiart, J. (1975a) Mise en evidence d'une fraction antigenique vaccinante de *Vibrio cholerae*. I. Isolement et caracterisation de la fraction. *Annals of Microbiology*, **126A**, 39–44.

Dodin, A. and Wiart, J. (1975b) Mise en evidence d'une fraction antigenique vaccinante de *Vibrio cholerae*. II. Proprietes biologiques de la fraction Ch1+2 comparees a celles du vaccin bacterien classique. *Annals of Microbiology*, **126A**, 45–56.

Dodin, A., Masengo, B. and Loucq, C. (1984) Resultats controles du vaccin anticholerique oral de l'Institut Pasteur au cours de l'epidemie du Shaba-Zaire en 1983. *Comptes Rendus de L'Academie des Sciences, Series III*, **299**, 205–7.

Enders, G.A., Delius, M., Ballhaus, S. and Brendel, W. (1987) Role of Peyer's patch in the intestinal immune response to cholera toxin in enterically immunized rats. *Infection and Immunity*, **55**, 1997–9.

Fishman, P.H. (1980) Mechanism of action of cholera toxin: events on the cell surface, in *Secretory Diarrhea* (eds M. Field, J.S. Fordtran and S.G. Schultz), American Physiological Society, Bethesda, pp. 85–106.

Flo, J., Massouh, E.J. and Roux, M.E. (1993) Long term effects of malnutrition during lactation on GALT. *Regional Immunology*, **5**, 100–5.

Forrest, B.D. (1988a) The identification of an intestinal immune response using peripheral blood lymphocytes. *Lancet*, **1**, 81–3.

Forrest, B.D. (1988b) The development of a bivalent vaccine against diarrhoeal disease. *Southeast Asian Journal of Tropical Medicine and Public Health*, **19**, 449–57.

Forrest, B.D. (1992) The indirect measurement of specific intestinal immune responses to an orally administered attenuated bacterial vaccine. *Infection and Immunity*, **60**, 2023–9.

Forrest, B.D. (1994) Clinical evaluation of attenuated *Salmonella* vaccines in human subjects, in *Oral Delivery Systems* (ed. D. O'Hagan), CRC Press, Boca Raton, pp. 59–85.

Forrest, B.D. and LaBrooy, J.T. (1991) *In vivo* evidence of immunological masking of the *Vibrio cholerae* O antigen of a hybrid *Salmonella typhi* Ty21a-*Vibrio cholerae* oral vaccine in humans. *Vaccine*, **9**, 515–20.

Forrest, B.D., LaBrooy, J.T., Attridge, S.R., Boehm, G., Beyer, L., Morona, R., Shearman, D.J.C. and Rowley, D. (1989) Immunogenicity of a candidate live oral typhoid/cholera hybrid vaccine in humans. *Journal of Infectious Diseases*, **159**, 145–6.

Freter, R., De, S.P., Mondal, A., Shrivastava, D.L. and Sunderman, F.W. (1965) Coproantibody and serum antibody in cholera patients. *Journal of Infectious Diseases*, **115**, 83–7.

Fujisawa, H., Tsuru, S., Taniguchi, M. and Zinnaki, Y. (1987) Oral immunization in adult mice to live and heat-killed *Vibrio cholerae*. *Microbiology and Immunology*, **31**, 763–70.

Fujita, K. and Finkelstein, R.A. (1972) Antitoxic immunity in experimental cholera: comparison of immunity induced perorally and parenterally in mice. *Journal of Infectious Diseases*, **125**, 647–55.

Gill, D.M. and Meren, R. (1978) ADP-ribosylation of membrane proteins catalyzed by cholera toxin: basis of the activation of adenylate cyclase. *Proceedings of the National Academy of Sciences, USA*, **75**, 3050–4.

Glass, R.I., Becker, S., Huq, M.I., Stoll, B.J., Khan, M.U., Merson, M.H., Lee, J.V. and Black, R.E. (1982) Endemic cholera in rural Bangladesh, 1966–1980. *American Journal of Epidemiology*, **116**, 959–70.

Gotuzzo, E., Butron, B., Seas, C. *et al.* (1993) Safety, immunogenicity, and excretion pattern of single-dose live oral cholera vaccine CVD 103-HgR in Peruvian adults of high and low socioeconomic levels. *Infection and Immunity*, **61**, 3994–7.

Hall, R.H., Losonsky, G., Silveira, A.P., Taylor, R.K., Mekalanos, J.J., Witham, N.D. and Levine, M.M. (1991) Immunogenicity of *Vibrio cholerae* O1 toxin-coregulated pili in experimental and clinical cholera. *Infection and Immunity*, **59**, 2508–12.

Hall, R.H., Khambaty, F.M., Kothary, M.H., Keasler, S.P. and Tall, B.D. (1994) *Vibrio cholerae* non-O1 serogroup associated with cholera gravis genetically and physiologically resembles O1 El Tor cholera strains. *Infection and Immunity*, **62**, 3859–63.

Halperin, S.A., Barreto, L., Eastwood, B.J., Law, B. and Roberts, E.A. (1994) Safety and immunogenicity of a five-component acellular pertussis vaccine with varying antigen quantities. *Archives of Pediatric and Adolescent Medicine*, **148**, 1220–4.

Haq, J.A. and Szewczuk, M.R. (1993) Regional dichotomy of interleukin-4 and -5 regulation of senescent B cell responses specific for cholera toxin in Peyer's patches, lamina propria, and mesenteric lymph nodes. *Regional Immunology*, **5**, 60–7.

Holmgren, J. and Svennerholm, A.-M. (1977) Mechanisms of disease and immunity in cholera. *Journal of Infectious Diseases*, **136** (Suppl.), 105–12.

Holmgren, J., Svennerholm, A.-M., Ouchterlony, O., Andersson, A., Wallerstrom, G. and Westerberg Berndtsson, U. (1975) Antitoxic immunity in experimental cholera: protection, and serum and local antibody responses in rabbits after enteral and parenteral immunization. *Infection and Immunity*, **12**, 1331–40.

Hornqvist, E., Goldschmidt, T.J., Holmdahl, R. and Lycke, N. (1991) Host defense against cholera toxin is strongly CD4+ T cell dependent. *Infection and Immunity*, **59**, 3630–8.

Jansen, W.H., Gielen, H., Rijpkema, S.G. and Guinee, P.A. (1988) Priming and boosting of the rabbit intestinal immune system with live and killed, smooth and rough *Vibrio cholerae* cells. *Microbial Pathogenesis*, **4**, 21–6.

Jertborn, M., Svennerholm, A.-M. and Holmgren, J. (1984) Gut mucosal, salivary and serum antitoxic and antibacterial antibody responses in Swedes after oral immunization with B subunit-whole cell cholera vaccine. *International Archives of Allergy and Applied Immunology*, **75**, 39–43.

Jertborn, M., Svennerholm, A.-M. and Holmgren, J. (1988) IgG and IgA subclass distribution of antitoxin antibody responses after oral cholera vaccination or cholera disease. *International Archives of Allergy and Applied Immunology*, **85**, 358–63.

Jertborn, M., Svennerholm, A.-M. and Holmgren, J. (1992) Safety and immunogenicity of an oral recombinant cholera B subunit-whole cell vaccine in Swedish volunteers. *Vaccine*, **10**, 130–2.

Johnson, J.A., Salles, C.A., Panigrahi, P., Albert, M.J., Wright, A.C., Johnson, R.J. and Morris, J.G. Jr. (1994) *Vibrio cholerae* O139 synonym bengal is closely related to *Vibrio cholerae* El Tor but has important differences. *Infection and Immunity*, **62**, 2108–10.

Kabir, S. (1980) Composition and immunochemical properties of outer membrane proteins of *Vibrio cholerae. Journal of Bacteriology*, **144**, 392–9.

Kabir, S. (1983) Immunochemical properties of the major outer membrane protein of *Vibrio cholerae. Infection and Immunity*, **39**, 452–6.

Kabir, S. (1987) Immunological responses of rabbits to various somatic and secreted antigens of *Vibrio cholerae* after intra-duodenal inoculation. *Journal of Medical Microbiology*, **24**, 29–40.

Klemperer, G. (1892) Recherches sur la vaccination antitoxique dans le cholera. *Gezette Hebondaire de Medicine*, **29**, 399.

Koch, R. (1884) An address on cholera and its bacillus. *British Medical Journal*, **2**, 405–7, 453–9.

Kurosky, A., Markel, D.E., Touchstone, B. and Peterson, W. (1975) Chemical characterization of the structure of cholera toxin and its natural toxoid. *Journal of Infectious Diseases*, **133** (Suppl.), S14–22.

LaBrooy, J.T., Forrest, B.D. and Bartholomeusz, R.C.A. (1987) Immunization against cholera with a Ty21a-cholera hybrid. *Annales Sclavo Collagano Monographia*, **1–2**, 143–7.

Lagos, R., Avendano, A., Prado, V., Horwitz, I., Wasserman, S., Losonsky, G., Cryz, S. Jr, Kaper, J.B. and Levine, M.M. (1995) Attenuated live cholera vaccine strain CVD 103-HgR elicits significantly higher serum vibriocidal antibody titers in persons of blood group O. *Infection and Immunity*, **63**, 707–9.

Lebens, M. and Holmgren, J. (1994) Mucosal vaccines based on the use of cholera toxin B subunit as immunogen and antigen carrier. *Developments in Biological Standardization*, **82**, 215–27.

Lee, C.K., Weltzin, R., Soman, G., Georgakopoulos, K.M., Houle, D.M. and Monath, T.P. (1994) Oral administration of polymeric immunoglobulin A prevents colonization with *Vibrio cholerae* in neonatal mice. *Infection and Immunity*, **62**, 887–91.

Levine, M.M. (1980) Immunity to cholera as evaluated in volunteers, in *Cholera and Related Diarrheas*, 43rd Nobel Symposium 1978, (eds O. Ouchterlony and J. Holmgren), Karger, Basel, pp. 195–203.

Levine, M.M. and Kaper, J.B. (1993) Live oral vaccines against cholera: an update. *Vaccine*, **11**, 207–12.

Levine, M.M., Nalin, D.R., Craig, J.P. *et al.* (1979) Immunity to cholera in man: relative role of antibacterial versus antitoxic immunity. *Transactions of the Royal Society for Tropical Medicine and Hygiene*, **73**, 3–9.

Levine, M.M., Black, R.E., Clements, M.L., Cisneros, L., Nalin, D.R. and Young, C.R. (1981) Duration of infection-derived immunity to cholera. *Journal of Infectious Diseases*, **143**, 818–20.

Levine, M.M., Kaper, J.B., Black, R.E. and Clements, M.L. (1983) New knowledge on pathogenesis of bacterial enteric infections as applied to vaccine development. *Microbiological Reviews*, **47**, 510–50.

Levine, M.M., Kaper, J.B., Herrington, D., Ketley, J., Losonsky, G., Tacket, C.O., Tall, B. and Cryz, S. (1988) Safety, immunogenicity, and efficacy of recombinant live oral cholera vaccines, CVD 103 and CVD 103-HgR. *Lancet*, **2**, 467–70.

Lewis, D.J., Novotny, P., Dougan, G. and Griffin, G.E. (1991) The early cellular and humoral immune response to primary and booster oral immunization with cholera toxin B subunit. *European Journal of Immunology*, **21**, 2087–94.

Lewis, D.J., Castello-Branco, L.R., Novotny, P., Dougan, G., Poulton, T.A. and Griffin, G.E. (1993) Circulating cellular immune response to oral immunization of humans with cholera toxin B-subunit. *Vaccine*, **11**, 119–21.

Lycke, N. and Holmgren, J. (1987) Long-term cholera antitoxin memory in the gut can be triggered to antibody formation associated with protection within hours of an oral challenge immunization. *Scandinavian Journal of Immunology*, **25**, 407–12.

Lycke, N. and Holmgren, J. (1989) Adoptive transfer of gut mucosal antitoxin memory by isolated B cells 1 year after oral immunization with cholera toxin. *Infection and Immunity*, **57**, 1137–41.

Lycke, N., Eriksen, L. and Holmgren, J. (1987a) Protection against cholera toxin after oral immunization is thymus-dependent and associated with intestinal production of neutralizing IgA antitoxin. *Scandinavian Journal of Immunology*, **25**, 413–19.

Lycke, N., Hellstrom, U. and Holmgren, J. (1987b) Circulating cholera antitoxin memory cells in the blood one year after oral cholera vaccination in humans. *Scandinavian Journal of Immunology*, **26**, 207–11.

Lycke, N., Bromander, A.K. and Holmgren, J. (1989) Role of local IgA antitoxin-producing cells for intestinal protection against cholera toxin challenge. *International Archives of Allergy and Applied Immunology*, **88**, 273–9.

Manning, P.A., Heuzenroeder, M.W., Yeadon, J., Leavesley, D.I., Reeves, P.R. and Rowley, D. (1986) Molecular cloning and expression in *Escherichia coli* K-12 of the O antigens of the Inaba and Ogawa serotypes of the *Vibrio cholerae* O1 lipopolysaccharides and their potential for vaccine development. *Infection and Immunity*, **53**, 272–7.

Marchant, C.D., Loughlin, A.M., Lett, S.M. *et al.* (1994) Pertussis in Massachusetts, 1981–1991: incidence, serologic diagnosis, and vaccine effectiveness. *Journal of Infectious Diseases*, **169**, 1297–305.

Masengo, B. (1984) Premiers essais controles du vaccin anticholerique oral au cours de l'epidemie de cholera dans la zone de Malemba-NKulu (Shaba-Zaire). *Bulletin de la Société Pathologique Exotique*, **77**, 13–16.

McDermott, M.R. and Bienenstock, J. (1979) Evidence for a common mucosal immunologic system. I. Migration of B-immunoblasts into intestinal, respiratory and genital tissues. *Journal of Immunology*, **122**, 1892–8.

McWilliams, M., Phillips-Quagliata, J.M. and Lamm, M.E. (1977) Mesenteric lymph node B lymphoblasts which home to the small intestine are precommitted to IgA synthesis. *Journal of Experimental Medicine*, **145**, 866–75.

Merson, M.H., Black, R.E., Khan, M. and Huq, I. (1980) Epidemiology of cholera and enterotoxigenic *Escherichia coli* diarrhea, in *Cholera and Related Diarrheas*, 43rd Nobel Symposium 1978, (eds O. Ouchterlony and J. Holmgren), Karger, Basel, pp. 34–45.

Mestecky, J. (1987) The common mucosal immune system and current strategies for induction of immune responses in external secretions. *Journal of Clinical Immunology*, **7**, 265–76.

Mestecky, J., Russell, M.W., Jackson, S. and Brown, T.A. (1986) The human IgA system: a reassessment. *Clinical Immunology and Immunopathology*, **40**, 105–14.

Michalski, J., Galen, J.E., Fasano, A. and Kaper, J.B. (1993) CVD110, an attenuated *Vibrio cholerae* O1 El Tor live oral vaccine strain. *Infection and Immunity*, **61**, 4462–8.

Mosley, W.H., Benenson, A.S. and Barui, R. (1968) A serological survey for cholera antibodies in rural East Pakistan. I. The distribution of antibody in the control population of a cholera vaccine field-trial area and the relation of antibody titre to the pattern of endemic cholera. *Bulletin of the World Health Organization*, **38**, 327–34.

Nair, G.B., Shimada, T., Kurazono, H. *et al.* (1994) Characterization of phenotypic, serological, and toxigenic traits of *Vibrio cholerae* O139 bengal. *Journal of Clinical Microbiology*, **32**, 2775–9.

Northrup, R.S. and Hossain, S.A. (1970) Immunoglobulins and antibody activity in the intestine and serum in cholera. II. Measurement of antibody activity in jejunal aspirates and sera of cholera patients by radioimmunodiffusion. *Journal of Infectious Diseases*, **121** (Suppl.), S142–6.

Northrup, R.S., Bienenstock, J. and Tomasi, T.B. (1970) Immunoglobulins and antibody activity in the intestine and serum in cholera. I. Analysis of immunoglobulins in cholera stool. *Journal of Infectious Diseases*, **121** (Suppl.), S137–41.

Owen, R.L. (1977) Sequential uptake of horseradish peroxidase by lymphoid follicle epithelium of Peyer's patches in the normal unobstructed mouse intestine: an ultrastructural study. *Gastroenterology*, **72**, 440–51.

Owen, R.L., Pierce, N.F., Apple, R.T. and Cray, W.C. Jr. (1986) M cell transport of *Vibrio cholerae* from the intestinal lumen into Peyer's patches: a mechanism for antigen sampling and for microbial transepithelial migration. *Journal of Infectious Diseases*, **153**, 1108–18.

Pierce, N.F. (1978) The role of antigen form and function in the primary and secondary intestinal immune responses to cholera toxin and toxoid in rats. *Journal of Experimental Medicine*, **148**, 195–206.

Pierce, N.F. and Reynolds, H.Y. (1974) Immunity to experimental cholera. I. Protective effect of humoral IgG antitoxin demonstrated by passive immunization. *Journal of Immunology*, **113**, 1017–23.

Pierce, N.F., Sack, R.B. and Sircar, B.K. (1977) Immunity to experimental cholera. III. Enhanced duration of protection after sequential parenteral administration of toxoid to dogs. *Journal of Infectious Diseases*, **135**, 888–96.

Pierre, P., Langendries, A. and Vaerman, J.P. (1988) Cholera toxin neutralization: a comparison of purified serum IgG and biliary secretory IgA antibodies. *Immunology Letters*, **18**, 51–5.

Plaut, A.G. and Keonil, P. (1969) Immunoglobulins in the small intestinal fluid. *Gastroenterology*, **56**, 522–30.

Podda, A., De Luca, E.C., Contu, B. *et al.* (1994) Comparative study of a whole-cell pertussis vaccine and a recombinant acellular pertussis vaccine. The Italian Multicenter Group for the Study of Recombinant Acellular Pertussis Vaccine. *Journal of Pediatrics*, **124**, 921–6.

Pollitzer, R. (1959) *Cholera*, World Health Organization, Geneva.

Rappaport, R.S., Bonde, G., McCann, T., Rubin, B.A. and Tint, H. (1974) Development of a purified cholera toxoid. II. Preparation of a stable antigenic toxoid by reaction of purified toxoid with glutaraldehyde. *Infection and Immunity*, **9**, 304–17.

Richardson, K., Kaper, J.B. and Levine, M.M. (1989) Human immune response to *Vibrio cholerae* O1 whole cells and isolated outer membrane antigens. *Infection and Immunity*, **57**, 495–501.

Rijpkema, S.G., Jansen, W.H., Gielen, H. and Guinee, P.A. (1987) Immunoglobulins in bile and serum of the rabbit associated with protection after *Vibrio cholerae* infection and vaccination. *Microbial Pathogenesis*, **3**, 365–75.

Rudzik, O., Clancy, R.L., Perey, D.Y.E., Day, R.P. and Bienenstock, J. (1975) Re-population with IgA-containing cells of bronchial and intestinal lamina propria after transfer of homologous Peyer's patch and bronchial lymphocytes. *Journal of Immunology*, **114**, 1599–604.

Sears, S.D., Richardson, K., Young, C., Parker, D. and Levine, M.M. (1984) Evaluation of the human immune response to outer membrane proteins of *Vibrio cholerae*. *Infection and Immunity*, **44**, 439–44.

Sengupta, D., Datta-Roy, K., Banerjee, K. and Ghose, A.C. (1989) Identification of some antigenically related outer-membrane proteins of strains of *Vibrio cholerae* O1 and non-O1 serovars involved in intestinal adhesion and the protective role of antibodies to them. *Journal of Medical Microbiology*, **29**, 33–9.

Sengupta, D.K., Sengupta, T.K. and Ghose, A.C. (1992a) Antibodies to outer membrane proteins of *Vibrio cholerae* induce protection by inhibition of intestinal colonization of vibrios. *FEMS Microbiology and Immunology*, **4**, 261–6.

Sengupta, D.K., Sengupta, T.K. and Ghose, A.C. (1992b) Major outer membrane proteins of *Vibrio cholerae* and their role in induction of protective immunity through inhibition of intestinal colonization. *Infection and Immunity*, **60**, 4848–55.

Serizawa, H., Miura, S., Tashiro, H. *et al.* (1993) Significant changes in intestinal lymphatic system and immune response elicited by Peyer's patch excision in adult rats. *Journal of Gastroenterology and Hepatology*, **8**, 63–9.

Svennerholm, A.-M. (1980) Nature of protective cholera immunity, in *Cholera and Related Diarrheas*, 43rd Nobel Symposium 1978, (eds O. Ouchterlony and J. Holmgren), Karger, Basel, pp. 171–84.

Svennerholm, A.-M. and Holmgren, J. (1976) Synergistic protective effect in rabbits of immunization with *Vibrio cholerae* lipopolysaccharide and toxin/toxoid. *Infection and Immunity*, **13**, 735–40.

Svennerholm, A.-M., Sack, D.A., Holmgren, J. and Bardhan, P.K. (1982) Intestinal antibody responses after immunisation with cholera B subunit. *Lancet*, **1**, 305–8.

Svennerholm, A.-M., Jertborn, M., Gothefors, L., Karim, A.M.M.M., Sack, D.A. and Holmgren, J. (1984) Mucosal antitoxic and antibacterial immunity after cholera disease and after immunization with a combined B subunit-whole cell vaccine. *Journal of Infectious Diseases*, **149**, 884–93.

Tacket, C.O., Forrest, B., Morona, R., Attridge, S.R., LaBrooy, J.T., Tall, B.D., Reymann, M., Rowley, D. and Levine, M.M. (1990) Safety, immunogenicity, and efficacy against cholera challenge in humans of a typhoid-cholera hybrid vaccine derived from *Salmonella typhi* Ty21a. *Infection and Immunity*, **58**, 1620–7.

Tacket, C.O., Losonsky, G., Nataro, J.P., Cryz, S.J., Edelman, R., Kaper, J.B. and Levine, M.M. (1992) Onset and duration of protective immunity in challenged volunteers after vaccination with live oral cholera vaccine CVD 103-HgR. *Journal of Infectious Diseases*, **166**, 837–41.

Tacket, C.O., Losonsky, G., Nataro, J.P., Cryz, S.J., Edelman, R., Fasano, A., Michalski, J., Kaper, J.B. and Levine, M.M. (1993) Safety and immunogenicity of live oral cholera vaccine candidate CVD 110, a delta *ctxA* delta *zot* delta *ace* derivative of El Tor Ogawa *Vibrio cholerae*. *Journal of Infectious Diseases*, **168**, 1536–40.

Targan, S.R. (1987) Immunologic mechanisms in intestinal diseases. *Annals of Internal Medicine*, **106**, 853–70.

Taylor, R., Shaw, C., Peterson, K., Spears, P. and Mekalanos, J. (1988) Safe, live *Vibrio cholerae* vaccines? *Vaccine*, **6**, 151–4.

Tomasi, T.B. and Bienenstock, J. (1968) Secretory immunoglobulins. *Advances in Immunology*, **9**, 1–96.

Vajdy, M., Kosco-Vilbois, M.H., Kopf, M., Kohler, G. and Lycke, N. (1995) Impaired mucosal immune responses in interleukin 4-targeted mice. *Journal of Experimental Medicine*, **181**, 41–53.

Veldkamp, J., van der Gaag, R. and Willers, J.M.N. (1973) The role of Peyer's patch cells in antibody formation. *Immunology*, **25**, 761–71.

Waldman, R.H., Bencic, Z., Sakazaki, R., Sinha, R., Ganguly, R., Deb, B.C. and Mukerjee, S. (1971) Cholera immunology. I. Immunoglobulin levels in serum, fluid from the small intestine, and feces from patients with cholera and noncholeraic diarrhea during illness and convalescence. *Journal of Infectious Diseases*, **123**, 579–86.

Waldman, R.H., Bencic, Z., Sinha, R., Deb, B.C., Sakazaki, R., Tamura, K., Mukerjee, S. and Ganguly, R. (1972) Cholera immunology. II. Serum and intestinal secretion antibody response after naturally occurring cholera. *Journal of Infectious Diseases*, **126**, 401–7.

Walker, W.A. and Isselbacher, K.J. (1977) Intestinal antibodies. *New England Journal of Medicine*, **297**, 767–73.

Watanabe, Y., Verwey, W.F., Guckian, J.C., Williams, H.R., Phillips, P.E. and Rocha, S.S. (1969) Some of the properties of mouse protective antigens derived from *Vibrio cholerae*. *Texas Reports in Biology and Medicine*, **27** (suppl. 1), 275–98.

Weintraub, A., Widmalm, G., Jansson, P.E., Jansson, M., Hultenby, K. and Albert, M.J. (1994) *Vibrio cholerae* O139 Bengal possesses a capsular polysaccharide which may confer increased virulence. *Microbial Pathogenesis*, **16**, 235–41.

Williams, R.C. and Gibbons, R.J. (1972) Inhibition of bacterial adherence by secretory immunoglobulin A: a mechanism of antigen disposal. *Science*, **177**, 697–9.

Winner, L. III, Mack, J., Weltzin, R., Mekalanos, J.J., Kraehenbuhl, J.P. and Neutra, M.R. (1991) New model for analysis of mucosal immunity: intestinal secretion of specific monoclonal immunoglobulin A from hybridoma tumors protects against *Vibrio cholerae* infection. *Infection and Immunity*, **59**, 977–82.

Wintermeyer, S.M., Nahata, M.C. and Kyllonen, K.S. (1994) Whole-cell and acellular pertussis vaccines. *Annals of Pharmacotherapeutics*, **28**, 925–39.

Wolf, J.L. and Bye, W.A. (1984) The membranous epithelial (M) cell and the

mucosal immune system. *Annual Reviews in Medicine*, **35**, 95–112.
World Health Organization Diarrhoeal Diseases Control Programme (1989) Programme for control of diarrhoeal diseases. Interim programme report 1988. WHO Report number WHO/CDD/89.31, Geneva.
World Health Organization Scientific Working Group (1980) Cholera and other vibrio-associated diarrhoeas. *Bulletin of the World Health Organization*, **58**, 353–74.
Xu-Amano, J., Jackson, R.J., Fujihashi, K., Kiyono, H., Staats, H.F. and McGhee, J.R. (1994) Helper Th1 and Th2 cell responses following mucosal or systemic immunization with cholera toxin. *Vaccine*, **12**, 903–11.
Yamamoto, T., Albert, M.J. and Sack, R.B. (1994) Adherence to human small intestines of capsulated *Vibrio cholerae* O139. *FEMS Microbiology Letters*, **119**, 229–35.
Yoshiyama, Y. and Brown, W.R. (1987) Specific antibodies to cholera toxin in rabbit milk are protective against *Vibrio cholerae*-induced intestinal secretion. *Immunology*, **61**, 543–7.

Cholera: pathogenesis and vaccine development

5

M.M. Levine and J.B. Kaper

5.1 AN OVERVIEW OF CHOLERA PATHOGENESIS

Cholera, which is one of the best-studied and understood of all infectious diseases from the purview of both clinical pathophysiology and molecular pathogenesis, commences with ingestion of a food or water vehicle containing pathogenic *Vibrio cholerae* O1 or O139. If the vibrios successfully transit the acid barrier of the stomach and pass through the pylorus, they may then initiate a series of adaptations that allow them to colonize the proximal small intestine, overcoming the peristaltic defence mechanism.

Fundamental to this process is the activation by the regulatory gene *toxR* of an array of genes encoding factors involved in pathogenesis. Yet other virulence attributes that are not part of the ToxR regulon also make critical contributions. By guided motility whereby the vibrios respond to chemotaxins and utilize mucus-dissolving enzymes to migrate through the mucous layer, the vibrios ultimately reach the enterocytes of the proximal small intestine to which they attach by means of toxin co-regulated pili (TCP) and other adhesins. Cholera enterotoxin elaborated by the adherent vibrios (and possibly other toxins) disrupts ion transport, causing diminished absorption by villus tip cells and outright secretion by crypt cells. The culminating, cumulative effect of this intestinal loss of water and electrolytes is the voluminous, electrolyte-rich, watery diarrhoea characteristic of cholera. The various steps in the pathogenesis of cholera and the participating virulence attributes are reviewed below.

Cholera and the Ecology of Vibrio cholerae. Edited by B.S. Drasar and B.D. Forrest.
Published in 1996 by Chapman & Hall, London. ISBN 0 412 61220 8.

5.2 STUDIES OF PATHOGENESIS IN THE EXPERIMENTAL CHOLERA MODEL IN VOLUNTEERS

Volunteer challenge studies, in which consenting informed adults are experimentally infected with *V. cholerae* under quarantine conditions, have yielded many insights into cholera pathogenesis and host immune response and have been invaluable in directing vaccine development strategies. Volunteer studies demonstrate the critical importance that the gastric barrier plays as a first-line non-specific host defence. If vibrios are ingested with water by a fasting normochlorhydric host, they are destroyed by gastric acid. Indeed, ingestion of as many as 10^9 vibrios in water, without buffer, by normochlorhydric North American volunteers results neither in infection nor diarrhoeal illness (Cash *et al.*, 1974; Levine *et al.*, 1981b). In contrast, ingestion by North American volunteers of 10^6 vibrios with buffer or with food results in an attack rate of 90–100%, a proportion of whom develop copious purges (Levine *et al.*, 1981b). Moreover, most volunteers who receive as few as 10^3 to 10^4 organisms with buffer develop diarrhoea (Levine *et al.*, 1981b), although with lower inocula the incubation period tends to be longer and the total diarrhoeal stool volume is less than with higher inocula (10^5 to 10^6) (Table 5.1). These data suggest that for certain highly susceptible hosts (e.g. of blood group O; see below) the effective inoculum in nature may be quite low. The marked variability in the severity of diarrhoea observed among volunteers who ingest the identical challenge inoculum, emphasizes the critical role that host factors play in determining the clinical consequences of cholera infection. Some of these host factors have been elucidated, whereas others remain enigmatic. For example, volunteer studies have incriminated blood group O as a risk factor for the development of cholera gravis (Levine *et al.*, 1979b; Tacket *et al.*, 1995b), corroborating correlations made in the field (Glass *et al.*, 1985a; Clemens *et al.*, 1989).

The crucial role of cholera toxin (CT) in causing the syndrome of cholera gravis was shown in clinical trials in which volunteers ingested varying doses of purified CT (Levine *et al.*, 1983). As little as 5 μg of pure CT resulted in diarrhoea in five of six volunteers who experienced 1–6 l of diarrhoea and ingestion of 25 μg of pure CT caused purging of over 20 l of rice-water stool and a clinical syndrome resembling cholera gravis. Subsequent studies with genetically engineered *V. cholerae* strains harbouring deletions in *ctx* genes encoding one or both subunits of CT demonstrated that, unlike their virulent parent strains, these CT deletion mutants could no longer cause severe diarrhoea. Nonetheless, ingestion of a number of these engineered vaccine strains resulted in mild diarrhoea, even in the

Table 5.1 Response of healthy North American adult volunteers following ingestion of various doses of *V. cholerae* O1 El Tor Inaba strain N16961 administered with buffer, food or plain water

Dose (CFU)	Vehicle for administering the oral inoculum	Attack rate for diarrhoea (%)		Mean incubation (h)	Mean diarrhoeal stool volume per ill volunteer (l)
10^6	Buffer + water*	9/10	(90)	26	3.2
10^5	Buffer + water*	3/5	(60)	18	3.1
10^4	Buffer + water*	4/5	(80)	37	1.1
10^3	Buffer + water*	3/5	(60)	33	0.9
10^6	Food$^+$	6/6	(100)	24	3.5
10^6	Plain water‡	0/7	(0)	–	–

*2.0 g of $NaHCO_3$ in 150 ml of water.
$^+$120 g of fish, 120 g of rice, 120 g of custard and 240 ml of skim milk.
‡300 ml.

absence of expression of CT (Levine *et al.*, 1988; Tacket *et al.*, 1993). Thus, although CT is responsible for the severe purging that results in dehydration and cholera gravis, there exist additional factors by which *V. cholerae* elicits mild diarrhoea and other signs and symptoms.

The critical roles played by the ToxR regulatory system and the TCP colonization factor in the pathogenesis of cholera were demonstrated in volunteer studies (Herrington *et al.*, 1988). Conversely, volunteer studies showed that *V. cholerae* O1 strains isolated from sewage water in Brazil which lacked CT and TCP could not readily colonize the intestine, cause diarrhoeal illness, or elicit strong vibriocidal responses (Levine *et al.*, 1982).

5.3 ANIMAL MODELS OF CHOLERA

Although *V. cholerae* O1 diarrhoeal disease does not occur in animals in nature, a few animal models have yielded useful information on cholera infection and immunity in humans (Richardson *et al.*, 1984; Sack and Spira, 1986). Extensively used models include the rabbit ligated ileal loop (De, 1959), the sealed adult mouse model (Richardson *et al.*, 1984; in which the anorectal canal is sealed with cyanoacrylate ester glue), the infant mouse model (Baselski *et al.*, 1977) and the infant rabbit model (Dutta *et al.*, 1959). The most popular intact animal model is the RITARD (removable intestinal tie-adult rabbit diarrhoea) model (Spira *et al.*, 1981), which utilizes a temporary slip knot tie of the small bowel that is subsequently removed 2 hours after inoculation of live *V. cholerae* O1 proximal to the tie. If a sufficiently large inoculum is used, this model allows massive, often fatal, diarrhoea to occur within 1–5 days. The RITARD model has been employed after oral immunization to test the protective capacity of attenuated *V. cholerae* strains. Successful colonization and immunization of adult rabbits by oral inoculation (usually without diarrhoea) can be accomplished with administration of tincture of opium to induce hypoperistalsis (Cray *et al.*, 1983).

5.4 MOTILITY AND FLAGELLA

V. cholerae achieve motility by means of a single, sheathed, polar flagellum and motility plays a fundamental role in virulence. Motility is directional as *V. cholerae* enter the mucous gel and move towards the mucosal surface in response to chemotaxins (Freter and O'Brien, 1981; Freter *et al.*, 1981a,b) achieving their objective within hours (Guentzel and Berry, 1975; Jones and Freter, 1976; Schrank and Verwey, 1976; Freter and O'Brien, 1981). Fully enterotoxinogenic but non-motile mutants are diminished in virulence (Schrank and Verwey, 1976;

Attridge and Rowley, 1983; Guentzel and Berry, 1975; Richardson, 1991). Since it has been suggested that the flagellum may serve as an adhesin (Eubanks *et al.*, 1977; Hranitzky *et al.*, 1980; Attridge and Rowley, 1983), an important question is whether the mere presence of a flagellum is sufficient for colonization or whether the organelle must manifest functional motility. Richardson *et al.* (1989), after testing isogenic strains in three different animal models, concluded that motility is indeed an important factor contributing to the pathogenicity and colonizing ability of *V. cholerae*.

The flagellar core consists of two proteins of M_r 47 and 49 kDa but antibodies against these proteins do not inhibit the motility of *V. cholerae* (Richardson and Parker, 1985). A membrane sheath, that appears to be contiguous with the outer membrane, covers the flagellar core (Follett and Gordon, 1963). Although antibodies against the 38 and 40 kDa proteins of the flagellar sheath inhibit motility of *V. cholerae*, immunization with these proteins did not induce protective immunity against disease in the rabbit (Richardson and Parker, 1985). *V. cholerae* O1 LPS antibodies bind to the sheathed flagella, suggesting that there exist shared antigens (Fuerst and Perry, 1988).

5.5 COLONIZATION FACTORS

Once the motile vibrios successfully breach the mucous gel and reach the mucosal surface, they must adhere in order to overcome the potent peristaltic defence mechanism. In recent years much has been learned about the mechanisms by which *V. cholerae* adheres to the small-bowel epithelium. Early work focused on agglutination of erythrocytes as a surrogate for intestinal colonization factors. *V. cholerae* produces at least four haemagglutinins (HA) (Hanne and Finkelstein, 1982): (i) a D-mannose-, D-fructose-sensitive HA expressed by El Tor strains (MFSHA); (ii) a L-fucose-sensitive HA expressed by classical strains (FSHA); (iii) a D-mannose-,L-fucose-resistant HA (MFRHA) expressed by both biotypes; and (iv) a soluble HA produced by both biotypes which also has protease activity. Some of these haemagglutinins, such as the MSHA, appear to play a role in intestinal colonization. However, it was with the discovery of the TCP that an intestinal colonization factor critical for disease was first and unequivocally demonstrated (Herrington *et al.*, 1988). Although TCP is critical for adherence to epithelial cells, intestinal colonization by *V. cholerae* requires the expression and interaction of many factors.

The identity of the intestinal cell receptors which are the target of *V. cholerae* adherence is not known. Jones and Freter (1976) found that L-fucose partially inhibited the attachment of *V. cholerae* to brush-border membranes and proposed that this sugar formed part of the intestinal receptor. Yamamoto *et al.* (1988) confirmed that L-fucose had the most prominent inhibitory effect in interfering with the attachment of *V. cholerae* to isolated human small-intestinal cells. These investigators found that the primary targets for *V. cholerae* adhesion in human small-intestinal cells were (in order of adherence efficiency): (i) mucus; (ii) the surface of the epithelial cells of ileal lymphoid follicles; and (iii) the surface of absorptive cells of jejunal or ileal villi (Yamamoto and Yokota, 1988; Yamamoto *et al.*, 1988). M cells were better adherence targets than other epithelial cells of the lymphoid follicles (Yamamoto and Yokota, 1988). The specificity of the targets varied according to the conditions used to grow the bacteria (Yamamoto *et al.*, 1988), indicating that the expression of *V. cholerae* adherence factors is highly regulated.

5.5.1 TCP

Taylor *et al.* (1987) used Tn*phoA* mutagenesis to discover a pilus colonization factor in *V. cholerae* O1 which morphologically is composed of long filaments 7 nm in diameter that form into bundles. One of their Tn*phoA* mutants exhibited decreased colonization in infant mice and was found to be defective in production of a pilus which mediated haemagglutination in the presence of L-fucose. The pilus was named Toxin Co-regulated Pilus (TCP) because its expression was correlated with that of CT.

The importance of TCP in colonization of the human intestine was proved by Herrington *et al.* (1988) who showed that intestinal colonization by volunteers by *V. cholerae* was abolished when the gene encoding the TCP subunit (*tcpA*) was inactivated in a derivative of classical Ogawa strain 395. Volunteers did not excrete vibrios, nor did they manifest good vibriocidal antibody responses despite ingesting large inocula of this mutant with buffer. Although functional TCP must be present in order for classical biotype *V. cholerae* O1 to elicit vibriocidal antibody responses, volunteers do not exhibit antibodies against TCP (Hall *et al.*, 1991).

The predicted amino aid sequence of the 20.5 kDa TcpA subunit shows significant homology to the Type IV pili of *Pseudomonas aeruginosa, Neisseria gonorrhoeae, Moraxella bovis* and *Bacteroides nodosus* (Shaw and Taylor, 1990). TCP is found in both El Tor and Classical strains and the predicted protein sequences share 82–83% homology between the biotypes (Iredell and Manning, 1994; Rhine and Taylor, 1994). Monoclonal antibodies against TCP show epitope differences in pili produced by Classical and El Tor strains (Sun *et al.*, 1990; Jonson *et al.*, 1991b). El Tor strains produce less TCP than Classical strains and culture conditions for optimal TCP expression differ between biotypes

(Sun *et al.*, 1990; Jonson *et al.*, 1992; Voss and Attridge, 1993). Sequences encoding TCP were not found in a collection of *V. cholerae* non-O1 strains nor in environmental isolates of *V. cholerae* O1 which do not produce cholera toxin (Taylor *et al.*, 1988). Synthesis of TCP is complex and up to 15 open reading frames are found in the *tcp* gene cluster (Ogierman and Manning, 1992b; Kaufman *et al.*, 1993) and at least one other unlinked gene, *tcpG* (*dsbA*), is also required for expression of a functional pilus (Peek and Taylor, 1992). TCP (identical to El Tor TCP) are also expressed by *V. cholerae* O139 when strains are grown under conditions that favour expression of TCP in El Tor *V. cholerae* O1 (Hall *et al.*, 1993).

5.5.2 ACCESSORY COLONIZATION FACTOR (ACF)

Tn*phoA* mutagenesis was also used to discover another potential colonization factor of *V. cholerae*. Peterson and Mekalanos (1988) isolated Tn*phoA* mutants which were regulated by ToxR (see below) and found four mutants whose colonization ability was diminished in infant mice. The reduction of colonization observed with insertions in the *acf* (for *a*ccessory *c*olonization *f*actor) locus, i.e. ca. 10 to 50-fold compared with the wild-type strain, was not as severe as that observed with the TCP mutant (ca. 1000-fold) (Taylor *et al.*, 1987). The exact nature of the ACF has not been reported, but one of the four open reading frames in this locus (*acfD*) encodes a lipoprotein (Parsot *et al.*, 1991). The nucleotide sequence of the *acf* locus has been determined and found to be located immediately adjacent to the *tcp* and *toxT* loci. Thus, there appears to be at least one other region of the *V. cholerae* chromosome, in addition to the region containing *ctx*, *zot* and *ace*, in which genes encoding virulence factors are clustered.

5.5.3 MANNOSE–FUCOSE-RESISTANT HAEMAGGLUTININ

Both biotypes of *V. cholerae* produce a cell-associated haemagglutinin which is not inhibited by mannose, fucose or other sugars. The gene encoding this haemagglutinin, termed mannose–fucose-resistant haemagglutinin (MFRHA), was cloned by Franzon and Manning (1986). The nucleotide sequence of a 693 bp open reading frame predicts a 26.9 kDa protein associated with this activity (Franzon *et al.*, 1993). An isogenic strain mutated in this gene was tested for virulence in infant mice and found to be markedly attenuated; whereas the parent strain had an LD_{50} of 9×10^5, the LD_{50} of the mutant was $>1 \times 10^8$ (Franzon *et al.*, 1993). Competitive colonization experiments in infant mice showed that *in vivo* colonization of the mutant was reduced 500 to 1300-fold relative to the parent strain.

5.5.4 MANNOSE-SENSITIVE HAEMAGGLUTININ

The mannose-sensitive haemagglutinin (MSHA) of *V. cholerae* is expressed by strains of the El Tor biotype but only rarely by strains of the Classical biotype (Jonson *et al.*, 1991a). Jonson *et al.* (1991a) have identified this factor as a thin, flexible pilus composed of subunits with a molecular mass of ca. 17 kDa. The sequence of *mshA* encoding the structural subunit suggests that MSHA is a type IV pilus (Jonson *et al.*, 1994). A monoclonal antibody capable of inhibiting El Tor vibrio-mediated agglutination of chicken erythrocytes bound specifically to the length of the pilus, suggesting that the receptor binding portion may reside in the actual structural subunit rather than a minor pilus protein (Jonson *et al.*, 1991a). The same monoclonal antibody protects against experimental El Tor cholera in the infant mouse and rabbit intestinal loop models (Osek *et al.*, 1992). The monoclonal antibody did not protect against challenge by *V. cholerae* O1 of the classical biotype, thus suggesting that there is an El Tor-specific protective antigen. Previous work by Finn *et al.* (1987) characterized Tn5-mutants of an El Tor strain which lacked MSHA. These mutants were defective in colonization of rabbit ileal tissue by a factor of four logs relative to the parent strain.

5.5.5 CORE-ENCODED PILUS

In addition to containing the toxin genes *ctx*, *zot* and *ace*, the 4.5-kb core element contains a gene encoding a factor that enhances colonization in infant mice. Pearson *et al.* (1993) showed that an 82-amino acid protein is encoded by the *cep* locus (core-encoded pilus). The predicted amino acid sequence of this protein shows homology to a flexible pilus of *Aeromonas hydrophila*. Deletion of the *cep* locus reduces colonization in infant mice 13 to 21-fold (Pearson *et al.*, 1993). Previous experiments showing that the 4.5-kb core region can be amplified *in vivo* (Mekalanos, 1983) may be explained by the selective colonization advantage conferred by the *cep* locus. The available evidence suggests that *cep* contributes little, if any, to intestinal colonization of humans. When volunteers were fed a derivative of El Tor strain E7946 which was deleted of the 4.5-kb core region and the *cep* genes (Tacket *et al.*, 1993) the mutant strain was excreted in large numbers and the immune responses observed were indistinguishable from that stimulated by the wild-type strain containing *cep*.

5.5.6 SOLUBLE HAEMAGGLUTININ/PROTEASE

The *V. cholerae* soluble haemagglutinin/protease (HA/protease) does not appear to be a colonization factor but is a zinc metalloenzyme that nicks cholera enterotoxin and also cleaves fibronectin, mucin and lactoferrin (Finkelstein *et al.*, 1983). The HA/protease also manifests a 'detachase' activity which allows the vibrios to detach from cultured human intestinal epithelial cells (Finkelstein *et al.*, 1992). El Tor strains produce higher levels of HA/protease than do Classical strains (Svennerholm *et al.*, 1983). The gene encoding the HA/protease (*hap*) has been cloned (Häse and Finkelstein, 1991) and sequenced. Surprisingly, an isogenic strain of *V. cholerae* specifically mutated in the gene encoding this protein was no less virulent in infant rabbits than the parent strain (Finkelstein *et al.*, 1992).

5.5.7 OTHER FIMBRIAE

Hall *et al.* (1988) described at least two other fimbrial types besides TCP and reported that multiple fimbrial types can be expressed by the same strain. Ehara *et al.* (1987) purified fimbriae of *V. cholerae* O1 which had a molecular mass of 16 kDa. The role of these fimbriae is unclear.

5.5.8 OUTER MEMBRANE PROTEINS

Numerous outer membrane proteins (OMPs) of *V. cholerae* have been identified and the genes encoding many of these proteins have been cloned. Antibodies to purified OMPs inhibit intestinal colonization in an infant mouse model (Sengupta *et al.*, 1992). OmpU, a 38 kDa OMP whose expression is positively regulated by ToxR (see below), has recently been shown to be a potential adherence factor for *V. cholerae* (Sperandio *et al.*, 1995). Antibodies raised against purified OmpU completely inhibited adhesion of several *V. cholerae* strains to HeLa, Hep-2, Caco-2, and Henle 407 epithelial cells and also inhibited intestinal colonization and conferred protection in newborn mice against both biotypes of *V. cholerae* O1. Another OMP which is important for virulence in infant mice is IrgA, a 77 kDa OMP the expression of which is regulated by iron. A *V. cholerae* O1 strain specifically mutated in the *irgA* gene showed a ca. 100-fold increase in LD_{50} compared with the parent strain (Goldberg *et al.*, 1990). The competitive colonization ability of the *irgA* mutant was reduced about 10-fold (Goldberg *et al.*, 1990), suggesting a colonization defect in this mutant. An OMP which is necessary for biogenesis of the TCP pilus is TcpC, which has a molecular mass of 55 kDa as predicted from the sequence of the *tcpC* gene (Ogierman and Manning, 1992b). TcpC is a lipoprotein that confers serum resistance upon *V. cholerae* (Parsot *et al.*, 1991).

The role of other OMPs in virulence has not been established in either animal or human studies. OmpV is a 25-kDa OMP, the gene for which has been cloned and sequenced (Pohlner *et al.*, 1986). Although this is a major immunogenic component of the *V. cholerae* surface, a Tn*phoA* insertion into this gene did not affect LD_{50} or competitive colonization in infant mice (Taylor *et al.*, 1987). Another immunogenic OMP is a 22-kDa protein called OmpW (Jalajakumari and Manning, 1990), for which no role in virulence has been shown. OmpS is a 44-kDa protein which is induced by maltose and shares homology with the bacteriophage λ receptor LamB of *E. coli* (Lang and Palva, 1987; unpublished data); in *E. coli*, the LamB protein functions as a pore and a similar role has been proposed for the OmpS protein of *V. cholerae* (Lang and Palva, 1987). *V. cholerae* also produces a 35-kDa OMP which cross-reacts with antisera against the *E. coli* OmpA protein. An additional OMP, OmpT (40 kDa) under the control of ToxR (see below) (Miller and Mekalanos, 1988) but no role in virulence has yet been shown for this protein.

5.5.8 LIPOPOLYSACCHARIDES (LPS)

LPS appears to play a role in adherence of *V. cholerae* O1 to the intestinal mucosa. Purified Inaba LPS significantly inhibits attachment of *V. cholerae* Inaba to rabbit mucosa (Chitnis *et al.*, 1982) and antibodies against Ogawa or Inaba LPS prevent adhesion of *V. cholerae* to intestinal mucosa (Freter and Jones, 1976; Chitnis *et al.*, 1982).

5.5.9 POLYSACCHARIDE CAPSULE

Although *V. cholerae* O1 is unencapsulated, strains of *V. cholerae* non-O1 have been shown to produce a polysaccharide capsule (Johnson *et al.*, 1992). Such a capsule could facilitate septicaemia in susceptible hosts, an outcome which is not uncommon with non-O1 strains. Johnson *et al.* (1992) found that an unencapsulated mutant derived by transposon mutagenesis was less virulent in a mouse model than the encapsulated parent. The encapsulated strain was protected from serum bactericidal activity while the unencapsulated mutant was readily killed. Capsular polysaccharide (CPS) may also help mediate adherence to epithelial cells. Notably, strains of *V. cholerae* O139 also produce a polysaccharide capsule (Johnson *et al.*, 1994; Preston *et al.*, 1995) which could possibly have contributed to a case of septicaemia reported with this serogroup (Jesudason *et al.*, 1993).

5.6 TOXINS

Once they are within a critical distance from epithelial cells, *V. cholerae* produces a variety of extracellular products that have deleterious effects on eukaryotic cells. The massive, dehydrating diarrhoea characteristic of cholera is induced by cholera enterotoxin, also referred to as cholera toxin (CT) or choleragen (Finkelstein, 1992). The volunteer studies described above show the pivotal role that CT plays in disease. However, *V. cholerae* strains deleted of genes encoding CT can still cause mild and even moderate diarrhoea in a proportion of volunteers and this has prompted a search for additional toxins produced by *V. cholerae* (Fasano *et al.*, 1991; Trucksis *et al.*, 1993).

5.6.1 CHOLERA ENTEROTOXIN

The existence of a toxin responsible for the symptoms of cholera was first proposed by Robert Koch (1884) and first demonstrated in 1959 independently by two groups in India (De, 1959; Dutta *et al.*, 1959). Finkelstein and LoSpalluto, (1969) purified the toxin, paving the way for other investigators to study its structure, receptor binding and mode of action.

(a) Structure

The structure of cholera toxin is typical of the A-B subunit group of toxins in which each of the subunits has a specific function. The B subunit binds the holotoxin to the eukaryotic cell receptor and the A subunit acts intracellularly in a specific enzymatic manner. Antibodies directed against B, the immunodominant subunit, are much more efficient at neutralizing toxin activity than are antibodies directed against the A subunit (Peterson *et al.*, 1979, 1991).

Binding of cholera toxin to epithelial cells is enhanced by a neuraminidase produced by *V. cholerae*. This enzyme, with a predicted molecular mass of 83 kDa (Galen *et al.*, 1992), enhances the effect of cholera toxin by catalysing the conversion of higher order gangliosides to GM_1 (Holmgren *et al.*, 1975).

(b) Enzymatic activity

The intracellular target of cholera toxin is adenylate cyclase, one of the most important regulatory systems of the eukaryotic cell. This enzyme mediates the transformation of ATP to cAMP, a crucial intracellular messenger for a variety of cellular pathways. Normally, adenylate cyclase is activated or inactivated in response to a variety of stimuli. Regulation of adenylate cyclase is mediated by G proteins, which serve to link many cell-surface receptors to effector proteins at the plasma membrane. G proteins are heterotrimers composed of three distinct subunits: α ($M_r = 39–46$ kDa), β ($M_r = 37$ kDa) and γ ($M_r = 8$ kDa) (Helper and Gilman, 1992). The specific G protein involved is the G_s protein, activation of which leads to increased adenylate cyclase activity.

Cholera toxin catalyses the transfer of the ADP-ribose moiety of NAD to a specific arginine residue in the $G_{s\alpha}$ protein, resulting in the activation of adenylate cyclase and subsequent increases in intracellular levels of cAMP. cAMP activates a cAMP-dependent protein kinase, leading to protein phosphorylation, alteration of ion transport, and ultimately to diarrhoea.

After CT binds to intact cells, there is a lag of 15–60 min before adenylate cyclase is activated (Gill and King, 1975). The lag time is necessary to allow the A_1 peptide to translocate through the membrane and to come into contact with the G proteins. The adenylate cyclase is located on the basolateral membrane of polarized intestinal epithelial cells and a variety of mechanisms has been proposed to explain how CT bound to the brush border (apical) membrane reaches the basolateral adenylate cyclase. The most likely explanation is that whole CT enters the cells by endocytosis and the low pH of the endosomal compartment facilitates the translocation of the A subunit across the endosomal membrane. The endosome then travels through the cell with the A_1 peptide still associated with the endosomal membrane. Ultimately, the A_1 peptide ADP-ribosylates $G_{s\alpha}$ located in the basolateral membrane. Several investigators have recently reported that the action of CT is completely inhibited by addition of brefeldin A, a fungal metabolite known to interfere with vesicular transport in endosomal and transcytotic pathways of many eukaryotic cells (Lencer *et al.*, 1992, 1993; Donta *et al.*, 1993; Nambiar *et al.*, 1993). These results show that an intact Golgi region is required for intracellular trafficking of CT and implicate toxin endocytosis in the entry of CT into cells.

The A_1 peptide catalyses the transfer of ADP-ribose from NAD to an arginine residue on $G_{s\alpha}$ in the following reaction:

$$NAD + G_{s\alpha} \rightarrow [\text{ADP-ribosyl } G_{s\alpha}] + \text{nicotinamide} + H^+$$ (Moss and Vaughan, 1977).

The α subunit of G_s contains a GTP binding site and an intrinsic GTPase activity (Hepler and Gilman, 1992). Binding of GTP to the α subunit leads to dissociation of the α and the β-γ subunits and subsequent increased affinity of α for adenylate cyclase. The resulting activation of adenylate cyclase continues until the intrinsic GTPase activity hydrolyses GTP to GDP, thereby inactivating the G protein and adenylate cyclase. ADP-ribosylation of the α subunit by the A_1 peptide of CT inhibits the hydrolysis of GTP to GDP, thus leaving adenylate cyclase constitutively activated, probably for the life of the cell (Kahn and Gilman, 1984; Cassel and Selinger, 1977).

The ADP-ribosylation activity of A_1 is stimulated *in vitro* by a family of proteins termed ARFs for ADP-ribosylation factors (reviewed in

Moss and Vaughan, 1991). The ARF proteins are ca. 20 kDa GTP-binding proteins and constitute a distinct family within the larger group of ca. 20 kDa guanine nucleotide-binding proteins which include the *ras* oncogene protein. In the presence of GTP, ARF appears to act directly on the A_1 peptide, independently of B subunits or $G_{s\alpha}$. At least *in vitro*, ARFs serve as allosteric activators of A_1 and increase the ADP-ribosyl transferase activity of the proteins.

(c) Cellular response

The increased intracellular cAMP concentrations resulting from the activation of adenylate cyclase by CT leads to increased Cl^- secretion by intestinal crypt cells and decreased NaCl coupled absorption by villus tip cells (Field *et al.*, 1989). The net movement of electrolytes into the lumen results in a transepithelial osmotic gradient which causes water flow into the lumen. The massive volume of water overwhelms the absorptive capacity of the intestine, resulting in diarrhoea. The steps between increased levels of cAMP and secretory diarrhoea are not known in their entirety. Pioneering work in this area by Field (Field *et al.*, 1969; Field, 1971) and subsequent investigators (reviewed in Finkelstein, 1992) demonstrated that CT could concomitantly increase cAMP formation and ion transport in isolated intestinal epithelium mounted in Ussing chambers. One crucial step resulting from increased cAMP levels is activation of protein kinase A which subsequently phosphorylates numerous substrates in the cell (Chang and Rao, 1991).

Although it is clear that chloride channels can be regulated by cAMP-dependent protein kinases, the actual ion channel or channels affected by the CT are not known with certainty. There are multiple types of Cl^- channels in apical membranes with different modes of activation and subcellular distribution. It is also not certain if protein kinase A directly phosphorylates the ion channel or if it phosphorylates an intermediate protein which then phosphorylates other proteins in a cascade. One attractive candidate for a relevant target protein is the cystic fibrosis (CF) gene product, CFTR. The CFTR protein is a Cl^- channel (Bear *et al.*, 1992) and has multiple potential substrate sequences for kinase A. Unlike normal intestinal tissue, tissues obtained from patients with cystic fibrosis (CF homozygotes) do not respond to either cAMP- or Ca-mediated secretagogues (Berschneider *et al.*, 1988). It has been postulated that individuals who are CF heterozygotes may have a selective advantage over 'normal' homozygotes in surviving cholera (Rodman and Zamudio, 1991). Heterozygotes presumably have only half the normal number of chloride channels responsive to kinase. After infection with *V. cholerae*, the CF heterozygote may have less intestinal chloride secretion, and therefore less diarrhoea, due to the smaller number of chloride channels.

(d) Alternate mechanisms of action

The activation of adenylate cyclase leading to increased cAMP and subsequent altered ion transport is the 'classic' mode of action of cholera toxin. However, increased levels of cAMP and subsequent protein kinase A activation may not explain entirely the secretory effects of cholera toxin. Prostaglandins and the enteric nervous system are apparently also involved in the response to cholera toxin.

Prostaglandins

The role of prostaglandins (PG), leukotrienes and other metabolites of arachidonic acid in causing intestinal secretion has been well documented and several reports have implicated prostaglandins in the pathogenesis of cholera (Bennett, 1971; Beubler *et al.*, 1989; Peterson and Ochoa, 1989). Cholera patients in the active secretory disease stage have elevated jejunal concentrations of PGE_2 compared with patients in the convalescent stage (Speelman *et al.*, 1985) and the prostaglandins PGE_1 and PGE_2 have been reported to stimulate adenylate cyclase and increase short circuit current (Isc) in Ussing chambers (Kimberg *et al.*, 1971). Peterson and Ochoa (1989) have reported that addition of cAMP induced only a small, transient fluid accumulation in rabbit intestinal loops whereas addition of PGE_2 caused a much stronger fluid accumulation. Addition of CT led to increases in both cAMP and PGE in rabbit loops and in CHO cells, resulting in the release of arachidonic acid from membrane phospholipids (Reitmeyer and Peterson, 1990). A model has been suggested in which cAMP levels increased by CT serve not only to activate protein kinase A but also to regulate transcription of a phospholipase or a phospholipase-activating protein. The activated phospholipase could act on membrane phospholipids to produce arachidonic acid, a precursor of prostaglandins and leukotrienes (Peterson *et al.*, 1991). Consistent with this model, de Jonge and colleagues (1991), using the human intestinal cell line HT29.c1.19A mounted in Ussing chambers, report that 40–60% of the Isc response to CT is inhibited by relatively low concentrations of the phospholipase A_2 inhibitor, mepacrine. The effect of CT on arachidonic acid metabolism might occur through ADP-ribosylation of G_s in the apical membrane, which would then directly activate a phospholipase or the effect may be mediated through protein kinase A which would activate the phospholipase.

Enteric nervous system

The enteric nervous system (ENS), part of the autonomic nervous system, plays an important role in intestinal secretion and absorption (Cooke, 1991). Two products elaborated by intestinal cells with well-established roles in causing secretion are vasoactive intestinal peptide (VIP) and serotonin (5-hydroxytryptamine or 5-HT) (Cooke, 1991). The ENS may play a crucial role in secretion due to cholera toxin. It has been hypothesized that cholera toxin binds to 'receptor cells', namely enterochromaffin cells, which release a substance such as serotonin which activates dendrite-like structures located beneath the intestinal epithelium. This leads to the release of VIP, resulting in electrolyte and fluid secretion.

This model is supported by a variety of evidence including: (i) CT-induced secretion is inhibited by a variety of ganglionic or neurotransmitter blockers (Cassuto *et al.*, 1981a, 1982, 1983); (ii) CT stimulates release of serotonin (5-HT) from enterochromaffin cells into the intestinal lumen (Nilsson *et al.*, 1983) and fluid secretion induced by CT is markedly diminished by 5-HT receptor antagonists (Sjöqvist *et al.*, 1992); and (iii) CT-induced secretion is accompanied by an increased release of VIP from the small bowel and the increased release of VIP is blocked by tetrodotoxin (Nilsson *et al.*, 1983). In addition, cholera patients also show increased levels of VIP in blood (Holmgren, 1992). Cassuto *et al.* (1981b) estimate that ca. 60% of the effect of CT on intestinal fluid transport could be attributed to nervous mechanisms. Besides the direct secretory effect, there is also evidence that CT increases intestinal motility and could thereby contribute to diarrhoea (Mathias *et al.*, 1976).

Genetics

The genes encoding cholera toxin (*ctxAB*) were initially cloned by exploiting their homology to the genes encoding the heat-labile enterotoxin (LT) of enterotoxigenic *E. coli* (Kaper and Levine, 1981; Gennaro *et al.*, 1982; Pearson and Mekalanos, 1982). DNA sequence analysis shows 78% overall nucleotide homology and 80% predicted protein homology between the two genes (Lockman and Kaper, 1983; Mekalanos *et al.*, 1983; Lockman *et al.*, 1984; Yamamoto and Yokota, 1983; Yamamoto *et al.*, 1984). The A and B subunits are encoded on two separate but overlapping open reading frames; the first two bases of the *ctxA* translation termination signal (TGA) are the last two of the *ctxB* translation initiation codon (ATG) (Lockman and Kaper, 1983; Mekalanos *et al.*, 1983). The A and B cistrons possess ribosomal binding sites immediately upstream of their start codons, with the site for *ctxB* being located

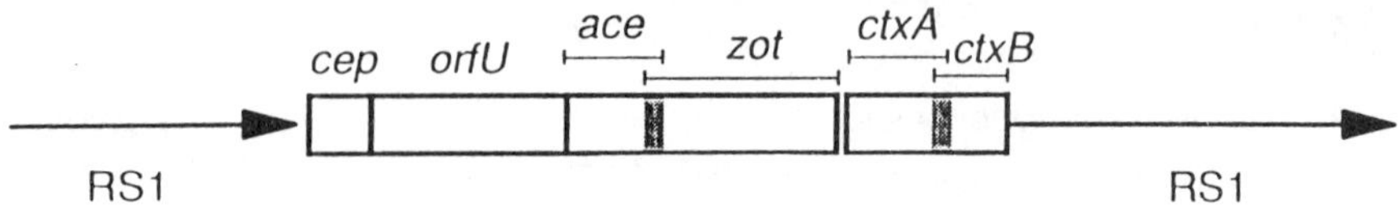

Fig. 5.1 The 'core region' of the *ctx* operon.

in the 3′ end of the *ctxA* sequence. The higher expression of the B subunit, leading to the 5:1 ratio of B:A subunits in the holotoxin, results from more efficient translation due to a stronger ribosomal binding site for the B subunit (Mekalanos *et al.*, 1983). Transcription of the *ctx* operon is regulated by the ToxR regulatory system.

Many strains of *V. cholerae* O1 contain multiple copies of the *ctx* operon. Classical strains contain two copies which are separated on the *V. cholerae* O1 chromosome by an unknown distance (Moseley and Falkow, 1980; Mekalanos, 1983). Most El Tor strains contain only a single copy of the *ctx* operon but about 30% of El Tor strains contain two or more adjacent gene copies (Mekalanos, 1983, 1985). The *ctx* operon, along with genes encoding the Zot and Ace toxins (see below) are located on a 4.5 kb region called the 'core region' (Goldberg and Mekalanos, 1986) (Figure 5.1). Flanking the 4.5 kb region are one or more copies of a 2.7-kb sequence called RS1. Recombination between RS1 sequences can lead to tandem duplication and amplification of the core region as well as deletion of the core region (Goldberg and Mekalanos, 1986). Serial passage of an El Tor strain through rabbit ligated intestinal loops led to *in vivo* amplification, resulting in strains expressing higher levels of CT and containing more *ctx* copies than the strain injected into the first loop (Mekalanos, 1983). Although cholera toxin by itself can aid colonization (Pierce *et al.*, 1985), the *in vivo* amplification of the entire 4.5 kb core region may be due to the presence in this region of a gene (*cep*) encoding an intestinal colonization factor (Pearson *et al.*, 1993). Classical strains and some El Tor strains possess RS1 sequences only on one side of the core region and therefore this gene duplication and amplification is not seen (Mekalanos, 1983, 1985). The RS1 sequence encodes a site-specific transposable element which can insert into a specific 18-bp sequence called attRS1 (Pearson *et al.*, 1993). Naturally occurring isolates of *V. cholerae* which do not produce CT lack sequences homologous to *ctx* and the rest of the core region (Kaper *et al.*, 1981; Miller and Mekalanos, 1984) but do contain attRS1 sequences (Pearson *et al.*, 1993). Pearson *et al.* (1993) recently demonstrated that when core region and RS1 sequences are cloned into a suicide plasmid incapable of replicating in *V. cholerae* and transferred from a conjugation-proficient *E. coli*, the RS1 and *ctx* sequences can insert into the attRS1 site in a *recA*-independent manner.

5.6.2 OTHER TOXINS PRODUCED BY *V. CHOLERAE*

(a) Zonula occludens toxin (Zot)

The zonula occludens toxin (Zot) of *V. cholerae* increases the permeability of the small intestinal mucosa by affecting the structure of the intercellular tight junction, or zonula occludens (ZO) (Fasano *et al.*, 1991). This activity was discovered by testing culture supernatants of *V. cholerae*, both wild-type and Δ*ctx*, in Ussing chambers, a classic technique for measuring transepithelial transport of electrolytes across intestinal tissue. When culture supernatants of *V. cholerae* strain CVD 101 (Δ*ctxA*) were added to rabbit ileal tissue mounted in Ussing chambers, an immediate increase in tissue conductivity (i.e. a decrease in tissue resistance) was observed (Fasano *et al.*, 1991). Unlike the increase in potential difference (PD) observed in response to CT, which reflects ion transport across the membrane, i.e. the transcellular pathway, variation in transepithelial conductance reflects primarily, although not exclusively, modification of tissue permeability through the intercellular space, i.e. the paracellular pathway. Examination by electron microscopy of the epithelial tight junctions, the major barrier in this paracellular pathway, revealed that exposure of ileal tissue to culture supernatants of CVD 101 resulted in 'loosening' of the tight junction so that an electron-dense marker could permeate into the paracellular space. In contrast, tissue treated with uninoculated broth control was not permeable to this marker. Freeze-fracture electron microscopy showed that the anastomosing network of strands comprising tight junctions suffered a striking decrease in strand complexity in tissue treated with supernatants (Fasano *et al.*, 1991). The toxin responsible for this striking effect on ZO was named Zot for zonula occludens toxin. By increasing intestinal permeability, Zot might cause diarrhoea by leakage of water and electrolytes into the lumen under the force of hydrostatic pressure (Fasano *et al.*, 1991).

The gene encoding Zot was cloned and found to be located immediately upstream of the *ctx* locus (Baudry *et al.*, 1992). *zot* gene sequences are present in both *V. cholerae* O1 and non-O1 strains and strains that contain *ctx* sequences almost always contain *zot* sequences and vice versa (Johnson *et al.*, 1993; Karasawa *et al.*, 1993; Faruque *et al.*, 1994). The *zot* gene consists of a 1.3 kb open reading frame which could potentially encode a 44.8-kDa polypeptide. The predicted amino acid sequence of the Zot protein shows no homology to any other bacterial toxin including the toxin A of *Clostridium difficile*, a large 300-kDa protein which also alters tight junctions (Hecht *et al.*, 1988).

(b) Accessory cholera enterotoxin (Ace)

The Accessory Cholera Enterotoxin (Ace) of *V. cholerae* was identified by Trucksis *et al.* (1993) who demonstrated that the gene product of an open reading frame located immediately upstream of *zot* can increase short circuit current in Ussing chambers. Like CT, and in contrast to Zot, this new toxin increases potential difference rather than tissue conductivity. Strains containing the cloned *ace* gene caused significant fluid accumulation in rabbit ligated ileal loops. The *ace* gene appears to encode a 96-residue peptide with a predicted M_r of 11.3 kDa (Trucksis *et al.*, 1993) that shows similarity to eukaryotic ion-transporting ATPases, including the human plasma membrane calcium pump, the calcium-transporting ATPase from rat brain and the product of the cystic fibrosis gene, the cystic fibrosis transmembrane regulator. The first two proteins are involved in transport of calcium ions across the membrane and CFTR functions as a chloride ion channel (Bear *et al.*, 1992). Ace also shows sequence similarity with a virulence protein of *Salmonella dublin*, SpvB, which is essential for virulence in mice (Krause *et al.*, 1991).

The predicted structure of the Ace protein suggests a model in which multimers of Ace insert into the eukaryotic membrane with hydrophobic surfaces facing the lipid bilayer and the hydrophilic sides facing the interior of a transmembrane pore. Such a structure is found for the 26-residue δ toxin of *Staphylococcus aureus* (Freer and Birkbeck, 1982) which has a variety of effects on eukaryotic tissue, including increasing vascular permeability in guinea pig skin and inhibiting water absorption and increasing cAMP concentration in the ileum (Kapral *et al.*, 1976). The C-terminal region of Ace shows 47% amino acid similarity with residues 2 to 20 of δ toxin (Trucksis *et al.*, 1993), thus lending some support to a model in which Ace acts by aggregating and inserting into the eukaryotic membrane to form an ion channel.

(c) Haemolysin/cytolysin

In the early years of the seventh cholera pandemic, haemolysis of sheep erythrocytes was used as one of the parameters to distinguish the El Tor from the Classical biotype of *V. cholerae*, although more recent El Tor isolates are only poorly haemolytic on sheep erythrocytes (Barrett and Blake, 1981). The haemolysin was initially purified by Honda and Finkelstein (1979) and shown to be cytolytic for a variety of erythrocytes and mammalian cells in culture and rapidly lethal for mice. The haemolysin/cytolysin is initially made as an 82-kDa protein and processed in two steps to a 65-kDa active cytolysin (Yamamoto *et al.*, 1990). An identical haemolysin, indistinguishable biologically, physicochemically, and antigenically, from haemolysin produced by El Tor strains is produced by *V. cholerae* non-O1 (Yamamoto *et al.*, 1990). Genes encoding this haemolysin, *hlyA*, are present in Classical, El Tor, and non-O1 strains of *V. cholerae* (Brown and Manning, 1985). Although Classical strains do not lyse sheep erythrocytes, *E. coli* containing cloned *hlyA* genes from Classical strain 395 can lyse rabbit and chicken, but not sheep, erythrocytes (Richardson *et al.*, 1986). The molecular basis for differences in haemolytic activity among Classical strains, early El Tor isolates, and recent El Tor isolates is not completely understood, but in one non-haemolytic Classical strain, 569B, an 11-bp deletion was found which would produce a truncated *hlyA* gene product of 27 kDa (Alm *et al.*, 1988). An oligonucleotide probe spanning this 11-bp region hybridized to all El Tor and non-O1 strains tested but did not hybridize to any Classical strains tested (Alm and Manning, 1990). Genes encoding a second haemolysin, genetically distinct from the El Tor haemolysin/cytolysin, have been cloned from Classical strain 395 (Richardson *et al.*, 1986) but no further characterization of this haemolysin is available.

The purified haemolysin can lead to the accumulation in ligated rabbit ileal loops of fluid tinged with blood and containing mucus (Ichinose *et al.*, 1987). McCardell *et al.* (1985) have described a cytolysin which is cytotoxin for Y-1 adrenal and CHO cells and causes fluid accumulation in rabbit ileal loops. This cytolysin was proposed to be identical to the El Tor haemolysin. To test the hypothesis that this haemolysin was responsible for the diarrhoea seen with Δ*ctx* strains of *V. cholerae*, Kaper *et al.* (1988) constructed derivatives of such strains which were mutated in the *hlyA* gene by deletion of an internal 400-bp *Hpa*I fragment. When fed to volunteers, the Δ*hlyA* strains CVD 104 and CVD 105 still caused mild diarrhoea in 33% of the subjects (Levine *et al.*, 1988b), indicating that the haemolysin/cytolysin was not the only explanation for the residual diarrhoea seen in recipients of Δ*ctx V. cholerae* strains.

(d) Miscellaneous toxins

In addition to CT, Zot, Ace, and haemolysin/cytolysin, which are widely distributed in *V. cholerae* and for which genes have been cloned and sequenced, a number of other toxins have been reported for this species.

Shiga-like toxin

A Shiga-like toxin, identified on the basis of cytotoxicity in HeLa cells neutralizable by antibody raised against Shiga toxin purified from *Shigella dysenteriae* 1, has been reported for *V. cholerae* O1 (O'Brien *et al.*, 1984). This toxin was hypothesized to be responsible for diarrhoea in volunteers who ingested genetically engineered CT^- strains. Consistent with this hypothesis, *V. cholerae* strain CVD 103-HgR (Δ*ctx*), which does not produce detectable Shiga-like activity, causes no reactogenicity (Levine *et al.*, 1988a; Levine and Kaper, 1993; Levine and Tacket, 1994). Genes encoding a Shiga-like toxin activity have not been cloned from *V. cholerae*, despite repeated efforts (Pearson *et al.*, 1990).

Heat-stable enterotoxin (ST)

Some strains of *V. cholerae* non-O1 produce a toxin that shares 50% protein sequence homology to the heat-stable enterotoxin (ST) of enterotoxigenic *E. coli* (Ogawa *et al.*, 1990). This toxin is only rarely found in *V. cholerae* (Takeda *et al.*, 1991). The genes encoding the ST enterotoxin in *V. cholerae* O1 and non-O1 are flanked by 123 bp direct repeats (Ogawa and Takeda, 1993), suggesting that this toxin has been introduced into *V. cholerae* via a transposon.

New cholera toxin (NCT)

Sanyal *et al.* (1983) reported that culture filtrates of environmental strains of *V. cholerae* O1 which lack genes encoding CT could cause fluid accumulation in ligated rabbit ileal loops and increase the capillary permeability of rabbit skin (but unlike CT would also cause blanching or necrosis along with the bluing reaction). The enterotoxic activity was subsequently found in CT^+ strains such as classical Inaba 569B (Saha and Sanyal, 1990). The toxin was termed 'new cholera toxin' (NCT) and proposed as the cause of diarrhoea in volunteers fed genetically engineered CT^- *V. cholerae* strains (Saha and Sanyal, 1990). Additional biochemical or genetic characterization of this toxin has not been reported.

Sodium-channel inhibitor

Tamplin *et al.* (1987) observed that strains of *V. cholerae* O1, including CT^- CVD 101, produce a factor that inhibits sodium channels. The initial characterization of this factor suggested that the active compound is tetrodotoxin or a related toxin. No further characterization of this sodium-channel inhibitor in *V. cholerae* O1 has been reported.

5.6.3 ROLE OF ADDITIONAL TOXINS IN DISEASE

The role of toxins other than CT in the pathogenesis of disease due to *V. cholerae* is unknown. These toxins clearly cannot cause cholera gravis because severe purging is never seen after ingestion of Δ*ctx* strains that retain the ability to elaborate these toxins, in contrast with the severe purging seen with the wild-type parent *V. cholerae* strains (Levine *et al.*, 1988b). However, toxins other than CT may contribute to the diarrhoea and other symptoms seen with wild-type *V. cholerae* strains and may be partly responsible for the symptoms seen with CT^- strains. Such toxins may serve as a secondary secretogenic mechanism when conditions for producing CT are not optimal. The five miscellaneous toxins reviewed above, Shiga-like toxin, ST, NCT, and the sodium-channel inhibitor, are either found only in a very small percentage of strains or are insufficiently characterized to assess the potential role in disease.

The three non-CT *V. cholerae* toxins that are the best characterized are the haemolysin/cytolysin, Zot, and Ace. Genes encoding the haemolysin/cytolysin are found in nearly all pathogenic and non-pathogenic strains of *V. cholerae* O1 and non-O1, with no correlation seen with the presence of *ctx* sequences. As noted above, a CT-negative strain specifically mutated in *hlyA* still caused mild diarrhoea in volunteers. The *ace* and *zot* genes are almost always found in strains containing *ctx* but are rarely found in strains lacking *ctx*. Thus, there is a strong epidemiological correlation of the presence of *zot* and *ace* with disease. It is intriguing that genes encoding three distinct toxins acting by different mechanisms are located in tandem on a dynamic region of the *V. cholerae* chromosome. The arrangement of the 4.5-kb core region flanked by RS1 elements can, in some strains, lead to amplification or deletion of all three toxin genes as a unit (Goldberg and Mekalanos, 1986). This region also contains a gene encoding an intestinal colonization factor, *cep* (core encoded pilin) (Pearson *et al.*, 1993). Thus, this region may be perceived as a 'virulence cassette' of *V. cholerae* (Trucksis *et al.*, 1993).

Recent volunteer studies tested *V. cholerae* strains specifically deleted of sequences encoding Zot, Ace and haemolysin/cytolysin as well as the CT A subunit (Tacket *et al.*, 1993; Taylor *et al.*, 1994). Several of these strains were well tolerated. However, two vaccine strains derived from wild-type strain El Tor Ogawa E7946, including strains CVD 110 (Michalski *et al.*, 1993) and Bahrain 3 (Taylor *et al.*, 1994), still caused mild to moderate diarrhoea in the majority of volunteers, as well as fever and abdominal cramps (Tacket *et al.*, 1993; Taylor *et al.*, 1994). While these volunteer studies do not exclude a role for Ace, Zot, and haemolysin/cytolysin in the pathogenesis of cholera, they clearly indicate that there are additional features of *V. cholerae* that result in diarrhoea.

5.7 REGULATION

There are multiple systems involved in regulation of virulence in *V. cholerae*. The ToxR regulon controls expression of several critical virulence factors and has been the most extensively characterized. Regulation in response to iron concentration is a distinct regulatory system that controls additional putative virulence factors. Other putative virulence factors such as neuraminidase and various haemagglutinins, are apparently not controlled by either regulatory system. There is also a set of genes that are expressed only *in vivo* and do not belong to the ToxR or iron regulatory systems. These different regulatory systems allow *V. cholerae* to vary expression of its genes to optimize survival in a variety of environments, from the human intestine to the estuarine environment.

5.7.1 THE TOXR REGULON

The expression of several virulence genes in *V. cholerae* O1 is coordinately regulated so that multiple genes respond in a similar fashion to environmental conditions (Miller and Mekalanos, 1984, 1988). For *V. cholerae* O1, the 'master switch' for control of these factors is ToxR, a 32-kDa transmembrane protein (Miller *et al.*, 1987) that binds to a tandemly repeated 7-bp DNA sequence found upstream of the *ctxAB* structural genes and increases transcription of *ctxAB*, resulting in higher levels of cholera toxin expression (Miller *et al.*, 1987; Miller and Mekalanos, 1984). The activity of ToxR is enhanced by the ToxS protein, a 19-kDa transmembrane protein that interacts with ToxR (DiRita and Mekalanos, 1991). A model has been proposed whereby ToxR dimers are active for DNA binding and transcriptional enhancement while ToxS serves to assemble or stabilize ToxR monomers into the dimeric form (DiRita and Mekalanos, 1991).

ToxR also controls expression of the TCP colonization factor (Taylor *et al.*, 1987), the ACF accessory colonization factor (Peterson and Mekalanos, 1988), the outer membrane proteins OmpT and OmpU (Miller and Mekalanos, 1988), and three lipoproteins (Parsot *et al.*, 1991). The effect of ToxR on expression of most of these factors is to increase expression, but expression of OmpT is decreased in the presence of ToxR (Miller and Mekalanos, 1988). A survey of genes whose expression is controlled by ToxR was undertaken by Peterson and Mekalanos (1988) who found at least 17 distinct genes that are regulated by ToxR. These genes comprise what is termed the 'ToxR regulon'. The importance of ToxR in human disease was demonstrated by Herrington *et al.* (1988) who fed volunteers a derivative of classical *V. cholerae* strain 395 in which the *toxR* gene was mutated. The ToxR mutant was severely compromised in its ability to colonize the intestine of the volunteers and elicited only meager immune responses in only a few subjects.

Binding of ToxR directly to the *ctx* operon appears to be the exception rather than the rule for the ToxR regulon. The repeated 7-bp target sequence for binding of ToxR is not present upstream of any other genes regulated by ToxR (DiRita, 1992). Expression of at least some of the genes of the ToxR regulon is controlled by another regulatory factor, ToxT (DiRita *et al.*, 1991). ToxT is a 32-kDa protein that shares significant sequence homology to the AraC family of transcriptional activators (Higgins *et al.*, 1992; Ogierman and Manning, 1992a). ToxR controls transcription of the *toxT* gene (DiRita *et al.*, 1991). The resulting increased expression of the ToxT protein then leads to activation of other genes in the ToxR regulon. Thus, a regulatory cascade controls expression of important virulence factors in *V. cholerae* where ToxR is at the top of the hierarchy, ToxT is at the next level, and a number of virulence genes controlled by ToxT are at the lowest level (DiRita *et al.*, 1991). This regulatory cascade has recently been reviewed by DiRita (1992).

The influence of environmental conditions on expression of CT has been known for many years (Richardson, 1969). Expression of other virulence factors such as TCP responds in similar fashion and many of these responses are mediated by the ToxR regulon. It has been proposed that the ToxR protein senses environmental conditions and transmits this information to other genes in the ToxR regulon by signal transduction (DiRita, 1992). For classical strains, increased expression of the ToxR regulon is found at pH 6.5 versus 8.5, 30 versus 37°C, and 66 mM NaCl versus lower or higher concentrations (Miller and Mekalanos, 1988). These conditions apply only to strains of the Classical biotype, in which the ToxR regulon has primarily been characterized. Expression of CT, TCP, and other potential virulence factors differs between the Classical and El Tor biotype (Jonson *et al.*, 1990) and control of the ToxR regulon in El Tor strains has not been thoroughly characterized. The decreased *in vitro* expression at 37°C of virulence factors known to be expressed at 37°C in the intestine is paradoxical. Perhaps this *in vitro* temperature effect may be due to the lack of other signals *in vitro* that are present at 37°C *in vivo* (DiRita, 1992). Additional regulatory factors may play a role in regulation of the ToxR regulon. Immediately upstream of the gene encoding ToxR is a gene for a heat shock protein, *htpG* (Parsot and Mekalanos, 1990). The *htpG* and *toxR* genes are transcribed in opposite directions and the promoters for the two genes are so close that only one RNA polymerase can bind in the intergenic region. In the model proposed by Parsot and Mekalanos (1990), the normal sigma-70 RNA polymerase can bind to the *toxR* promoter and transcribe the *toxR* gene only at low temperature. At elevated temperatures (>37°C), the RNA polymerase sigma subunit involved in transcription of heat shock genes, sigma-32, binds to the *htpG* promoter, thus repressing the *toxR* promoter. It has been suggested that upon entry into the body and passage through the stomach, the heat shock response is first induced, thereby decreasing expression of the ToxR regulon (Parsot and Mekalanos, 1990). Once the organism is in the small bowel, the heat shock system would no longer be expressed and the ToxR regulon would then be expressed.

5.7.2 IRON REGULATION

Growth of *V. cholerae* under low-iron conditions induces the expression of several outer membrane proteins that are not seen when the vibrios are grown in iron-rich media (Sigel and Payne, 1982). Many of these proteins are similar to proteins induced by *in vivo* growth of *V. cholerae,* indicating that the intestinal site of *V. cholerae* is a low-iron environment (Sciortino and Finkelstein, 1983). In addition, expression of some OMPs decreases under iron-limiting conditions (Sigel and Payne, 1982; Sciortino and Finkelstein, 1983). The expression of the haemolysin/cytolysin, vibriobactin, and IrgA (Stoebner and Payne, 1988; Goldberg *et al.*, 1990) are increased under low-iron conditions.

V. cholerae has at least two high-affinity systems for acquiring iron. The first involves a phenolate-like siderophore, vibriobactin, which is produced under low-iron conditions (Payne and Finkelstein, 1978; Griffiths *et al.*, 1984). Vibriobactin binds iron extracellularly and transports it into the cell through a specific receptor (Stoebner and Payne, 1988). The vibriobactin receptor is a 74-kDa outer membrane protein encoded by the *viuA* gene (Butterton *et al.*, 1992). Mutation of *viuA* prevents growth in low-iron media but does not affect synthesis of vibriobactin or transport of other iron complexes including ferrichrome, haemin and ferric citrate (Stoebner *et al.*, 1992). Vibriobactin production is not essential for virulence, suggesting that other iron transport systems must be operative. Indeed, a second system for acquiring iron utilizes heme and haemoglobin (Stoebner and Payne, 1988). Henderson and Payne (1993) have recently cloned genes encoding a 26-kDa inner membrane protein and a 77-kDa OMP (distinct from IrgA) which allow transport of heme into the cell.

Iron-regulation gene expression in *V. cholerae* involves the protein 'Fur', which shares 76% homology to the *E. coli* Fur protein (Butterton *et al.*, 1992). Regulation by Fur involves binding of the Fur protein in the presence of iron to a 21-bp operator sequence found in the promoter of iron-regulated genes, thereby repressing transcription (Calderwood and Mekalanos, 1987). In *V. cholerae*, Fur acts as a repressor for the *irgA* and *viuA* genes (Butterton *et al.*, 1992). Regulation of *irgA* also requires a second protein, IrgB, which acts as a positive transcriptional activator (Goldberg *et al.*, 1991). The *irgB* gene is located immediately upstream of *irgA* and transcription of *irgB* itself is repressed by Fur in the presence of iron.

5.7.3 *IN VIVO* REGULATION

In addition to the genes whose expression is regulated by ToxR or by iron, there are additional genes that are expressed *in vivo* but not *in vitro*. Sciortino and Finkelstein (1983) first demonstrated that *V. cholerae* grown in ligated rabbit ileal loops expresses a different protein profile than *V. cholerae* grown *in vitro*. Many of these proteins also appeared to be induced under iron-limiting conditions. Subsequent studies by Richardson *et al.* (1989) and Jonson *et al.* (1989) also reported novel proteins expressed *in vivo*. Additional antigens that were expressed *in vitro* but were down-regulated during *in vivo* growth were also noted. Jonson and colleagues reported *in vivo* proteins ranging from 29 to 200 kDa and most were not induced during culture in iron-depleted media (Jonson *et al.*, 1989). These investigators did not observe any increase in expression of ToxR-regulated proteins such as OmpU and TCP in the bacteria grown *in vivo*. The human immune response to proteins expressed *in vivo* was examined by Richardson *et al.* (1989) using serum IgG and jejunal fluid IgA from convalescent volunteers experimentally infected with *V. cholerae*. In addition to containing antibodies reacting to antigens expressed *in vitro*, both serum and jejunal fluid samples reacted to several antigens found only in cells grown *in vivo*.

5.8 CHOLERA VACCINES

Recognizing that *V. cholerae* O1 and O139 are non-invasive pathogens and that intestinal immunity is important in mediating protection, most modern approaches to developing new cholera vaccines have involved oral vaccines aimed to stimulate more effectively the intestinal immune system. Non-living vaccines oral vaccines consisting of inactivated *V. cholerae* (a mixture of serotypes and biotypes) in combination with the B subunit of cholera toxin were the first to validate the underlying rationale for oral immunization. Since these non-living antigen vaccines require that multiple (at least two) spaced doses be given in order to elicit protection, much research has been undertaken using recombinant DNA technology to construct attenuated strains of *V. cholerae* that can be used as single-dose live oral vaccines. Before 1992, research efforts were directed exclusively towards the development of vaccines to prevent disease caused by *V. cholerae* O1. In addition to summarizing progress on O1 vaccines, recent work on development of vaccines against O139 cholera will also be reviewed in this chapter.

5.8.1 NON-LIVING ORAL VACCINES

Two non-living oral vaccines that have been extensively tested in volunteer trials and in controlled field trials include: (i) killed whole vibrios (WCV) alone; or (ii) killed whole vibrios in combination with the B subunit of cholera toxin (the immunogenic non-toxic portion of cholera toxin) (BS/WCV) (Black *et al.*, 1987; Clemens *et al.*, 1987). The inactivated whole-cell component of these vaccines contains, per dose, a total of 1×10^{11} bacteria representing a mixture of *V. cholerae* O1 biotypes (Classical and El Tor) and serotypes (Inaba and Ogawa) and acts to stimulate antibacterial immunity. In addition, each dose of the combination vaccine contains 1 mg of B subunit to stimulate antitoxic immunity.

In extensive clinical trials in Sweden, the USA and Bangladesh, these vaccines proved to be well tolerated in adults and children as young as 2 years of age (Jertborn *et al.*, 1984; Black *et al.*, 1987; Clemens *et al.*, 1987; Sack *et al.*, 1991). Vibriocidal and other antibacterial immune responses were modest but the combination vaccine elicited prominent antitoxic responses in serum and intestine (Black *et al.*, 1987; Clemens *et al.*, 1987; Sack *et al.*, 1991; Jertborn *et al.*, 1984, 1993). On average these vaccines elicit a two-fold rise in geometric mean titre of vibriocidal antibody in subjects given two or three spaced doses of vaccine (Jertborn *et al.*, 1984, 1986, 1993; Black *et al.*, 1987; Clemens *et al.*, 1987; Sack *et al.*, 1991).

The protective effect of three spaced doses of the BS-WC vaccine and the WCV was initially assessed in clinical trials in adult American volunteers (Black *et al.*, 1987). Overall, a protective efficacy of 64% was observed for BS-WCV and 56% for WCV. However, both vaccines conferred complete (100%) protection against moderate (total purge $\geqslant$3.0 l) and severe cholera (total purge $\geqslant$5.0 l) (Black *et al.*, 1987).

These vaccines were evaluated in a large-scale, randomized, placebo-controlled field trial carried out in Bangladesh from 1985–1988 (Clemens *et al.*, 1986, 1990). During the initial 6 months of the trial the combination vaccine conferred 85% protection, a higher level than the WCV provided during the same period (56% protection) (Clemens *et al.*, 1986). However, this difference disappeared during the remaining 24 months of follow-up so that over the 3 years of surveillance the two vaccines provided essentially the same level of moderate protection, 50% vaccine efficacy for the combination and 53% for the killed whole-cell alone vaccine (Clemens *et al.*, 1990) (Table 5.2). These data suggest that for a period of several months following immunization with the BS/WCV antitoxic immunity and antibacterial immunity work synergistically to confer a high level of protection but the intestinal antitoxic immunity then appears to wane and the level of protection drops.

The original BS/WCV formulation has been modified to incorporate B subunit made by a recombinant vibrio expressing high levels of B subunit. Results of a recent field trial in Peru with the new formulation have corroborated the finding that a high level of protection (86%) is evident during the first 4 months after oral immunization with two spaced doses (2 weeks apart) of killed whole-cell/B subunit combination vaccine (Sanchez *et al.*, 1994). Long-term protective results with this vaccine have not yet been reported, although a large-scale, randomized, double-blind field trial is underway in Arequipa, Peru.

The inactivated oral vaccines evaluated in the field trials in Bangladesh and Peru represent a notable improvement in protection against cholera. They confirm the rationale for the oral route of immunization and demonstrate that improved vaccines can achieve a moderate level of protection against cholera that endures for several years (at least in certain segments of the population).

Recently, the Swedish investigators have further modified their vaccine to include inactivated *V. cholerae* O139 (Sack and Albert, 1994). This multivalent O1/O139 oral BS/WCV is presently undergoing Phase 1 and 2 clinical trials in Swedish adults.

While representing a major advance in immunization against cholera, nevertheless, the oral BS/WCV and WCV suffer from certain notable drawbacks: (i) young children are poorly protected (Clemens *et al.*, 1990); and (ii) multiple (at least two) spaced doses must be administered to elicit protection (Clemens *et al.*, 1990). It is hoped that these drawbacks can be overcome by use of certain types of live oral cholera vaccines.

ATTENUATED *V. CHOLERAE* O1 STRAINS AS LIVE ORAL CHOLERA VACCINES

(a) Attributes of an ideal cholera vaccine

An ideal cholera vaccine would be given orally (for practicality and to stimulate mucosal immunity in the intestine) in a single-dose regimen, and would confer high level, long-term protection that would already be operative within a few days of vaccination. An ideal cholera vaccine would satisfactorily protect high-risk groups (including young children and persons of blood group O) that have not been well protected by earlier cholera vaccines such as the parenteral or oral inactivated

Table 5.2 Overall occurrence of cholera after 6 and 36 months of surveillance in recipients of oral B subunit/whole-cell combination vaccine (BS/WCV), oral whole-cell vaccine (WCV) alone, or placebo, in a large-scale randomized, controlled field trial in Bangladesh. (From Clemens *et al.*, 1986, 1990.)

Group	First 6 months			Full 36 months		
	No. of cases	Incidence per 10^4	Mean (range) vaccine efficacy (%)	No. of cases	Incidence per 10^4	Mean (range) vaccine efficacy (%)
BS/WCV	4*	1.89	85(62–94)[a]	131⁺	63.3	50(39–60)
WCV	11†	5.20	58(14–79)	127‡	61.2	52(41–61)
Placebo	26§	12.25	–	266¶	127.7	–

[a] (95% confidence intervals of vaccine efficacy)

*vs §, $P < 0.0001$; †vs §, $P < 0.01$; *vs †, $P < 0.04$; ⁺vs ¶, $P < 0.0001$.

vaccines (Clemens *et al.*, 1989, 1990; Levine and Pierce, 1992). An ideal cholera vaccine would be inexpensive and would be available in a simple formulation that could facilitate mass vaccination, including young children.

In recent years much progress has been made in engineering recombinant *V. cholerae* O1 (and more recently, O139) and adapting them for use as live oral cholera vaccines. In extensive clinical studies carried out so far, one such recombinant strain, CVD 103-HgR, has been found to exhibit many of the features of an ideal cholera vaccine (Levine and Kaper, 1993; Levine and Tacket, 1994). Yet other attractive O1 and O139 vaccine candidates are entering early phase clinical trials (Taylor *et al.*, 1994; Coster *et al.*, 1995; Tacket *et al.*, 1995a).

(b) Rationale for live cholera vaccines

The rationale for using attenuated *V. cholerae* O1 prepared by recombinant DNA methods as live oral cholera vaccines is based on a series of fundamental observations:

(a) An initial clinical infection caused by wild-type *V. cholerae* O1 confers significant protection against cholera upon subsequent exposure to wild-type *V. cholerae* O1 (Levine *et al.*, 1979a, 1981a, 1983; Glass *et al.*, 1982; Clemens *et al.*, 1991b; Levine and Pierce, 1992; Levine and Kaper, 1993)

(b) While multiple virulence properties contribute to the pathogenesis of cholera, the *in vivo* expression of cholera enterotoxin is an absolute necessity for the profuse purging of rice-water stools that is characteristic of cholera gravis (Levine *et al.*, 1983)

(c) The fundamental protective immunity to cholera is antibacterial rather than antitoxic in nature (Levine *et al.*, 1979a; Levine and Pierce, 1992), although in the short term antitoxic immunity may synergistically enhance antibacterial immunity (Levine *et al.*, 1983; Clemens *et al.*, 1986)

(d) The degree of stimulation of serum vibriocidal antibody following ingestion of a live oral *V. cholerae* O1 vaccine or following infection with wild-type *V. cholerae* O1 constitutes the best correlate for the elicitation of antibacterial immunity in the intestine (Mosley *et al.*, 1968a,b; Glass *et al.*, 1985b; Clemens *et al.*, 1991a; Levine and Pierce, 1992; Levine and Kaper, 1993)

(e) Whereas many antigens on the surface of *V. cholerae* O1 have been identified, the precise combination of antigens that constitutes the protective repertoire remains the subject of debate.

(c) Infection-derived immunity

The level of protection that a prior clinical infection with wild-type enterotoxigenic *V. cholerae* O1 stimulates against subsequent cholera challenge is measured by volunteer studies and by epidemiologic studies in endemic areas. Re-challenge studies in volunteers shown in Table 5.3 demonstrate that an initial clinical infection with either classical or El Tor biotype *V. cholerae* O1 confers 90% (El Tor) to 100% (Classical) protection against clinical illness when the volunteers are re-challenged with *V. cholerae* O1 of the same biotype (Levine *et al.*, 1981b, 1983; Levine and Kaper, 1993). Protection is equal against re-challenge with either the homologous or heterologous serotype (i.e. either Inaba or Ogawa) within the same biotype. In studies with Classical biotype vibrios, protection was seen to persist as long as 3 years, the longest interval tested (Levine *et al.*, 1981a). Unfortunately, no data are available from cross-biotype re-challenge studies in volunteers with wild-type vibrios to ascertain the degree of cross-biotype immunity evident in this model. One obvious difference observed in relation to biotype was the extent of excretion of *V. cholerae* O1 (Table 5.3). Excretion of *V. cholerae* O1 was so low in volunteers re-challenged with classical biotype vibrios that direct cultures of stool onto thiosulphate-citrate-bile salt-sucrose (TCBS) medium were all negative. In contrast, *V. cholerae* O1 biotype El Tor could be recovered on direct coprocultures from approximately 30% of individuals who were re-challenged with that biotype. These data suggest that in some manner Classical biotype vibrios stimulate a more potent immunity than El Tor vibrios.

Natural infection-derived immunity has been studied among persons living in Matlab Bazaar, Bangladesh, an area where cholera is highly endemic. Glass *et al.* (1982), working mainly with data from Classical biotype infections concluded that an initial clinical infection diminishes by approximately 90% the risk of experiencing a subsequent episode of clinical cholera (Table 5.4). Notably, all individuals who experienced recurrent episodes of cholera had their first illness when they were young children 5 years of age or less.

Clemens *et al.* (1991b) reviewed the frequency of repeat clinical cholera infections in Matlab during a period when infections of both biotypes were prevalent (Table 5.4). They observed that an initial clinical infection due to *V. cholerae* O1 of Classical biotype conferred 100% protection against subsequent cholera due to either biotype. In contrast, an initial infection with El Tor biotype was seen to be unprotective against subsequent cholera due to the heterologous Classical biotype and to be poorly protective (29% efficacy) even against subsequent cholera due to the homologous El Tor biotype.

Table 5.3 Protective efficacy in volunteers against biotype homologous challenge conferred by prior clinical infection with pathogenic *Vibrio cholerae* O1 of Classical or El Tor biotype

Immunizing *V. cholerae* biotype	Attack rate for diarrhoea*		Protective efficacy (%)	Isolation of *V. cholerae* from direct coprocultures	
	Controls	Veterans⁺		Controls	Veterans
Classical	24/27 (89%)	0/16 (0%)	100	26/27 (96%)	0/16 (0%) $P = 0.012$
El Tor	32/37 (86%)	2/22 (9%)	90	34/37 (92%)	8/22 (36%)

* Challenge with 10^6 pathogenic *V. cholerae* O1 given with $NaHCO_3$.
⁺ Includes both serotype-homologous and serotype-heterologous challenges. Volunteers who developed diarrhoea following ingestion of *V. cholerae* O1 on initial challenge were rechallenged 4–6 weeks later with *V. cholerae* O1 of either the same or the heterologous serotype within the identical biotype.

Table 5.4 Infection-derived immunity conferred by clinical infection with different biotypes of *Vibrio cholerae* O1. Observations from field studies in a cholera-endemic area of Bangladesh

Study	Biotype of initial infection	Biotype of subsequent infection	Protective efficacy (%)
Glass *et al.* 1982	Mostly classical	Mostly classical	90
Clemens *et al.* 1991b	Classical	Classical	100
	Classical	El Tor	100
	El Tor	El Tor	29
	El Tor	Classical	0

The volunteer model and the field studies concur in demonstrating that superior protective immunity follows clinical infection with Classical biotype *V. cholerae* O1. The discrepancy over the degree of immunity following biotype El Tor clinical infections may be explained by the duration of follow-up. The high level of protection following El Tor cholera infection in volunteers was based on re-challenge after only 2 months. It is conceivable that the protective immunity conferred by El Tor wanes over time. In the epidemiologic studies in Bangladesh, the El Tor infection-derived immunity was examined over years rather than several months. Under any circumstances, after reviewing the above data, one might conclude that if the intention is to utilize recombinant *V. cholerae* O1 strains as live oral vaccines, it would be wise to include strains of Classical biotype; arguably, attenuated classical strains alone might suffice.

(d) Cholera enterotoxin is a prerequisite for cholera gravis

The well-studied molecular pathogenesis of cholera infection is characterized by cascades of coordinately regulated virulence properties as *V. cholerae* O1 converts itself from an environmental niche in brackish water to adapt to survival in the human small intestine (DiRita *et al.*, 1991). Of the many recognized virulence properties of *V. cholerae* O1 (Levine *et al.*, 1983), one that **must** be expressed for the syndrome of cholera gravis (i.e. severe, rapidly dehydrating, potentially fatal cholera) to ensue is expression *in vivo* of cholera enterotoxin. Indeed, ingestion of minute quantities of purified cholera enterotoxin by adult North American volunteers resulted in severe purging and a syndrome resembling cholera gravis (Levine *et al.*, 1983). These data emphasize the fundamental importance of inactivating cholera holotoxin in developing candidate live vaccine strains.

(e) Antibacterial immunity

That the main mechanism of protection against *V. cholerae* O1 involves antibacterial rather than antitoxic immunity, is supported by the following observations:

1. Parenteral whole-cell inactivated vaccines that elicit serum vibriocidal antibody but not antitoxin confer significant protection, albeit for only short periods (several months) (Levine and Pierce, 1992).
2. Parenteral toxoids that stimulate high levels of serum antitoxin do not confer credible protection, even short-term (Noriki, 1976; Curlin *et al.*, 1974).
3. Over 3 years of surveillance in a controlled field trial in Bangladesh, three spaced doses (6 weeks apart) of an oral vaccine consisting of only inactivated *V. cholerae* O1 conferred virtually identical protection (52% efficacy) as three doses of the same inactivated *V. cholerae* O1 vaccine given in combination with B subunit (50% efficacy) (Clemens *et al.*, 1990). Although the large doses of B subunit stimulated strong antitoxic serologic responses in the intestine and in serum, this was not accompanied by any long-term enhancement of the protection conferred by the inactivated bacteria alone (Clemens *et al.*, 1990).
4. A single oral dose of a recombinant A^-B^- *V. cholerae* O1 El Tor candidate vaccine strain (JBK 70) conferred on volunteers 89% protection against experimental challenge with wild-type *V. cholerae* El Tor (Levine *et al.*, 1988b). Cholera antitoxin did not contribute to the high level of protection since this strain elaborated neither the A nor the B subunit of CT and did not stimulate an antitoxic immune response.

(f) Serum vibriocidal antibody as a correlate of protection

Vibriocidal antibody is measured by bacterial lysis when serial dilutions of serum are incubated with a large standardized inoculum of *V. cholerae* O1 in the presence of guinea pig complement (Benenson *et al.*, 1968). Following natural or experimental infection of humans, the titre of serum vibriocidal antibody rise many-fold. In cholera endemic areas, the repeated ingestion of *V. cholerae* O1 gives rise to long-lived elevated titres of IgG vibriocidal antibody. Thus, in endemic areas the prevalence of geometric mean titre of vibriocidal antibody increase with age (Mosley *et al.*, 1968a,b; Glass *et al.*, 1985b; Clemens *et al.*, 1991a) and such elevated titres are correlated with protection (Mosley *et al.*, 1968a,b; Glass *et al.*, 1985b; Clemens *et al.*, 1991a). Vibriocidal responses following experimental challenge of adult volunteers tend to be mainly

IgM and are short-lived, typically falling to a level two-fold over baseline within 6–12 months. Nevertheless, the initial short-lived serum vibriocidal response is a marker for the elicitation of long-lived intestinal immunity since such volunteers are protected for as long as 3 years after initial challenge, despite the fact that they no longer have elevated vibriocidal titres (Levine *et al.*, 1981a). The potent serum vibriocidal antibody response that appears following the ingestion of live oral antigens, be they wild-type or attenuated vibrios, serves as a marker for the stimulation of a potent intestinal immunity that endures long after the serum vibriocidal antibody titres have returned towards baseline levels (Levine *et al.*, 1981a; Levine and Kaper, 1993; Losonsky *et al.*, 1993).

For the above reasons, serum vibriocidal antibody is monitored as the proxy for elicitation of a protective intestinal immune response following the administration of candidate live oral vaccines (Levine and Kaper, 1993). In general, the more potent the serum vibriocidal immune response following oral vaccination, the greater the protective immunity induced.

5.9 PROTECTIVE SURFACE ANTIGENS OF *V. CHOLERAE* O1

Many surface antigens have been proposed as protective antigens (Levine *et al.*, 1983; Levine and Pierce, 1992), including lipopolysaccharide (LPS) O antigen, pili (fimbriae), outer membrane proteins and certain haemagglutinins, but their relative importance is disputed. Moreover, there exist antigens that are expressed *in vivo* but that are not readily observed when *V. cholerae* O1 are cultured *in vitro*, despite attempts to mimic *in vivo* conditions (Sigel and Payne, 1982; Jonson *et al.*, 1989).

While approximately 85–90% of the vibriocidal antibodies stimulated by wild-type infection are directed against the LPS O antigen (Levine *et al.*, 1983; Chongsa-nguan *et al.*, 1986; Sharma *et al.*, 1987; Hall *et al.*, 1991), important protective antibodies also reside among the vibriocidal antibodies directed against antigens other than LPS (Levine *et al.*, 1983; Chongsa-nguan *et al.*, 1986; Sharma *et al.*, 1987). TCP must be expressed for Classical biotype *V. cholerae* O1 to elicit strong vibriocidal responses (Herrington *et al.*, 1988), yet the immune response to the pili themselves is minimal (Hall *et al.*, 1991).

5.9.1 EARLY GENERATIONS OF ATTENUATED *V. CHOLERAE* O1 RECOMBINANT VACCINE CANDIDATES

The main approach followed in developing live recombinant vaccines against cholera has been to disarm known pathogenic strains of their ability to express certain specific virulence attributes, thereby presumably rendering them incapable of causing cholera yet leaving intact the various surface antigens (known and unknown) involved in protection. This approach has been pioneered by research groups at the Center for Vaccine Development of the University of Maryland School of Medicine (Kaper *et al.*, 1984a,b; Kaper and Levine, 1990; Ketley *et al.*, 1993; Levine and Kaper, 1993) and at Harvard Medical School (Mekalanos *et al.*, 1983; Pearson *et al.*, 1990; Taylor *et al.*, 1994). A radically different approach was taken by Australian investigators in the 1980s who utilized attenuated *Salmonella typhi* vaccine strain Ty21a as a live vector to express a putative protective antigen of *V. cholerae* O1 and deliver it to the human intestinal immune system (Forrest *et al.*, 1989). These approaches are discussed below.

5.9.2 FIRST- AND SECOND-GENERATION ATTENUATED *V. CHOLERAE* O1 RECOMBINANT VACCINE CANDIDATES

Early vaccine candidates were constructed from wild-type *V. cholerae* strains known to be pathogenic for volunteers by introducing deletions in the chromosomal genes encoding the A subunit (Mekalanos *et al.*, 1983; Kaper *et al.*, 1984b) or both the A and B subunits (Kaper *et al.*, 1984a) of cholera enterotoxin. These first generation vaccine strains, such as JBK 70, CVD 101 and 395N1, were markedly attenuated compared with their wild-type parents (Levine *et al.*, 1988b), as they were no longer capable of causing severe diarrhoea. Nevertheless, some residual reactogenicity remained (Herrington *et al.*, 1988; Levine *et al.*, 1988b) as approximately one-half of the volunteers suffered adverse reactions consisting of combinations of malaise, nausea, vomiting, abdominal cramps, low-grade fever, headache and mild diarrhoea. Although these first-generation strains were unacceptably reactogenic, a single oral dose containing as few as 10^3 colony forming units (CFU), given with sodium bicarbonate buffer, elicited vigorous vibriocidal responses (Levine *et al.*, 1988b). The objective of subsequent research was to diminish reactogenicity, while retaining immunogenicity with a single oral dose.

Two second-generation vaccine strains provided invaluable information in directing further live cholera vaccine development (Herrington *et al.*, 1988). JJM43 is a further derivative of strain 395N1 (*ctxA* mutant of Ogawa 395) in which the regulatory gene *toxR* is inactivated, making the vaccine candidate unable to respond to environmental signals and unable to coordinate the regulation of virulence gene expression (DiRita *et al.*, 1991). Strain TCP2 is another derivative of 395N1 in which *tcpA*, the gene encoding the structural subunit of cholera (TCP), has been inactivated (Taylor *et al.*, 1984). Each of these strains exhibited greatly diminished ability to colonize the intestine of volunteers and to elicit vibriocidal antibody responses (Herrington *et al.*, 1988). Results of these studies emphasize the importance of retaining intact the fimbrial colonization factors and the master regulatory gene (*toxR*) of *V. cholerae* O1 in attenuated *V. cholerae* candidate live vaccines.

5.9.3 ATTENUATED *V. CHOLERAE* O1 VACCINE STRAINS CVD 103 AND CVD 103-HGR

The first engineered vaccine strain to be well tolerated yet highly immunogenic and protective was strain CVD 103, derived from pathogenic classical Inaba strain 569B (Levine *et al.*, 1988a; Ketley *et al.*, 1993) by deletion of 94% of the gene encoding the A subunit of CT (*ctxA*), leaving intact expression of B subunit. A single dose of CVD 103 vaccine elicited seroconversions of vibriocidal antibody in 90% of vaccinees and of antitoxin in >80% and conferred upon volunteers significant protection against challenge with pathogenic *V. cholerae* O1 of either serotype or biotype (Levine *et al.*, 1988a,b).

A gene encoding resistance to Hg^{++} (*mer*) was introduced into the *hlyA* locus of the chromosome of CVD 103 to provide a marker to differentiate readily the vaccine strain from wild-type vibrios (Ketley *et al.*, 1993). This derivative encoding resistance to Hg^{++} was designated CVD 103-HgR (Levine *et al.*, 1988a; Kaper and Levine, 1990; Ketley *et al.*, 1993).

(a) Overview of clinical and field studies with CVD 103-HgR

To date, approximately 6000 subjects ranging in age from 12 months to 65 years have participated in placebo-controlled clinical trials establishing the safety, immunogenicity and excretion pattern following immunization with CVD 103-HgR and the ability to co-administer CVD 103-HgR with other live oral vaccines. These trials have been carried out in industrialized countries (USA, Switzerland, Italy, Austria) (Levine *et al.*, 1988a; Cryz *et al.*, 1990, 1992, 1995; Kotloff *et al.*, 1992; Tacket *et al.*, 1992b; Levine and Kaper, 1993) and in developing countries with endemic cholera (Indonesia, Thailand) (Migasena *et al.*, 1989; Su-Arehawaratana *et al.*, 1992; Suharyono *et al.*, 1992; Simanjuntak *et al.*, 1993), epidemic cholera (Peru, Colombia) (Gotuzzo *et al.*, 1993), or little or no cholera (Chile, Costa Rica) (Lagos *et al.*, 1993, 1995). A practical formulation of the vaccine was used in these studies consisting of two aluminum foil sachets, one containing lyophilized vaccine (and aspartame as sweetener) and the other containing buffer (to protect the vaccine strain from gastric acid). The two sachets are mixed in a cup containing 100 ml of water and the resultant suspension is ingested by the subject. In 1993, a large-scale field trial involving approximately 67 000 subjects was initiated in North Jakarta, Indonesia to assess the efficacy of a single dose of CVD 103-HgR in preventing cholera under natural conditions of challenge.

(b) Safety of CVD 103-HgR

Since CVD 103-HgR was derived from a pathogenic parent strain (Classical 569B) known to be capable of causing cholera gravis in volunteers, considerable attention has been directed to ascertaining whether CVD 103-HgR causes diarrhoea. Table 5.5 summarizes the results of a number of clinical studies that have assessed the safety of CVD 103-HgR, most of which are randomized, double-blind, placebo-controlled trials (Migasena *et al.*, 1989; Cryz *et al.*, 1990, 1992; Kotloff *et al.*, 1992; Su-Arehawaratana *et al.*, 1992; Suharyono *et al.*, 1992; Tacket *et al.*, 1992b; Gotuzzo *et al.*, 1993; Lagos *et al.*, 1993, 1995; Simanjuntak *et al.*, 1993; Wasserman *et al.*, 1994). A large clinical experience attests to the safety of this vaccine strain in all age groups and shows that diarrhoeal adverse reactions have not been observed more often in vaccinees than in placebo recipients in controlled trials. Placebo-controlled studies in infants and toddlers 6–23 months of age have recently begun.

(c) Immunogenicity of CVD 103-HgR

Since vibriocidal antibody is currently considered to be the best correlate of protection and the best measure of the successful stimulation of antibacterial immunity, the serum vibriocidal antibody response has been used to assess the immunogenicity of CVD 103-HgR (Levine *et al.*, 1988a; Migasena *et al.*, 1989; Cryz *et al.*, 1990, 1992; Kotloff *et al.*, 1992; Su-Arehawaratana *et al.*, 1992; Suharyono *et al.*, 1992; Tacket *et al.*, 1992b; Gotuzzo *et al.*, 1993; Lagos *et al.*, 1993; Levine and Kaper, 1993; Losonsky *et al.*, 1993; Simanjuntak *et al.*, 1993; Wasserman *et al.*, 1994); four-fold or greater rise is considered significant (i.e. seroconversion).

Immunogenicity in populations in industrialized countries

In clinical studies in North Americans, Swiss, Italian and Austrian subjects, a single 5×10^8 CFU dose of CVD 103-HgR has consistently elicited four-fold or greater rises in vibriocidal antibody in approximately 85–92% of vaccinees with an 80- to 160-fold rise in geometric mean titre (Levine *et al.*, 1988a; Cryz *et al.*, 1990, 1992, 1995; Kotloff *et al.*, 1992; Tacket *et al.*, 1992b; Levine and Kaper, 1993; Losonsky *et al.*, 1993). Two explanations have been advanced to account for the vigorous immunogenicity of this live oral vaccine. The first relates to

Table 5.5 Rate of diarrhoeal adverse reactions in recipients of CVD 103-HgR live oral cholera vaccine versus placebo recipients in randomized, double-blind trials in adults and children

Age group and site	Vaccine dose (CFU)	Rate of diarrhoea				*P*	Reference
		Vaccinees		Controls			
Adults							
USA	5×10^8	1/200	(0.5%)	–			Levine and Kaper (1993)
USA	5×10^8	1/94	(1.1%)	0/94	(0%)	NS	Levine and Kaper (1993)
Switzerland	5×10^8	1/25	(4%)	2/25	(8%)	NS	Cryz *et al.* (1990)
Thailand	5×10^8	0/12	(05)	0/12	(0%)	NS	Migasena *et al.* (1989)
Thailand	5×10^8	11/102	(11%)	13/104	(13%)	NS	Su-Arehawatana *et al.* (1992)
Thailand	5×10^8	0/103	(0%)	–		–	Su-Arehawatana *et al.* (1992)
Thailand	5×10^9	5/119	(4%)	2/89	(2%)	NS	Su-Arehawatana *et al.* (1992)
Peru	5×10^8	2/41	(4.9%)	3/40	(7.5%)	NS	Gotuzzo *et al.* (1993)
	5×10^9	5/41	(12.2%)	3/40	(7.5%)	NS	Gotuzzo *et al.* (1993)
Chile	5×10^9	0/40	(0%)	1/41	(2.4%)	NS	Lagos *et al.* (1993)
5–9-year-olds							
Indonesia	$5 \times 10^{6-8}$	10/209	(4.8%)	4/65	(6.2%)	NS	Suharyone *et al.* (1992)
	$5 \times 10^{9-10}$	4/124	(3.2%)	4/32	(1%)	NS	Suharyone *et al.* (1992)
Chile	5×10^9	2/178	(1.1%)	1/171	(0.6%)	NS	Lagos *et al.* (1993)
Costa Rica	5×10^9	2/196	(1.0%)	4/193	(2.1%)	NS	
Peru	5×10^8	1/40	(2.5%)	1/40	(2.5%)	NS	
	5×10^9	1/40	(2.5%)	1/40	(2.5%)	NS	
24–59-month-olds							
Indonesia	5×10^9	18/155	(11.6%)	12/148	(8.1%)	NS	Simanjuntak *et al.* (1993)
Costa Rica	5×10^9	2/118	(1.7%)	2/118	(1.7%)	NS	
Chile	5×10^9	4/100	(4.0%)	4/100	(4.0%)	NS	

CFU, colony forming units; NS, not significant.

the fact that initiation of the intestinal immune response begins with uptake of antigen by the dome-like epithelial cells (so-called M cells) that cover the Peyer's patches and other organized lymphoid tissue of the gut (McGhee *et al.*, 1992). Owen *et al.* (1986) have shown that in rabbit ileal loops containing a Peyer's patch, live *V. cholerae* O1 are taken up more readily by the M cells than are inactivated bacteria. The second is based on the observation that *V. cholerae* O1 infection requires coordinate regulation of multiple tightly controlled virulence properties that are activated or depressed by regulatory genes that respond to environmental signals (DiRita *et al.*, 1991). With inactivated vaccine, the surface antigens present during growth in the fermentor are the surface antigens that will be exposed to the intestinal immune system; the inactivated vibrios cannot change in response to environmental stimuli in the intestine. In contrast, in live vaccines such as CVD 103-HgR the *toxR* regulatory gene is intact; thus, this attenuated vaccine strain can modify and express the new proteins that are required to survive in and colonize the intestine. Some of these *in vivo*-activated gene products are immunogenic and may play a role in protection.

Immunogenicity in populations in less-developed countries

In the course of performing community studies of adults and children in less-developed countries where cholera was endemic or epidemic, it was found that the 5×10^8 CFU dosage level, which is highly immunogenic in subjects in industrialized countries (90% seroconversion), elicited seroconversions of vibriocidal antibody in only ca. 25% of Thai soldiers (Su-Arehawaratana *et al.*, 1992) and in only 16% of Indonesian children (Suharyono *et al.*, 1992; Simanjuntak *et al.*, 1993). It was subsequently found that many individuals in endemic areas already have elevated vibriocidal titres and are presumably at least partially immune (Su-Arehawaratana *et al.*, 1992; Suharyono *et al.*, 1992; Simanjuntak *et al.*, 1993). In several studies the baseline vibriocidal geometric mean titre in subjects who did not seroconvert (i.e. at least a four-fold rise in titre) was significantly higher than the baseline geometric mean titre of individuals who did seroconvert (Su-Arehawaratana *et al.*, 1992; Suharyono *et al.*, 1992; Simanjuntak *et al.*, 1993). In such persons with elevated baseline titres, a further rise in serum vibriocidal antibody is not a good measure of vaccine 'take' or boosting. In these populations it was found that the vibriocidal seroconversion rate could be markedly increased simply by administering a single dose of vaccine containing one log higher number of vaccine organisms (i.e. 5×10^9 CFU). With a single 5×10^9 CFU dose of vaccine, high rates of seroconversion of vibriocidal antibody (usually 75–85%) have been achieved in the less-developed country populations studied so far (Su-Arehawaratana *et al.*,

1992; Suharyono *et al.*, 1992; Simanjuntak *et al.*, 1993; Gotuzzo *et al.*, 1993; Lagos *et al.*, 1993, 1995).

It is well recognized that persons of blood group O are at greatly increased risk of developing cholera gravis. While assessing the safety and immunogenicity of CVD 103-HgR in 5- to 9-year-old Chilean children, Lagos *et al.* (1995) observed that whereas the rates of seroconversion of vibriocidal antibody were similar for subjects of blood group O versus non-O, the former manifested significantly higher geometric mean titres of vibriocidal antibody.

(d) Efficacy of CVD 103-HgR in experimental challenge studies

A series of challenge studies have been carried out in the volunteer model of experimental cholera. A total of ten such challenges have been performed, three with volunteers who received CVD 103 and seven with volunteers who received CVD 103-HgR (Levine *et al.*, 1988a; Tacket *et al.*, 1992b; Levine and Kaper, 1993). In all but one study only a single dose of vaccine was administered; in one study (El Tor Ogawa challenge) the volunteers ingested two doses of vaccine 1 week apart. Subsequent immunogenicity studies have detected no difference in the vibriocidal antibody response between groups of subjects who receive a single dose of CVD 103-HgR versus those who get two doses 1 week apart (Su-Arehawaratana *et al.*, 1992).

In each of these ten challenge studies, CVD 103-HgR (or its parent CVD 103) conferred significant protection against challenge with fully enterotoxigenic wild-type *V. cholerae* O1 (strains which caused cholera diarrhoea in 78–100% of unimmunized control volunteers). The overall protection was quite high against challenge with Classical biotype (82–100%, irrespective of serotype) and moderate with El Tor biotype (49–67%, irrespective of serotype). However, a more relevant way to consider the challenge data is from the perspective of prevention of severe and moderate diarrhoea, since it is the syndrome of cholera gravis that establishes cholera as a public health problem. Table 5.6 summarizes the ten challenge studies with respect to protection against diarrhoea of different severity determined by total diarrhoeal stool volume. CVD 103 and CVD 103-HgR provide complete protection against severe (⩾5.0 l total purge) and moderate (⩾3.0 l total purge) volume diarrhoea. In fact, of the 19 vaccinees who manifested diarrhoea following challenge with wild-type *V. cholerae* O1, only two subjects purged a total of diarrhoeal stool volume of ⩾2.0 l. Thus, CVD 103-HgR is highly protective against severe and moderate cholera of the type that would lead to dehydration.

Tacket *et al.* (1992b) showed that a single dose of CVD 103-HgR provided 100% protection in volunteers who were challenged with

Table 5.6 Efficacy of CVD 103 and CVD 103-HgR live oral cholera vaccines against experimental challenge with wild-type *V. cholerae* O1

Severity of diarrhoea*	Controls	Vaccinees	Protective efficacy (%)	*P*
⩾ 5.0 l	9/88	0/101	100	< 0.001
⩾ 3.0 l	19/88	0/101	100	< 0.0001
> 2.0 l	28/88	2/101	94	< 0.0001
⩾ 1.0 l	42/88	7/101	86	< 0.0001
Any diarrhoea	70/88	19/101	76	< 0.0001

* Total diarrhoeal stool volume in litres during episode of experimental cholera.

wild-type *V. cholerae* O1 (classical Inaba) 6 months after immunization (the longest interval so far tested). Moreover, when a group of volunteers were challenged with classical Inaba a mere 8 days after ingesting a single dose of CVD 103-HgR, they were also 100% protected against cholera (Tacket *et al.*, 1992b).

(e) Field trials of efficacy of CVD 103-HgR in endemic areas

In 1993, a large-scale, randomized, placebo-controlled field trial was initiated in 67 000 subjects aged 2–42 years in North Jakarta to assess the efficacy of a single oral dose of CVD 103-HgR in protecting against cholera in a population exposed to natural challenge in an endemic area. The aims of this study include measuring vaccine efficacy in persons less than 6 years of age (an age group not well protected by earlier vaccines) and comparing efficacy in persons of blood group O versus other blood groups. Surveillance will be maintained for at least 3 years before data are analysed.

5.9.4 NEWER ATTENUATED *V. CHOLERAE* O1 EL TOR VACCINE STRAINS

The *ctx* virulence cassette region of the chromosome in El Tor strains (that contains the genes encoding CT, Zot and Ace toxins and core-encoded pili) is found between two insertion-like sequences, RS, that make this transposon-like region of the chromosome unstable and subject to duplication or deletion (Goldberg and Mekalanos, 1986). Accordingly, teams of investigators at the University of Maryland and at Harvard University have each been able to construct new vaccine candidate strains from wild-type El Tor parent strains by deleting the entire CT virulence cassette that falls between the two RS elements (Roberts *et al.*, 1992; Michalski *et al.*, 1993; Taylor *et al.*, 1993, 1994; Levine and Tacket, 1994). Thus, these new candidates lack Zot and Ace, as well as CT and Cep.

(a) Construction of CVD 110

In constructing CVD 110, following deletion of the *ctx* virulence cassettes, the Maryland investigators (Michalski *et al.*, 1993) then introduced the gene encoding B subunit under control of its native promoter into another chromosomal locus, *hlyA*, along with a gene encoding resistance to Hg^{++}.

(b) Construction of Bah-3, Bang-3, Peru-3 and Peru-5

The Harvard investigators (Roberts *et al.*, 1992; Taylor *et al.*, 1994) prepared a series of El Tor vaccine candidates generated from wild-type El Tor strains, including several (El Tor Inaba P27459 from Bangladesh and El Tor Ogawa E7946 from Bahrain) of established virulence based on volunteer studies (Levine *et al.*, 1981b, 1982; Taylor *et al.*, 1994). The various vaccine constructs have in common a deletion of the virulence cassette that contains *ctxAB*, *zot*, *ace* and *cep*, as well as factors RS1 and attRS1 which are involved in site-specific and homologous recombination. These vaccine candidates, except Peru-5, have an inactivation of *recA* as a consequence of the insertion into that gene of *ctxB* fused to a promoter from the heat shock gene *htpG*. In the Peru-5 strain, *ctxB* under its native promoter is inserted into *lacZ*.

(c) Clinical trials with CVD 110, Bah-3, Bang-3, Peru-3 and Peru-5

Somewhat surprising results were obtained when Phase 1 clinical trials were carried out with the new generation of El Tor candidate strains, particularly CVD 110 (Michalski *et al.*, 1993) and Bah-3 (Roberts *et al.*, 1992; Taylor *et al.*, 1994). Candidate vaccine strains CVD 110 and Bahr-3, which were both derived from volunteer-tested wild-type strain El Tor Ogawa E7946 (isolated from an epidemic in Bahrain), each caused mild to moderate diarrhoea in most of the vaccinees, accompanied by headache, malaise, abdominal cramps or low-grade fever (Tacket *et al.*, 1993; Taylor *et al.*, 1994). The other vaccine strains, Bang-3, Peru-3 and Peru-5, also caused adverse reactions in volunteers but at a lower frequency and milder in nature. The attack rate for such adverse reactions and the total diarrhoeal stool volumes (ca. 300–500 ml) appeared to be similar to what was observed with earlier generation vaccine candidate strain CVD 104 (Levine *et al.*, 1988b).

These are important findings because they demonstrate that *V. cholerae* O1 candidate vaccine strains that are derived from known strains of high virulence by deleting the capacity to express CT, Zot, Ace, El Tor haemolysin and Cep, still retain the propensity to cause definite (albeit not severe) diarrhoeal illness. The pathophysiologic explanation for the residual diarrhoeagenic potential of these strains has yet to be elucidated. One possibility is that additional, so far unidentified, enterotoxins are elaborated *in vivo*. An alternative plausible explanation is that the act of adherence of vaccine strains to enterocytes in the proximal small intestine *per se* results in net secretion, perhaps consequent to the release of pro-inflammatory cytokines such as interleukin-6.

(d) CVD 111 and Peru-14

Based on the results of clinical observations made with strains CVD 110, Bah-3, Bang-3, Peru-3 and Peru-5, the Maryland and Harvard groups prepared additional El Tor vaccine candidates intended to be diminished in reactogenicity, yet maintaining good immunogenicity and ability to protect against experimental challenge. CVD 111 was constructed in a manner identical to CVD 110, except that the starting wild-type parent (strain El Tor Ogawa N16117) was one that exhibited inherently diminished pathogenicity in volunteers, despite possessing a full array of virulence genes. Peru-14 was developed by selecting a motility-deficient variant of Peru-3 (Taylor *et al.*, 1994).

In preliminary Phase 1 clinical trials, both CVD 111 and Peru-14 have proven to be well tolerated and highly immunogenic (Taylor *et al.*, 1994; Tacket *et al.*, 1996), a stark contrast to the results of Phase 1 trials with CVD 110, Bah-3 and the other vaccine strains. Moreover, in experimental challenge studies independently carried out with each vaccine candidate, a single oral dose of CVD 111 or Peru-14 conferred ca. 80–85% protection against cholera (Taylor *et al.*, 1994; Tacket *et al.*, 1996).

5.10 LIVE VACCINES AGAINST *V. CHOLERAE* O139

Investigators at the University of Maryland documented the virulence of a prototype wild-type O139 strain, AI1837, by means of experimental challenge studies in volunteers (Morris *et al.*, 1995). The attack rates and severity of diarrhoea following ingestion of *V. cholerae* O139 strain AI1837 with buffer resemble those observed in volunteers challenged with *V. cholerae* O1 of the El Tor biotype. Seven volunteers who experienced O139 cholera were re-challenged 2 months later, along with 13 naive control subjects. Eleven of 13 controls developed O139 cholera versus only one of seven 'veterans' (14%) (who manifested only mild diarrhoea) ($P = 0.004$) (Morris *et al.*, 1995). Thus, in the volunteer model an initial clinical infection caused by wild-type *V. cholerae* O139 elicits immune responses that confer 83% protection against a challenge inoculum capable of causing a high attack rate in naive volunteers. These observations encourage the expectation that moderate levels of protective immunity may also be achieved by means of O139 vaccine candidates.

5.10.1 CVD 112

Comstock *et al.* (unpublished results) at the Center for Vaccine Development engineered an attenuated *V. cholerae* O139 vaccine candidate by deleting from volunteer-tested wild-type strain AI1837 the entire 'virulence cassette' region of the chromosome that includes the genes encoding cholera toxin, Zot, Ace and the Cep. A gene encoding resistance to Hg^{++} and the gene encoding B subunit under control of its native promoter were introduced into the chromosome in the *hlyA* locus, thereby also inactivating that haemolysin/cytotoxin/enterotoxin. The resultant vaccine candidate is designated CVD 112.

In Phase 1 clinical trials, CVD 112 has proven to be well tolerated and immunogenic (Tacket *et al.*, 1995a). In contrast with O1 strains, vibriocidal antibody responses to O139 strains are meager and somewhat non-specific. The quantification of gut-derived antibody-secreting cells that make IgA antibody to the O139 capsule appears to be the best measure of vaccine immunogenicity. In experimental challenge studies, a single oral dose of CVD 112 has conferred 84% protection against challenge with wild-type *V. cholerae* O139. Phase 1 and 2 clinical trials are commencing with bivalent and trivalent live vaccines consisting of mixtures of lyophilized CVD 111, CVD 112 and CVD 103-HgR.

5.10.2 BENGAL-15

The Harvard group have also constructed attenuated O139 vaccine candidates, starting from wild-type strain MO10 which was shown to be pathogenic in volunteer studies (Waldor and Mekalanos, 1994; Coster *et al.*, 1995). Attenuated derivative designated Bengal-15 has deletion of the virulence cassette that contains *ctxAB*, *zot*, *ace* and *cep* as well as RS1 and attRS1, a recombinant gene encoding B subunit under control of the promoter from heat shock gene *htpG* ($_{htpG}P$) has been inserted into *recA*, and the strain has an undefined spontaneous mutation characterized by diminished motility. In Phase 1 clinical trials, single oral doses of Bengal-15 were well tolerated and immunogenic and conferred 83% protection against challenge with wild-type *V. cholerae* O139 (Coster *et al.*, 1995).

5.10.3 CVD 103-HGR EXPRESSING O139 ANTIGENS

In view of the exceptional safety, immunogenicity and efficacy (in volunteer challenge studies) of single-dose live oral *V. cholerae* O1 classical Inaba vaccine strain CVD 103-HgR and the fact that has already been licensed by regulatory agencies in several countries, attempts are being made to express O139 LPS and capsular polysaccharide antigens in CVD 103-HgR. Following this strategy CVD 103-HgR would serve as a live vector vaccine to deliver O139 antigens to the human immune system. A bivalent vaccine could then be prepared consisting of CVD 103-HgR (to prevent O1 disease) and CVD 103-HgR expressing O139 antigens (to prevent O139 disease). This combination may diminish the chance of possible interference between the two live vaccine strains since they share a common background.

5.11 ATTENUATED *S. TYPHI* EXPRESSING *V. CHOLERAE* O1 ANTIGENS

Investigators in Adelaide, Australia cloned the genes required for expression of the Inaba serotype O antigen of *V. cholerae* O1 onto a plasmid which also carried *thyA* encoding thymine independence and introduced this plasmid into a rifampicin-resistant variant of attenuated *S. typhi* strain Ty21a (a licensed live oral typhoid vaccine) carrying a deletion mutation in *thyA*. The Ty21a was further altered in that the *rfa* chromosomal region involved with LPS expression was replaced with *rfa* from *E. coli* K-12. The resultant transconjugant, *S. typhi* strain EX645, stably expressed the O antigen of *V. cholerae* O1 Inaba (Forrest *et al.*, 1989).

In Phase 1 studies with EX645 (Forrest *et al.*, 1989; Tacket *et al.*, 1990), the vaccine was well tolerated when subjects were administered three large doses (alternate day schedule) each containing approximately 5×10^{10} CFU. EX645 was modestly immunogenic in eliciting serum or intestinal secretory immune responses to cholera (Forrest *et al.*, 1989; Tacket *et al.*, 1990), whereas immune responses to *S. typhi* antigens were more prominent.

One month following vaccination, eight subjects who received three doses of EX645 participated in an experimental challenge study with wild-type El Tor Inaba to assess vaccine efficacy, along with 13 unvaccinated controls (Tacket *et al.*, 1990). Following challenge, the attack rate for diarrhoea overall in the vaccinees (6/8, 75%) was not significantly lower than observed in the controls (13/13, 100%) (overall, 25% vaccine efficacy). However, the severity of the diarrhoea was significantly less in the vaccinees (mean 867 ml per ill volunteer) than in the controls (mean 2603 ml per ill volunteer) ($P < 0.05$). The excretion of wild-type *V. cholerae* O1 following challenge was also significantly diminished in the vaccinees than in the controls ($P < 0.05$).

Although the EX645 vaccine was notably less protective than a number of other live and inactivated oral vaccines evaluated in the experimental challenge model, it nevertheless clearly demonstrated a significant biologic effect by ameliorating the severity of illness and by curtailing excretion of vibrios. This protective effect, albeit modest, was accomplished by a live vector vaccine that expressed only a single *V. cholerae* O1 antigen and that was only modestly immunogenic. Presumably, if multiple additional *V. cholerae* O1 antigens could be expressed with greater efficiency in a new generation attenuated *S. typhi* live vector that is inherently more immunogenic (Hone *et al.*, 1991; Tacket *et al.*, 1992a), this strategy to elicit protective immunity against cholera might yet prove fruitful.

5.12 SUMMARY AND CONCLUSIONS

One hundred years have elapsed since clinical trials were initiated with the first cholera vaccine in 1885, which consisted of live, putatively attenuated, vibrios administered parenterally. During the past decade, recombinant DNA techniques have been applied to construct attenuated *V. cholerae* O1 strains to serve as live oral vaccines. One such strain, CVD 103-HgR (deleted of the CT A subunit gene and harbouring a gene encoding resistance to Hg^{++}), has proven to be particularly attractive following extensive testing. In randomized, placebo-controlled, double-blind clinical trials involving thousands of subjects from 2 years to 65 years of age, this live oral vaccine has been well tolerated and appears highly immunogenic. A single oral dose of CVD 103-HgR has conferred upon adult volunteers a high level of protection against experimental challenge with wild-type *V. cholerae* O1 of either biotype or serotype. Based on the extensive data on safety, immunogenicity and efficacy in volunteers, CVD 103-HgR is currently licensed in several countries in Europe (e.g. Switzerland) and South America (e.g. Argentina) for prevention of cholera in travellers. A large-scale field trial is underway in Indonesia to determine the efficacy of a single dose of CVD 103-HgR in preventing cholera in an endemic area.

Beginning in 1992, epidemic cholera caused by a new serogroup, O139, was observed in Asia. The subsequent spread of O139 cholera throughout much of Asia suggests that this new variant has pandemic potential. Epidemiologic data strongly indicate that prior immunity to *V. cholerae* O1 does not confer notable protection against serogroup

O139. Therefore, recent knowledge about the pathogenesis of cholera and techniques of biotechnology have been applied in accelerated programmes to develop attenuated strains of *V. cholerae* O139 for use as live oral vaccines. These candidate vaccine strains are presently undergoing preliminary clinical trials to assess their reactogenicity, immunogenicity and efficacy.

5.13 REFERENCES

Alm, R.A. and Manning, P.A. (1990) Biotype-specific probe for *Vibrio cholerae* serogroup O1. *Journal of Clinical Microbiology*, **28**, 823–4.

Alm, R.A., Stroeher, U.H. and Manning, P.A. (1988) Extracellular proteins of *Vibrio cholerae*: nucleotide sequence of the structural gene (*hlyA*) for the haemolysin of the haemolytic El Tor strain O17 and characterization of the *hlyA* mutation in the non-haemolytic classical strain 569B. *Molecular Microbiology*, **2**, 481-8.

Attridge, S.R. and Rowley, D. (1983) The role of the flagellum in the adherence of *Vibrio cholerae*. *Journal of Infectious Diseases*, **147**, 864–72.

Barrett, T.J. and Blake, P.A. (1981) Epidemiological usefulness of changes in hemolytic activity of *Vibrio cholerae* biotype El Tor during the seventh pandemic. *Journal of Clinical Microbiology*, **13**, 126–9.

Baselski, V., Briggs, R. and Parker, C. (1977) Intestinal fluid accumulation induced by oral challenge with *Vibrio cholerae* or cholera toxin in infant mice. *Infection and Immunity*, **15**, 704–12.

Baudry, B., Fasano, A., Ketley, J. and Kaper, J.B. (1992) Cloning of a gene (*zot*) encoding a new toxin produced by *Vibrio cholerae*. *Infection and Immunity*, **60**, 428–34.

Bear, C.E., Li, C., Kartner, N., Bridges, R.J., Jensen, T.J., Ramjeesingh, M. and Riordan, J.R. (1992) Purification and functional reconstitution of the cystic fibrosis transmembrane conductance regulator (CFTR). *Cell*, **68**, 809–18.

Benenson, A.S., Saad, A., Mosley, W.H. and Ahmed, A. (1968) Serological studies in cholera. 3. Serum toxin neutralization-rise in titre in response to infection with *Vibrio cholerae*, and the level in the 'normal' population in East Pakistan. *Bulletin of the World Health Organization*, **38**, 287–95.

Bennett, A. (1971) Cholera and prostaglandins. *Nature*, **231**, 536.

Berschneider, H.M., Knowles, M.R., Azizkhan, R.G., Boucher, R.C., Tobey, N.A., Orlando, R.C. and Powell, D.W. (1988) Altered intestinal chloride transport in cystic fibrosis. *FASEB Journal*, **2**, 2625–9.

Beubler, E., Kollar, G., Saria, A., Bukhave, K. and Rask-Madsen, J. (1989) Involvement of 5-hydroxytryptamine, prostaglandin E_2, and cyclic adenosine monophosphate in cholera toxin-induced fluid secretion in the small intestine of the rat in vivo. *Gastroenterology*, **96**, 368–76.

Black, R.E., Levine, M.M., Clements, M.L., Young, C.R., Svennerholm, A.M. and Holmgren, J. (1987) Protective efficacy in humans of killed whole-vibrio oral cholera vaccine with and without the B subunit of cholera toxin. *Infection and Immunity*, **55**, 116–20.

Brown, M.H. and Manning, P.A. (1985) Haemolysin genes of *Vibrio cholerae*: presence of homologous DNA in non-haemolytic O1 and haemolytic non-O1 strains. *FEMS Microbiology Letters*, **30**, 197–201.

Butterton, J.R., Stoebner, J.A., Payne, S.M. and Calderwood, S.B. (1992) Cloning, sequencing, and transcriptional regulation of *viuA*, the gene encoding the ferric vibriobactin receptor of *Vibrio cholerae*. *Journal of Bacteriology*, **174**, 3729–38.

Calderwood, S.B. and Mekalanos, J.J. (1987) Iron regulation of Shiga-like toxin expression in *Escherichia coli* is mediated by the fur locus. *Journal of Bacteriology*, **169**, 4759–64.

Cash, R.A., Music, S.I., Libonati, J.P., Snyder, M.J., Wenzel, R.P. and Hornick, R.B. (1974) Response of man of infection with *Vibrio cholerae*. I. Clinical, serologic, and bacteriologic responses to a known inoculum. *Journal of Infectious Diseases*, **129**, 45–52.

Cassel, D. and Selinger, Z. (1977) Mechanism of adenylate cyclase activation by cholera toxin: inhibition of GTP hydrolysis at the regulatory site. *Proceedings of the National Academy of Sciences, USA*, **74**, 3307–11.

Cassuto, J., Fahrenkrug, J., Jodal, M., Tuttle, R. and Lundgren, O. (1981a) The release of vasoactive intestinal polypeptide from the cat small intestine exposed to cholera toxin. *Gut*, **22**, 958–63.

Cassuto, J., Jodal, M., Tuttle, R. and Lundgren, O. (1981b) On the role of intramural nerves in the pathogenesis of cholera toxin-induced intestinal secretion. *Scandinavian Journal of Gastroenterology*, **16**, 377–84.

Cassuto, J., Jodal, M. and Lundgren, O. (1982) The effect of nicotinic and muscarinic receptor blockade on cholera toxin induced intestinal secretion in rats and cats. *Acta Physiologica Scandinavica*, **114**, 573–7.

Cassuto, J., Siewert, A., Jodal, M. and Lundgren, O. (1983) The involvement of intramural nerves in cholera toxin induced intestinal secretion. *Acta Physiologica Scandinavica*, **117**, 195–202.

Chang, E.B. and Rao, M.C. (1991) Intracellular mediators of intestinal electrolyte transport, in *Diarrheal Diseases*, (ed. M. Field), Elsevier, New York, pp. 49–72.

Chitnis, D.S., Sharma, K.D. and Kamat, R.S. (1982) Role of somatic antigen of *Vibrio cholerae* in adhesion to intestinal mucosa. *Journal of Medical Microbiology*, **5**, 53–61.

Chongsa-nguan, M., Chaicumpa, W., Kalambaheti, T., Soejedi, H., Luxananil, P., Swasdikosa, S., Thanasiri, P. and Mayurasakorn, S. (1986) Vibriocidal antibody and antibodies to *Vibrio cholerae* lipopolysaccharide, cell-bound haemagglutinin and toxin in Thai population. *Southeast Asian Journal of Tropical Medicine and Public Health*, **17**, 558–66.

Clemens, J.D., Sacks, D.A., Harris, J.R. *et al.* (1986) Field trial of oral cholera vaccines in Bangladesh. *Lancet*, **ii**, 124–7.

Clemens, J.D., Stanton, B.F., Chakraborty, J. *et al.* (1987) B subunit-whole cell and whole cell-only oral vaccines against cholera: studies on reactogenicity and immunogenicity. *Journal of Infectious Diseases*, **155**, 79–85.

Clemens, J.D., Sack, D.A., Harris, J.R. *et al.* (1989) ABO blood groups and cholera: new observations on specificity of risk and modification of vaccine efficacy. *Journal of Infectious Diseases*, **159**, 770–3.

Clemens, J.D., Sack, D.A., Harris, J.R. *et al.* (1990) Field trial of cholera vaccines in Bangladesh: results from three year follow-up. *Lancet*, **335**, 270–3.

Clemens, J.D., van Loon, F., Sack, D.A. *et al.* (1991a) Field trial of oral cholera vaccines in Bangladesh: serum vibriocidal and antitoxic antibodies as markers of the risk of cholera. *Journal of Infectious Diseases*, **163**, 1235–42.

Clemens, J.D., van Loon, F., Sack, D.A. *et al.* (1991b) Biotype as determinant of natural immunising effect of cholera. *Lancet*, **337**, 883–4.

Comstock, L.E., Michalski, J.M., Levine, M.M. and Kaper, J.B. (1995) CVD112, a live oral attenuated vaccine to *Vibrio cholerae* serogroup O139 synonym Bengal. Unpublished data.

Cooke, H.J. (1991) Hormones and neurotransmitters regulating intestinal ion transport, in *Diarrheal Diseases*, (ed. M. Field), Elsevier, New York, pp. 23–48.

Coster, T.S., Killeen, K.P., Waldor, M.K. *et al.* (1995) Safety, immunogenicity and efficacy of live attenuated *Vibrio cholerae* O139 vaccine prototype. *Lancet*, **345**, 949–52.

Cray, W.C. Jr, Tokunaga, E. and Pierce, N.F. (1983) Successful colonization and immunization of adult rabbits by oral inoculation with *Vibrio cholerae* O1. *Infection and Immunity*, **41**, 735–41.

Cryz, S.J., Levine, M.M., Kaper, J.B., Furer, E. and Althaus, B. (1990) Randomized double-blind placebo controlled trial to evaluate the safety and immunogenicity of the live oral cholera vaccine strain CVD 103-HgR in Swiss adults. *Vaccine*, **8**, 577–80.

Cryz, S.J. Jr, Levine, M.M., Losonsky, G., Kaper, J.B. and Althaus, B. (1992) Safety and immunogenicity of a booster dose of *Vibrio cholerae* CVD 103-HgR live oral cholera vaccine in Swiss adults. *Infection and Immunity*, **60**, 3916–17.

Cryz, S.J. Jr, Que, J.U., Levine, M.M., Wiedermann, G. and Kollaritsch, H. (1995) Safety and immunogenicity of a live oral bivalent typhoid fever (*Salmonella typhi* Ty21a)-cholera (*Vibrio cholerae* CVD 103-HgR) vaccine in healthy adults. *Infection and Immunity*, **63**, 1336–9.

Curlin, G., Levine, R., Aziz, K.M.A., Rahman, A.S.M.M. and Verwey, W.F. (1974) Field trial of cholera toxoid, in *Proceedings of the 11th Joint Conference on Cholera, U.S.–Japan Cooperative Medical Science Program*, US HEW, Washington, DC, pp. 314–29.

de Jonge, H.R. (1991) Intracellular mechanisms regulating intestinal secretion, in *Molecular Pathogenesis of Gastrointestinal Infections*, (eds T. Wadström, P.H. Mäkelä, A.-M. Svennerholm and H. Wolf-Watz), Plenum Press, New York, pp. 107–14.

De, S.N. (1959) Enterotoxicity of bacteria-free culture filtrate of *Vibrio cholerae*. *Nature*, **183**, 1533–4.

DiRita, V.J. (1992) Co-ordinate expression of virulence genes by ToxR in *Vibrio cholerae*. *Molecular Microbiology*, **6**, 451–8.

DiRita, V.J. and Mekalanos, J.J. (1991) Periplasmic interaction between two membrane regulatory proteins, ToxR and ToxS, results in signal transduction and transcriptional activation. *Cell*, **64**, 29–37.

DiRita, V.J., Parsot, C., Jander, G. and Mekalanos, J.J. (1991) Regulatory cascade controls virulence in *Vibrio cholerae*. *Proceedings of the National Academy of Sciences, USA*, **88**, 5403–7.

Donta, S.T., Beristain, S. and Tomicic, T.K. (1993) Inhibition of heat-labile cholera and *Escherichia coli* enterotoxins by brefeldin A. *Infection and Immunity*, **61**, 3282–6.

Dutta, N.K., Panse, M.W. and Kulkarni, D.R. (1959) Role of cholera toxin in experimental cholera. *Journal of Bacteriology*, **78**, 594–5.

Ehara, M., Ishibashi, M., Ichinose, Y., Iwanaga, M., Shimotori, S. and Naito, T.

(1987) Purification and partial characterization of fimbriae of *Vibrio cholerae* O1. *Vaccine*, **5**, 283–8.

Eubanks, E.R., Guentzel, M.N. and Berry, L.J. (1977) Evaluation of surface components of *Vibrio cholerae* as protective immunogens. *Infection and Immunity*, **15**, 533–8.

Faruque, S.M., Comstock, L., Kaper, J.B. and Albert, M.J. (1994) Distribution of *zonula occludens* toxin (*zot*) gene among clinical isolates of *Vibrio cholerae* O1 from Bangladesh and Africa. *Journal of Diarrhoeal Diseases Research*, **12**, 222–4.

Fasano, A., Baudry, B., Pumplin, D.W., Wasserman, S.S., Tall, B.D., Ketley, J.M. and Kaper, J.B. (1991) *Vibrio cholerae* produces a second enterotoxin, which affects intestinal tight junctions. *Proceedings of the National Academy of Sciences, USA*, **88**, 5242–6.

Field, M. (1971) Intestinal secretion: effect of cyclic AMP and its role in cholera. *New England Journal of Medicine*, 284, 1137–44.

Field, M., Fromm, D., Wallace, C.K. and Greenough, W.B. III (1969) Stimulation of active chloride secretion in small intestine by cholera exotoxin. *Journal of Clinical Investigation*, **48**, 24a (abstract).

Field, M., Rao, M.C. and Chang, E.B. (1989) Intestinal electrolyte transport and diarrheal disease – Part 2. *New England Journal of Medicine*, **321**, 879–83.

Finkelstein, R.A. (1992) Cholera enterotoxin (choleragen) – A historical perspective, in *Cholera*, (eds D. Barua and W.B. Greenough III), Plenum Medical Book Co., New York, pp. 155–87.

Finkelstein, R.A. and LoSpalluto, J.J. (1969) Pathogenesis of experimental cholera: preparation and isolation of choleragen and choleragenoid. *Journal of Experimental Medicine*, **130**, 185–202.

Finkelstein, R.A., Boesman-Finkelstein, M. and Holt, P. (1983) *Vibrio cholerae* hemagglutinin/lectin/protease hydrolyzes fibronectin and ovomucin: F.M. Burnet revisited. *Proceedings of the National Academy of Sciences, USA*, **80**, 1092–5.

Finkelstein, R.A., Boesman-Finkelstein, M., Chang, Y. and Häse, C.C. (1992) *Vibrio cholerae* hemagglutinin/protease, colonial variation, virulence, and detachment. *Infection and Immunity*, **60**, 472–8.

Finn, T.M., Reiser, J., Germanier, R. and Cryz, S.J. Jr (1987) Cell-associated hemagglutinin-deficient mutant of *Vibrio cholerae*. *Infection and Immunity*, **55**, 942–6.

Follett, E.A.C. and Gordon, J. (1963) An electron microscope study of vibrio flagella. *Journal of General Microbiology*, **32**, 235–9.

Forrest, B., LaBrooy, J.T., Attridge, S.R., Boehm, G., Beyer, L., Morona, R., Shearman, D.D.J.C. and Rowley, D. (1989) A candidate live oral typhoid/cholera hybrid vaccine is immunogenic in man. *Journal of Infectious Diseases*, **159**, 145–6.

Franzon, V.L. and Manning, P.A. (1986) Molecular cloning and expression in *Escherichia coli* K-12 of the gene for a hemagglutinin from *Vibrio cholerae*. *Infection and Immunity*, **52**, 279–84.

Franzon, V.L., Barker, A. and Manning, P.A. (1993) Nucleotide sequence encoding the mannose-fucose-resistant hemagglutinin of *Vibrio cholerae* O1 and construction of a mutant. *Infection and Immunity*, **61**, 3032–7.

Freer, J.H. and Birkbeck, T.H. (1982) Possible conformation of delta-lysin, a membrane-damaging peptide of *Staphylococcus aureus*. *Journal of Theoretical Biology*, **94**, 535–40.

Freter, R. and Jones, G.W. (1976) Adhesive properties of *Vibrio cholerae*: nature of the interaction with intact mucosal surfaces. *Infection and Immunity*, **14**, 246–56.

Freter, R. and O'Brien, P.C.M. (1981) Role of chemotaxis in the association of motile bacteria with intestinal mucosa: fitness and virulence of nonchemotactic *Vibrio cholerae* mutants in infant mice. *Infection and Immunity*, **34**, 222–33.

Freter, R., Allweiss, B., O'Brien, P.M.C., Halstead, S.A. and Macsai, M.S. (1981a) Role of chemotaxis in the association of motile bacteria with intestinal mucosa: in vitro studies. *Infection and Immunity*, **34**, 241–9.

Freter, R., O'Brien, P.C.M. and Macsai, M.S. (1981b) Role of chemotaxis in the association of motile bacteria with intestinal mucosa: in vivo studies. *Infection and Immunity*, **34**, 234–40.

Fuerst, J.A. and Perry, J.W. (1988) Demonstration of lipopolysaccharide on sheathed flagella of *Vibrio cholerae* O:1 by protein A-gold immunoelectron microscopy. *Journal of Bacteriology*, **170**, 1488–94.

Galen, J.E., Ketley, J.M., Fasano, A., Richardson, S.H., Wasserman, S.S. and Kaper, J.B. (1992) Role of *Vibrio cholerae* neuraminidase in the function of cholera toxin. *Infection and Immunity*, **60**, 406–15.

Gennaro, M.L., Greenaway, P.J. and Broadbent, D.A. (1982) The expression of biologically active cholera toxin in *Escherichia coli*. *Nucleic Acids Research*, **10**, 4883–90.

Gill, D.M. and King, C.A. (1975) The mechanism of action of cholera toxin in pigeon erythrocyte lysates. *Journal of Biological Chemistry*, **250**, 6424–32.

Glass, R.I., Becker, S., Huq, M.I., Stoll, B.J., Khan, M.U., Merson, M.H., Lee, J.V. and Black, R.E. (1982) Endemic cholera in rural Bangladesh, 1966–1980. *American Journal of Epidemiology*, **116**, 959–70.

Glass, R.I., Holmgren, J., Haley, C.E. *et al.* (1985a) Predisposition for cholera of individuals with O blood group. Possible evolutionary significance. *American Journal of Epidemiology*, **121**, 791–6.

Glass, R.I., Svennerholm, A.M., Khan, M.R., Huda, S., Huq, M.I. and Holmgren, J. (1985b) Seroepidemiological studies of El Tor cholera in Bangladesh: association of serum antibody levels with protection. *Journal of Infectious Diseases*, **151**, 236–42.

Goldberg, I. and Mekalanos, J.J. (1986) Effect of a *recA* mutation on cholera toxin gene amplification and deletion events. *Journal of Bacteriology*, **165**, 723–31.

Goldberg, M.B., DiRita, V.J. and Calderwood, S.B. (1990) Identification of an iron-regulated virulence determinant in *Vibrio cholerae*, using Tn*phoA* mutagenesis. *Infection and Immunity*, **58**, 55–60.

Goldberg, M.B., Boyko, S.A. and Calderwood, S.B. (1991) Positive transcriptional regulation of an iron-regulated virulence gene in *Vibrio cholerae*. *Proceedings of the National Academy of Sciences, USA*, **88**, 1125–9.

Gotuzzo, E., Butron, B., Seas, C. *et al.* (1993) Safety, immunogenicity, and excretion pattern of single-dose live oral cholera vaccine CVD 103-HgR in Peruvian adults of high and low socioeconomic levels. *Infection and Immunity*, **61**, 3994–7.

Griffiths, G.L., Sigel, S.P., Payne, S.M. and Neilands, J.B. (1984) Vibriobactin, a siderophore from *Vibrio cholerae*. *Journal of Biological Chemistry*, **259**, 393–5.

Guentzel, M.N. and Berry, L.J. (1975) Motility as a virulence factor for *Vibrio cholerae*. *Infection and Immunity*, **11**, 890–7.

Hall, R.H., Vial, P.A., Kaper, J.B., Mekalanos, J.J. and Levine, M.M. (1988) Morphological studies on fimbriae expressed by *Vibrio cholerae* O1. *Microbial Pathogenesis*, **4**, 257–65.

Hall, R.H., Losonsky, G., Silveira, A.P., Taylor, R.K., Mekalanos, J.J., Witham, N.D. and Levine, M.M. (1991) Immunogenicity of *Vibrio cholerae* O1 toxin-coregulated pili in experimental and clinical cholera. *Infection and Immunity*, **59**, 2508–12.

Hall, R.H., Khambaty, F.M., Kothary, M. and Keasler, S.P. (1993) Non-O1 *Vibrio cholerae*. *Lancet*, **342**, 430.

Hanne, L.F. and Finkelstein, R.A. (1982) Characterization and distribution of the hemagglutinins produced by *Vibrio cholerae*. *Infection and Immunity*, **36**, 209–14.

Häse, C.C. and Finkelstein, R.A. (1991) Cloning and nucleotide sequence of the *Vibrio cholerae* hemagglutinin/protease (HA/protease) gene and construction of an HA/protease-negative strain. *Journal of Bacteriology*, **173**, 3311–17.

Hecht, G., Pothoulakis, C., LaMont, J.T. and Madara, J.L. (1988) *Clostridium difficile* toxin A perturbs cytoskeletal structure and tight junction permeability of the cultured human intestinal epithelial monolayers. *Journal of Clinical Investigation*, **82**, 1516–24.

Henderson, D.P. and Payne, S.M. (1993) Cloning and characterization of the *Vibrio cholerae* genes encoding the utilization of iron from haemin and haemoglobin. *Molecular Microbiology*, **7**, 461–9.

Helper, J.R. and Gilman, A.G. (1992) G proteins. *Trends in Biochemical Science*, **17**, 393–7.

Herrington, D.A., Hall, R.H., Losonsky, G., Mekalanos, J.J., Taylor, R.K. and Levine, M.M. (1988) Toxin, toxin-coregulated pili, and the toxR regulon are essential for *Vibrio cholerae* pathogenesis in humans. *Journal of Experimental Medicine*, **168**, 1487–92.

Higgins, D.E., Nazareno, E. and DiRita, V.J. (1992) The virulence gene activator ToxT from *Vibrio cholerae* is a member of the AraC family of transcriptional activators. *Journal of Bacteriology*, **174**, 6974–80.

Holmgren, J. (1992) Pathogenesis, in *Cholera*, (eds D. Barua and W.B. Greenough III), Plenum Medical Book Co., New York, pp. 199–208.

Holmgren, J., Lonnroth, I., Mansson, J. and Svennerholm, L. (1975) Interaction of cholera toxin and membrane G_{M1} ganglioside of small intestine. *Proceedings of the National Academy of Science, USA*, **72**, 2520–4.

Honda, T. and Finkelstein, R.A. (1979) Purification and characterization of a hemolysin produced by *Vibrio cholerae* biotype El Tor: another toxic substance produced by cholera vibrios. *Infection and Immunity*, **26**, 1020–7.

Hone, D.M., Harris, A.M., Chatfield, S., Dougan, G. and Levine, M.M. (1991) Construction of genetically defined double *aro* mutants of *Salmonella typhi*. *Vaccine*, 9, 810–16.

Hranitzky, K.W., Mulholland, A., Larson, A.D., Eubanks, E.R. and Hart, L.T. (1980) Characterization of a flagellar sheath protein of *Vibrio cholerae*. *Infection and Immunity*, **27**, 597–603.

Ichinose, Y., Yamamoto, K., Nakasone, N., Tanabe, M.J., Takeda, T., Miwatani,

T. and Iwanaga, M. (1987) Enterotoxicity of El Tor-like hemolysin of non-O1 *Vibrio cholerae*. *Infection and Immunity*, **55**, 1090–3.

Iredell, J.R. and Manning, P.A. (1994) Biotype-specific *tcpA* genes in *Vibrio cholerae*. *FEMS Microbiology Letters*, **121**, 47–54.

Jalajakumari, M.B. and Manning, P.A. (1990) Nucleotide sequence of the gene, *ompW*, encoding a 22 kDa immunogenic outer membrane protein of *Vibrio cholerae*. *Nucleic Acids Research*, **18**, 2180.

Jertborn, M., Svennerholm, A.M. and Holmgren, J. (1984) Gut mucosal, salivary and serum antitoxic and antibacterial antibody responses in Swedes after oral immunization with B subunit-whole cell cholera vaccine. *International Archives of Allergy Applied Immunology*, **75**, 38–43.

Jertborn, M., Svennerholm, A.M. and Holmgren, J. (1986) Saliva, breast milk, and serum antibody responses as indirect measures of intestinal immunity after oral cholera vaccination or natural disease. *Journal of Clinical Microbiology*, **24**, 203–9.

Jertborn, M., Svennerholm, A.-M. and Holmgren, J. (1993) Evaluation of different immunization schedules for oral cholera B subunit-whole cell vaccine in Swedish volunteers. *Vaccine*, **11**, 1007–12.

Jesudason, M.V., Cherian, A.M. and John, T.J. (1993) Blood stream invasion by *Vibrio cholerae* O139. *Lancet*, **342**, 431.

Johnson, J.A., Panigrahi, P. and Morris, J.G. Jr (1992) Non-O1 *Vibrio cholerae* NRT36S produces a polysaccharide capsule that determines colony morphology, serum resistance, and virulence in mice. *Infection and Immunity*, **60**, 864–9.

Johnson, J.A., Morris, J.G. Jr and Kaper, J.B. (1993) Gene encoding zonula occludens toxin (*zot*) does not occur independently from cholera enterotoxin genes (*ctx*) in *Vibrio cholerae*. *Journal of Clinical Microbiology*, **31**, 732–3.

Johnson, J.A., Salles, C.A., Panigrahi, P., Albert, M.J., Wright, A.C., Johnson, R.J. and Morris, J.G. Jr (1994) *Vibrio cholerae* O139 synonym Bengal is closely related to *Vibrio cholerae* El Tor but has important differences. *Infection and Immunity*, **62**, 2108–10.

Jones, G.W. and Freter, R. (1976) Adhesive properties of *Vibrio cholerae*: nature of the interaction with isolated rabbit brush border membranes and human erythrocytes. *Infection and Immunity*, **14**, 240–5.

Jonson, G., Svennerholm, A.M. and Holmgren, J. (1989) *Vibrio cholerae* expresses cell surface antigens during intestinal infection which are not expressed during in vitro culture. *Infection and Immunity*, **57**, 1809–15.

Jonson, G., Svennerholm, A.-M. and Holmgren, J. (1990) Expression of virulence factors by classical and El Tor *Vibrio cholerae* in vivo and in vitro. *FEMS Microbiology and Ecology*, **74**, 221–8.

Jonson, G., Holmgren, J. and Svennerholm, A.-M. (1991a) Identification of a mannose-binding pilus on *Vibrio cholerae* El Tor. *Microbial Pathogenesis*, **11**, 433–41.

Jonson, G., Holmgren, J. and Svennerholm, A.M. (1991b) Epitope differences in toxin-coregulated pili produced by classical and El Tor *Vibrio cholerae* O1. *Microbial Pathogenesis*, **11**, 179–88.

Jonson, G., Holmgren, J. and Svennerholm, A.-M. (1992) Analysis of expression of toxin-coregulated pili in classical and El Tor *Vibrio cholerae* O1 in vitro and in vivo. *Infection and Immunity*, **60**, 4278–84.

Jonson, G., Lebens, M. and Holmgren, J. (1994) Cloning and sequencing of

Vibrio cholerae mannose-sensitive haemagglutinin pilin gene: Localization of *mshA* within a cluster of type 4 pilin genes. *Molecular Microbiology*, **13**, 109–18.

Kahn, R.A. and Gilman, A.G. (1984) ADP-ribosylation of G_s promotes dissociation of its a and b subunits. *Journal of Biological Chemistry*, **259**, 6235–40.

Kaper, J.B. and Levine, M.M. (1981) Cloned cholera enterotoxin genes in study and prevention of cholera. *Lancet*, **2**, 1162–3.

Kaper, J.B. and Levine, M.M. (1990) Recombinant attenuated *Vibrio cholerae* strains used as live oral vaccines. *Research in Microbiology*, **141**, 901–6.

Kaper, J.B., Moseley, S.L. and Falkow, S. (1981) Molecular characterization of environmental and nontoxigenic strains of *Vibrio cholerae*. *Infection and Immunity*, **32**, 661–7.

Kaper, J.B., Lockman, H., Baldini, M.M. and Levine, M.M. (1984a) Recombinant nontoxinogenic *Vibrio cholerae* strains as attenuated cholera vaccine candidates. *Nature*, **308**, 655–8.

Kaper, J.B., Lockman, H., Baldini, M.M. and Levine, M.M. (1984b) A recombinant live oral cholera vaccine. *Bio/Technology*, **2**, 345–9.

Kaper, J.B., Mobley, H.L.T., Michalski, J.M., Herrington, D.A. and Levine, M.M. (1988) Recent advances in developing a safe and effective live oral attenuated *Vibrio cholerae* vaccine, in *Advances in Research on Cholera and Related Diarrheas*, vol. 6, (eds N. Ohtomo and R.B. Sack), KTK Scientific Publishers, Tokyo, pp. 161–7.

Kapral, F.A., O'Brien, A.D., Ruff, P.D. and Drugan, W.J. Jr (1976) Inhibition of water absorption in the intestine by *Staphylococcus aureus* delta-toxin. *Infection and Immunity*, **13**, 140–5.

Karasawa, T., Mihara, T., Kurazono, H., Nair, G.B., Garg, S., Ramamurthy, T. and Takeda, Y. (1993) Distribution of the *zot* (zonula occludens toxin) gene among strains of *Vibrio cholerae* O1 and non-O1. *FEMS Microbiology Letters*, **106**, 143–6.

Kaufman, M.R., Shaw, C.E., Jones, I.D. and Taylor, R.K. (1993) Biogenesis and regulation of the *Vibrio cholerae* toxin-coregulated pilus: analogies to other virulence factor secretory systems. *Gene*, **126**, 43–9.

Ketley, J.M., Michalski, J., Galen, J., Levine, M.M. and Kaper, J.B. (1993) Construction of genetically-marked *Vibrio cholerae* O1 vaccine strains. *FEMS Microbiology Letters*, **111**, 15–22.

Kimberg, D.K., Field, M., Johnson, J., Henderson, E. and Gershon, E. (1971) Stimulation of intestinal mucosal adenyl cyclase by cholera enterotoxin and prostaglandins. *Journal of Clinical Investigation*, **50**, 1218–30.

Koch, R. (1884) An address on cholera and its bacillus. *British Medical Journal*, **2**, 403–7.

Kotloff, K.L., Wasserman, S.S., O'Donnell, S., Losonsky, G.A., Cryz, S.J. and Levine, M.M. (1992) Safety and immunogenicity in North Americans of a single dose of live oral cholera vaccine CVD 103-HgR: results of a randomized, placebo-controlled, double-blind crossover trial. *Infection and Immunity*, **60**, 4430–2.

Krause, M., Roudier, C., Fierer, J., Harwood, J. and Guiney, D. (1991) Molecular analysis of the virulence locus of the *Salmonella dublin* plasmid pSDL2. *Molecular Microbiology*, **5** 307–16.

Lagos, R., Avendaño, A., Horwitz, I. *et al.* (1993) Tolerance and immunogenicity of an oral dose of CVD 103-HgR: a double blind study in Chilean adults. *Revista Mévista de Chile*, **121**, 857–63.

Lagos, R., Avendaño, A., Prado, V., Horwitz, I., Wasserman, S., Losonsky, G., Cryz, S., Jr, Kaper, J.B. and Levine, M.M. (1995) Attenuated live cholera vaccine strain CVD 103-HgR elicits significantly higher serum vibriocidal antibody titers in persons of blood group O. *Infection and Immunity*, **63**, 707–9.

Lang, H.A. and Palva, E.T. (1987) A major outer membrane protein in *Vibrio cholerae* is maltose-inducible. *Microbial Pathogenesis*, **3**, 143–7.

Lencer, W.I., Delp, C., Neutra, M.R. and Madara, J.L. (1992) Mechanism of cholera toxin action on a polarized human intestinal epithelial cell line: role of vesicular traffic. *Journal of Cell Biology*, **117**, 1197–209.

Lencer, W.I., de Almeida, J.B., Moe, S., Stow, J.L., Ausiello, D.A. and Madara, J.L. (1993) Entry of cholera toxin into polarized human epithelial cells: identification of an early brefeldin-A sensitive event required for A_1 peptide generation. *Journal of Clinical Investigation*, **92**, 2941–51.

Levine, M.M. and Kaper, J.B. (1993) Live oral vaccines against cholera: an update. *Vaccine*, **11**, 207–12.

Levine, M.M. and Pierce, N.F. (1992) Immunity and vaccine development, in *Cholera*, (eds D. Barua and W.B. Greenough III), Plenum Medical Book Co., New York, pp. 285–327.

Levine, M.M. and Tacket, C.O. (1994) Recombinant live cholera vaccines, in *Vibrio cholerae and Cholera: Molecular to Global Perspectives*, (eds I.K. Wachsmuth, P.A. Blake and Ø. Olsvik), American Society for Microbiology, Washington, DC, pp. 395–413.

Levine, M.M., Nalin, D.R., Craig, J.P. *et al.* (1979a) Immunity of cholera in man: relative role of antibacterial versus antitoxic immunity. *Transactions of the Royal Society of Tropical Medicine and Hygiene*, **73**, 3–9.

Levine, M.M., Nalin, D.R., Rennels, M.B., Hornick, R.B., Sotman, S., Van Blerk, G., Hughes, T.P., O'Donnell, S. and Barua, D. (1979b) Genetic susceptibility to cholera. *Annals of Human Biology*, **6**, 369–74.

Levine, M.M., Black, R.E., Clements, M.L., Cisneros, L., Nalin, D.R. and Young, C.R. (1981a) Duration of infection-derived immunity to cholera. *Journal of Infectious Diseases*, **143**, 818–20.

Levine, M.M., Black, R.E., Clements, M.L., Nalin, D.R., Cisneros, L. and Finkelstein, R.A. (1981b) Volunteer studies in development of vaccines against cholera and enterotoxigenic *Escherichia coli*: a review, in *Acute Enteric Infections in Children. New Prospects for Treatment and Prevention*, (eds T. Holme, J. Holmgren, M.H. Merson and R. Mollby), Elsevier/North-Holland Biomedical Press, Amsterdam, pp. 443–59.

Levine, M.M., Black, R.E., Clements, M.L. *et al.* (1982) The pathogenicity of nonenterotoxigenic *Vibrio cholerae* serogroup O1 biotype El Tor isolated from sewage water in Brazil. *Journal of Infectious Diseases*, **145**, 296–9.

Levine, M.M., Kaper, J.B., Black, R.E. and Clements, M.L. (1983) New knowledge on pathogenesis of bacterial enteric infections as applied to vaccine development. *Microbiological Reviews*, **47**, 510–50.

Levine, M.M., Kaper, J.B., Herrington, D., Ketley, J., Losonsky, G., Tacket, C.O., Tall, B. and Cryz, S. (1988a) Safety, immunogenicity, and efficacy of recombinant live oral cholera vaccines, CVD 103 and CVD 103-HgR. *Lancet*, **ii**, 467–70.

Levine, M.M., Kaper, J.B., Herrington, D., Losonsky, G., Morris, J.G., Clements, M.L., Black, R.E., Tall, B. and Hall, R. (1988b) Volunteer studies of deletion mutants of *Vibrio cholerae* O1 prepared by recombinant techniques. *Infection and Immunity*, **56**, 161–7.

Lockman, H. and Kaper, J.B. (1983) Nucleotide sequence analysis of the A2 and B subunits of *Vibrio cholerae* enterotoxin. *Journal of Biological Chemistry*, **258**, 13722–6.

Lockman, H.A., Galen, J.E. and Kaper, J.B. (1984) *Vibrio cholerae* enterotoxin genes: nucleotide sequence analysis of DNA encoding ADP-ribosyltransferase. *Journal of Bacteriology*, **159**, 1086–9.

Losonsky, G.A., Tacket, C.O., Wasserman, S.S., Kaper, J.B. and Levine, M.M. (1993) Secondary *Vibrio cholerae*-specific cellular antibody responses following wild-type homologous challenge in people vaccinated with CVD 103-HgR live oral cholera vaccine: changes with time and lack of correlation with protection. *Infection and Immunity*, **61**, 729–33.

Mathias, J.R., Carlson, G.M., DiMarino, A.J., Bertiger, G., Morton, H.E. and Cohen, S. (1976) Intestinal myoelectric activity in response to live *Vibrio cholerae* and cholera enterotoxin. *Journal of Clinical Investigation*, **58**, 91–6.

McCardell, B., Madden, J.M. and Shah, D.B. (1985) Isolation and characterization of a cytolysin produced by *Vibrio cholerae* serogroup non-O1. *Canadian Journal of Microbiology*, **31**, 711–20.

McGhee, J.R., Mestecky, J., Dertzbaugh, M.T., Eldridge, J.H., Hirasawa, M. and Kiyono, H. (1992) The mucosal immune system: from fundamental concepts to vaccine development. *Vaccine*, **10**, 75–88.

Mekalanos, J.J. (1983) Duplication and amplification of toxin genes in *Vibrio cholerae*. *Cell*, **35**, 253–63.

Mekalanos, J.J. (1985) Cholera toxin: genetic analysis, regulation, and role in pathogenesis. *Current Topics in Microbiology and Immunology*, **118**, 97–118.

Mekalanos, J.J., Swartz, D.J., Pearson, G.D., Harford, N., Groyne, F. and de Wilde, M. (1983) Cholera toxin genes: nucleotide sequence, deletion analysis and vaccine development. *Nature*, **306**, 551–7.

Michalski, J., Galen, J.E., Fasano, A. and Kaper, J.B. (1993) *Vibrio cholerae* CVD110: a live oral attenuated El Tor vaccine strain. *Infection and Immunity*, **61**, 4462–8.

Migasena, S., Pitisuttitham, P., Prayurahong, P. *et al.* (1989) Preliminary assessment of the safety and immunogenicity of live oral cholera vaccine strain CVD 103-HgR in healthy Thai adults. *Infection and Immunity*, **57**, 3261–4.

Miller, V.L. and Mekalanos, J.J. (1984) Synthesis of cholera toxin is positively regulated at the transcriptional level by toxR. *Proceedings of the National Academy of Sciences, USA*, **81**, 3471–5.

Miller, V.L. and Mekalanos, J.J. (1988) A novel suicide vector and its use in construction of insertion mutations: osmoregulation of outer membrane proteins and virulence determinants in *Vibrio cholerae* requires toxR. *Journal of Bacteriology*, **170**, 2575–83.

Miller, V.L., Taylor, R.K. and Mekalanos, J.J. (1987) Cholera toxin transcriptional activator toxR is a transmembrane DNA binding protein. *Cell*, **48**, 271–9.

Morris, J.G. Jr, Losonsky, G.E., Johnson, J.A., Tacket, C.O., Nataro, J.P., Panigrahi, P. and Levine, M.M. (1995) Clinical and immunologic characteristics of *Vibrio cholerae* O139 Bengal infection in North American volunteers. *Journal of Infectious Diseases*, **171**, 903–8.

Moseley, S.L. and Falkow, S. (1980) Nucleotide sequence homology between the heat-labile enterotoxin gene of *Escherichia coli* and *Vibrio cholerae. Journal of Bacteriology*, **144**, 444–6.

Mosley, W.H., Ahmad, S., Benenson, A.S. and Ahmed, A. (1968a) The relationship of vibriocidal antibody titre to susceptibility of cholera in family contacts of cholera patients. *Bulletin of the World Health Organization*, **38**, 777–85.

Mosley, W.H., Benenson, A.S. and Barui, R. (1968b) A serological survey for cholera antibodies in rural east Pakistan. I. The distribution of antibody in the control population of a cholera vaccine field-trial area and the relation of antibody titer to the pattern of endemic cholera. *Bulletin of the World Health Organization*, **39**, 327–34.

Moss, J. and Vaughan, M. (1977) Mechanism of activation of choleragen. Evidence for ADP-ribosyltransferase activity with arginine as an acceptor. *Journal of Biological Chemistry*, **252**, 2455–7.

Moss, J. and Vaughan, M. (1991) Activation of cholera toxin and *Escherichia coli* heat-labile enterotoxins by ADP-ribosylation factors, a family of 20-kDa guanine nucleotide-binding proteins. *Molecular Microbiology*, **5**, 2621–7.

Nambiar, M.P., Oda, T., Chen, C., Kuwazuru, Y. and Wu, H.C. (1993) Involvement of the Golgi region in the intracellular trafficking of cholera toxin. *Journal of Cellular Physiology*, **154**, 222–8.

Nisson, O., Cassuto, J., Larsson, P.-A., Jodal, M., Lidberg, P., Ahlman, H., Dahlström, A. and Lundgren, O. (1983) 5-Hydroxytryptamine and cholera secretion: a histochemical and physiological study in cats. *Gut*, **24**, 542–8.

Noriki, H. (1976) Evaluation of toxoid field trial in the Philippines, in *Proceedings of the 12th Joint Conference on Cholera, U.S.–Japan Cooperative Medical Science Program*, Tokyo: Fuji, pp. 302–10.

O'Brien, A.D., Chen, M.E., Holmes, R.K., Kaper, J. and Levine, M.M. (1984) Environmental and human isolates of *Vibrio cholerae* and *Vibrio parahaemolyticus* produce a *Shigella dysenteriae* 1 (Shiga)-like cytotoxin. *Lancet*, **i**, 77–8.

Ogawa, A. and Takeda, T. (1993) The gene encoding the heat-stable enterotoxin of *Vibrio cholerae* is flanked by 123-base pair direct repeats. *Microbiology and Immunology*, **37**, 607–16.

Ogawa, A., Kato, J.-I., Watanabe, H., Nair, B.G. and Takeda, T. (1990) Cloning and nucleotide sequence of a heat-stable enterotoxin gene from *Vibrio cholerae* non-O1 isolated from a patient with traveler's diarrhea. *Infection and Immunity*, **58**, 3325–9.

Ogierman, M.A. and Manning, P.A. (1992a) Homology of TcpN, a putative regulatory protein of *Vibrio cholerae*, to the AraC family of transcriptional activators. *Gene*, **116**, 93–7.

Ogierman, M.A. and Manning, P.A. (1992b) TCP pilus biosynthesis in *Vibrio cholerae* O1: gene sequence of *tcpC* encoding an outer membrane lipoprotein. *FEMS Microbiology Letters*, **97**, 179–84.

Osek, J., Svennerholm, A.-M. and Holmgren, J. (1992) Protection against *Vibrio cholerae* El Tor infection by specific antibodies against mannose-binding hemagglutinin pili. *Infection and Immunity*, **60**, 4961–4.

Owen, R.L., Pierce, N.F., Apple, R.T. and Cray, W.C. Jr (1986) M cell transport of *Vibrio cholerae* from the intestinal lumen into Peyer's patches: a mechanism for antigen sampling and for microbial transepithelial migration. *Journal of Infectious Diseases*, **153**, 1108–18.

Parsot, C. and Mekalanos, J.J. (1990) Expression of ToxR, the transcriptional activator of the virulence factors in Vibrio-cholerae, is modulated by the heat shock response. *Proceedings of the National Academy of Sciences, USA*, **87**, 9898–902.

Parsot, C., Taxman, E. and Mekalanos, J.J. (1991) ToxR regulates the production of lipoproteins and the expression of serum resistance in *Vibrio cholerae*. *Proceedings of the National Academy of Sciences, USA*, **88**, 1641–5.

Payne, S.M. and Finkelstein, R.A. (1978) Siderophore production by *Vibrio cholerae*. *Infection and Immunity*, **20**, 310–11.

Pearson, G.D. and Mekalanos, J.J. (1982) Molecular cloning of *Vibrio cholerae* enterotoxin genes in *Escherichia coli* K-12. *Proceedings of the National Academy of Sciences, USA*, **79**, 2976–80.

Pearson, G.D., DiRita, V.J., Goldberg, M.B., Boyko, S.A., Calderwood, S.B. and Mekalanos, J.J. (1990) New attenuated derivatives of *Vibrio cholerae*. *Research in Microbiology*, **141**, 893–9.

Pearson, G.D.N., Woods, A., Chiang, S.L. and Mekalanos, J.J. (1993) CT genetic element encodes a site-specific recombination system and an intestinal colonization factor. *Proceedings of the National Academy of Sciences, USA*, **90**, 3750–4.

Peek, J.A. and Taylor, R.K. (1992) Characterization of a periplasmic thiol:disulfide interchange protein required for the functional maturation of secreted virulence factors of *Vibrio cholerae*. *Proceedings of the National Academy of Sciences, USA*, **89**, 6210–14.

Peterson, J.W. and Ochoa, L.G. (1989) Role of prostaglandins and cAMP in the secretory effects of cholera toxin. *Science*, **245**, 857–9.

Peterson, J.W., Hejtmancik, K.E., Markel, D.E., Craig, J.P. and Kurosky, A. (1979) Antigenic specificity of neutralizing antibody to cholera toxin. *Infection and Immunity*, **24**, 774–9.

Peterson, J.W., Reitmeyer, J.C., Jackson, C.A. and Ansari, G.A.S. (1991) Protein synthesis is required for cholera toxin-induced stimulation of arachidonic acid metabolism. *Biochimica et Biophysica Acta*, **1092**, 79–84.

Peterson, K.M. and Mekalanos, J.J. (1988) Characterization of the *Vibrio cholerae* ToxR regulon: identification of novel genes involved in intestinal colonization. *Infection and Immunity*, **56**, 2822–9.

Pierce, N.F., Kaper, J.B., Mekalanos, J.J. and Cray, W.C. (1985) Role of cholera toxin in enteric colonization by *Vibrio cholerae* O1 in rabbits. *Infection and Immunity*, **50**, 813–16.

Pohlner, J., Meyer, T.F. and Manning, P.A. (1986) Serological properties and processing in Escherichia coli K12 of OmpV fusion proteins of Vibrio cholerae. *Molecular and General Genetics*, **205**, 501–6.

Preston, L.M., Xu, Q., Johnson, J.A., Joseph, A., Maneval, D.R. Jr, Husain, K., Reddy, G.P., Bush, C.A. and Morris, J.G. Jr (1995) Preliminary structure determination of the capsular polysaccharide of *Vibrio cholerae* O139 Bengal Al1837. *Journal of Bacteriology*, **177**, 835–8.

Reitmeyer, J.C. and Peterson, J.W. (1990) Stimulatory effects of cholera toxin on arachidonic acid metabolism in Chinese hamster ovary cells. *Proceedings of the Society for Experimental Biology and Medicine*, **193**, 181–4.

Rhine, J.A. and Taylor, R.K. (1994) TcpA pilin sequences and colonization requirements for O1 and O139 Vibrio cholerae. *Molecular Microbiology*, **13**, 1013–20.

Richardson, K. (1991) Roles of motility and flagellar structure in pathogenicity of *Vibrio cholerae*: analysis of motility mutants in three animal models. *Infection and Immunity*, **59**, 2727–36.
Richardson, K. and Parker, C.D. (1985) Identification and occurrence of *Vibrio cholerae* flagellar core proteins in isolated outer membrane. *Infection and Immunity*, **47**, 674–9.
Richardson, K., Michalski, J. and Kaper, J.B. (1986) Hemolysin production and cloning of two hemolysin determinants from classical *Vibrio cholerae*. *Infection and Immunity*, **54**, 415–20.
Richardson, K., Kaper, J.B. and Levine, M.M. (1989) Human immune response to *Vibrio cholerae* O1 whole cells and isolated outer membrane antigens. *Infection and Immunity*, **57**, 495–501.
Richardson, S.H. (1969) Factors influencing in vitro skin permeability factor production by *Vibrio cholerae*. *Journal of Bacteriology*, **100**, 27–34.
Richardson, S.H., Giles, J.C. and Kruger, K.S. (1984) Sealed adult mice: new model for enterotoxin evaluation. *Infection and Immunity*, **43**, 482–6.
Roberts, A., Pearson, G. and Mekalanos, J. (1992) Cholera vaccine strains derived from a 1991 Peruvian isolate of Vibrio cholerae and other El Tor strains. *Abstracts of the 28th Joint US–Japan Cooperative Medical Science Program Cholera and Related Diarrheal Diseases Panel*, **28**, 43–7 (abstract).
Rodman, D.M. and Zamudio, S. (1991) The cystic fibrosis heterozygote – advantage in surviving cholera? *Medical Hypotheses*, **36**, 253–8.
Sack, D.A., Clemens, J.D., Huda, S. *et al.* (1991) Antibody response after immunization with killed oral cholera vaccines during the 1985 vaccine field trial in Bangladesh. *Journal of Infectious Diseases*, **164**, 407–11.
Sack, R.B. and Albert, M.J. (1994) Summary of a cholera vaccine workshop. *Journal of Diarrhoeal Diseases Research*, **12**, 138–43.
Sack, R.B. and Spira, W.M. (1986) Animal models of acute diarrheal diseases of bacterial etiology, in *Experimental Models in Antimicrobial Chemotherapy*, Vol. 2, (ed. O. Zak), Academic Press, London, p. 43.
Saha, S. and Sanyal, S.C. (1990) Immunobiological relationships of the enterotoxins produced by cholera toxin gene-positive (CT^+) and -negative (CT^-) strains of *Vibrio cholerae* O1. *Journal of Medical Microbiology*, **32**, 33–7.
Sanchez, J.L., Vasquez, B., Begue, R.E. *et al.* (1994) Protective efficacy of oral whole-cell/recombinant-B-subunit cholera vaccine in Peruvian military recruits. *Lancet*, **344**, 1273–6.
Sanyal, S.C., Alam, K., Neogi, P.K.B., Huq, M.I. and Al-Mahmud, K.A. (1983) A new cholera toxin. *Lancet*, **i**, 1337.
Schrank, G.D. and Verwey, W.F. (1976) Distribution of cholera organisms in experimental *Vibrio cholerae* infections: proposed mechanisms of pathogenesis and antibacterial immunity. *Infection and Immunity*, **13**, 195–203.
Sciortino, C.V. and Finkelstein, R.A. (1983) *Vibrio cholerae* expresses iron-regulated outer membrane proteins in vivo. *Infection and Immunity*, **42**, 990–6.
Sengupta, D.K., Sengupta, T.K. and Ghose, A.C. (1992) Major outer membrane proteins of *Vibrio cholerae* and their role in induction of protective immunity through inhibition of intestinal colonization. *Infection and Immunity*, **60**, 4848–55.
Sharma, D.P., Attridge, S., Hackett, J. and Rowley, D. (1987) Non-lipopoly-

saccharide protective antigens shared by classical and El Tor biotypes of Vibrio cholerae. *Journal of Infectious Diseases*, **155**, 716–23.

Shaw, C.E. and Taylor, R.K. (1990) *Vibrio cholerae* O395 *tcpA* pilin gene sequence and comparison of predicted protein structural features to those of type 4 pilins. *Infection and Immunity*, **58**, 3042–9.

Sigel, S.P. and Payne, S.M. (1982) Effect of iron limitation on growth, siderophore production, and expression of outer membrane proteins of *Vibrio cholerae*. *Journal of Bacteriology*, **150**, 148–55.

Simanjuntak, C.H., O'Hanley, P., Punjabi, N.H. *et al.* (1993) Safety, immunogenicity, and transmissibility of single-dose live oral cholera vaccine strain CVD 103-HgR in 24- to 59-month-old Indonesian children. *Journal of Infectious Diseases*, **168**, 1169–76.

Sjöqvist, A., Cassuto, J., Jodal, M. and Lundgren, O. (1992) Actions of serotonin antagonists on cholera-toxin-induced intestinal fluid secretion. *Acta Physiologica Scandinavica*, **145**, 229–37.

Speelman, P., Rabbani, G.H., Bukhave, K. and Rask-Madsen, J. (1985) Increased jejunal prostaglandin E_2 concentrations in patients with acute cholera. *Gut*, **26**, 188–93.

Sperandio, V., Giron, J.A., Silveira, W.D. and Kaper, J.B. (1995) The OmpU outer membrane protein, a potential adherence factor of *Vibrio cholerae*. *Infection and Immunity* **63**, 4433–8.

Spira, W.M., Sack, R.B. and Froehlich, J.L. (1981) Simple adult rabbit model for *Vibrio cholerae* and enterotoxigenic *Escherichia coli* diarrhea. *Infection and Immunity*, **32**, 739–47.

Stoebner, J.A. and Payne, S.M. (1988) Iron-regulated hemolysin production and utilization of heme and hemoglobin by *Vibrio cholerae*. *Infection and Immunity*, **56**, 2891–5.

Stoebner, J.A., Butterton, J.R., Calderwood, S.B. and Payne, S.M. (1992) Identification of the vibriobactin receptor of *Vibrio cholerae*. *Journal of Bacteriology*, **174**, 3270–4.

Su-Arehawaratana, P., Singharaj, P., Taylor, D.N. *et al.* (1992) Safety and immunogenicity of different immunization regimens of CVD 103-HgR live oral cholera vaccine in soldiers and civilians in Thailand. *Journal of Infectious Diseases*, **165**, 1042–8.

Suharyono Simanjuntak, C., Witham, N., Punjabi, N. *et al.* (1992) Safety and immunogenicity of single-dose live oral cholera vaccine CVD 103-HgR in 5–9 year old Indonesian children. *Lancet*, **340**, 689–94.

Sun, D., Tillman, D.M., Marion, T.N. and Taylor, R.K. (1990a) Production and characterization of monoclonal antibodies to the toxin coregulated pilus (TCP) of *Vibrio cholerae* that protect against experimental cholera in infant mice. *Serodiagnosis and Immunotherapy in Infectious Disease*, **4**, 73–81.

Sun, D.X., Mekalanos, J.J. and Taylor, R.K. (1990b) Antibodies directed against the toxin-coregulated pilus isolated from *Vibrio cholerae* provide protection in the infant mouse experimental cholera model. *Journal of Infectious Diseases*, **161**, 1231–6.

Svennerholm, A.M., Stromberg, G.J. and Holmgren, J. (1983) Purification of *Vibrio cholerae* soluble hemagglutinin and development of enzyme-linked immunosorbent assays for antigen and antibody quantitations. *Infection and Immunity*, **41**, 237–43.

Tacket, C.O., Forrest, B., Morona, R., Attridge, S.R., LaBrooy, J., Tall, B.D.,

Reymann, M., Rowley, D. and Levine, M.M. (1990) Safety, immunogenicity, and efficacy against cholera challenge in humans of a typhoid–cholera hybrid vaccine derived from *Salmonella typhi* Ty21a. *Infection and Immunity*, **58**, 1620–7.

Tacket, C.O., Hone, D.M., Losonsky, G.A., Guers, L., Edelman, R. and Levine, M.M. (1992a) Clinical acceptability and immunogenicity of CVD 908 *Salmonella typhi* vaccine strain. *Vaccine*, **10**, 443–6.

Tacket, C.O., Losonsky, G., Nataro, J.P., Cryz, S.J., Edelman, R., Kaper, J.B. and Levine, M.M. (1992b) Onset and duration of protective immunity in challenged volunteers after vaccination with live oral cholera vaccine CVD 103-HgR. *Journal of Infectious Diseases*, **166**, 837–41.

Tacket, C.O., Losonsky, G., Nataro, J.P., Cryz, S.J., Edelman, R., Fasano, A., Michalski, J., Kaper, J.B. and Levine, M.M. (1993) Safety and immunogenicity of live oral cholera vaccine candidate CVD 110, a *ΔctxA Δzot Δace* derivative of El Tor Ogawa *Vibrio cholerae. Journal of Infectious Diseases*, **168**, 1536–40.

Tacket, C.O., Losonsky, G., Nataro, J.P., Wasserman, S.S., Cryz, S.J., Edelman, R. and Levine, M.M. (1995a) Extension of the volunteer challenge model to study South American cholera in a population of volunteers with predominantly blood group antigen O. *Transactions of the Royal Society of Tropical Medicine and Hygiene*, **89**, 75–7.

Tacket, C.O., Losonsky, G., Nataro, J.P., Comstock, L., Michalski, J., Edelman, R., Kaper, J.B. and Levine, M.M. (1995b) Initial clinical studies of CVD 112 *Vibrio cholerae* O139 live oral vaccine: safety and efficacy against experimental challenge. *Journal of Infections Diseases*, **172**, 883–6.

Tacket, C.O., Losonsky, G., Nataro, J.P., Edelman, R., Kaper, J.B. and Levine, M.M. (1996) Safety, immunogenicity, and protective efficacy of attenuated *Vibrio cholerae* El Tor Ogawa strain CVD 111. Submitted publication.

Takeda, T., Peina, Y., Ogawa, A., Dohi, S., Abe, H., Nair, G.B. and Pal, S.C. (1991) Detection of heat-stable enterotoxin in a cholera toxin gene-positive strain of *Vibrio cholerae* O1. *FEMS Microbiology Letters*, **80**, 23–8.

Tamplin, M.L., Colwell, R.R., Hall, S., Kogure, K. and Strichartz, G.R. (1987) Sodium-channel inhibitors produced by enteropathogenic *Vibrio cholerae* and *Aeromonas hydrophila. Lancet*, **i**, 975.

Taylor, D.N., Hack, D.C., Killeen, K.P. *et al.* (1993) Development of a live, oral, attenuated vaccine against El Tor cholera. *Abstracts of the Twenty-ninth Joint Conference on Cholera and Related Diarrheal Diseases*, Asilomer, Pacific Grove, California, December 1–3, 1993, 189–193.

Taylor, D.N., Killeen, K.P., Hack, D.C. *et al.* (1994) Development of a live, oral, attenuated vaccine against El Tor cholera. *Journal of Infectious Diseases*, **170**, 1518–23.

Taylor, R., Shaw, C., Peterson, K., Spears, P. and Mekalanos, J. (1988) Safe, live *Vibrio cholerae* vaccines? *Vaccine*, **6**, 151–4.

Taylor, R.K., Miller, V.L., Furlong, D.B. and Mekalanos, J.J. (1987) Use of phoA gene fusions to identify a pilus colonization factor coordinately regulated with cholera toxin. *Proceedings of the National Academy of Sciences, USA*, **84**, 2833–7.

Trucksis, M., Galen, J.E., Michalski, J., Fasano, A. and Kaper, J.B. (1993) Accessory cholera enterotoxin (Ace), the third toxin of a *Vibrio cholerae* virulence cassette. *Proceedings of the National Academy of Sciences, USA*, **90**, 5267–71.

Voss, E. and Attridge, S.R. (1993) *In vitro* production of toxin-coregulated pili by *Vibrio cholerae* El Tor. *Microbial Pathogenesis*, **15**, 255–68.

Waldor, M.K. and Mekalanos, J.J. (1994) Emergence of a new cholera pandemic: molecular analysis of virulence determinants in *Vibrio cholerae* O139 and development of a live vaccine prototype. *Journal of Infectious Diseases*, **170**, 278–83.

Wasserman, S.S., Losonsky, G.A., Noriega, F., Tacket, C.O., Castañada, E. and Levine, M.M. (1994) Kinetics of the vibriocidal antibody response to live oral cholera vaccine. *Vaccine*, **12**, 1000–3.

Yamamoto, K., Ichinose, Y., Shinagawa, H., Makino, K., Nakata, A., Iwanaga,, M., Honda, T. and Miwatani, T. (1990) Two-step processing for activation of the cytolysin/hemolysin of *Vibrio cholerae* O1 biotype El Tor: nucleotide sequence of the structural gene (*hlyA*) and characterization of the processed products. *Infection and Immunity*, **58**, 4106–16.

Yamamoto, T. and Yokota, T. (1983) Sequence of heat-labile enterotoxin of *Escherichia coli* pathogenic for humans. *Journal of Bacteriology*, **155**, 728–33.

Yamamoto, T. and Yokota, T. (1988) Electron microscopic study of *Vibrio cholerae* O1 adherence to the mucus coat and villus surface in the human small intestine. *Infection and Immunity*, **56**, 2753–9.

Yamamoto, T., Tamura, T. and Yokota, T. (1984) Primary structure of heat-labile enterotoxin produced by *Escherichia coli* pathogenic for humans. *Journal of Biological Chemistry*, **259**, 5037–44.

Yamamoto, T., Kamano, T., Uchimura, M., Iwanaga, M. and Yokota, T. (1988) *Vibrio cholerae* O1 adherence to villi and lymphoid follicle epithelium: in vitro model using formalin-treated human small intestine and correlation between adherence and cell-associated hemagglutinin levels. *Infection and Immunity*, **56**, 3241–50.

Ecology of *Vibrio cholerae*: role of aquatic fauna and flora 6

M.S. Islam, B.S. Drasar and R.B. Sack

6.1 INTRODUCTION

Cholera is endemic in many parts of the world. Substantial advances in our understanding of treatment, pathophysiology and immunology of cholera have been made but many aspects of cholera epidemiology remain obscure. In endemic areas of Bangladesh, cholera epidemics occur twice a year and have a regular seasonal pattern (Martin *et al.*, 1969; McCormack *et al.*, 1969; Merson *et al.*, 1980; Glass *et al.*, 1982; Samadi *et al.*, 1983). During epidemics, *Vibrio cholerae* O1 is isolated from both patients and from surface water, but disappears from the environment during inter-epidemic seasons (Khan *et al.*, 1981). The reservoirs or sites of survival and multiplication of *V. cholerae* O1 between epidemics are not completely known (Islam and Aziz, 1981; Islam *et al.*, 1988, 1989, 1990a,b).

6.2 HISTORICAL ASPECTS OF CHOLERA

Cholera has been known as a killing disease from time immemorial. The total number of deaths reported to be caused by cholera from 1909–1948 in India was 788 000 (Pollitzer, 1959). Cholera still takes a toll of lives in many parts of the world. If one goes back to the history of cholera, a description of the disease was found in India dating from the time of Alexander The Great (Pollitzer, 1959). Subsequently, various Portuguese, Dutch, French and British observers have described cholera in India (Pollitzer, 1959).

However, from 1817 onwards, the literature and study of cholera are

Cholera and the Ecology of Vibrio cholerae. Edited by B.S. Drasar and B.D. Forrest.
Published in 1996 by Chapman & Hall, London. ISBN 0 412 61220 8.

more systematic. The first pandemic started in 1817 in India and spread to other parts of the world. After this time cholera began to gain more attention because it was no longer confined to India but rather was becoming a disease of most serious concern to the world.

Seven pandemics have been recorded so far. The seventh pandemic started in 1961 and is still continuing. The origin of all these pandemics and their spread to various countries of the world have been extensively reviewed by Pollitzer (1959), Kamal (1974) and Barua (1992).

It has been observed that most of the countries of the world have been visited by cholera at one time or another. Therefore, it would be simpler to list the countries which have not been invaded. Cholera did not penetrate generally into the northernmost and southernmost parts of the globe. Accordingly in Asia, northern Siberia and Chamchatka were spared. Similarly, the most northern parts of western Europe (Iceland, the Faroe Islands, Shetland and Orkney Islands) as well as North American regions including Newfoundland and Greenland were also spared. In South America, cholera was absent from the southernmost parts of Chile, Argentina and from the Falkland Islands (Figure 6.1).

One important aspect of the presence pandemic is its place of origin. All the previous pandemics started from the Ganges delta of Bengal but the seventh started from the island of Sulawesi (Celebes) in Indonesia. However, during this period of the seventh pandemic, endemic cholera continued in Bengal and it is still there. After outbreaks, cholera has sometimes disappeared from many countries but continues to be present in Bengal. It has, therefore, been endemic in Bengal from the

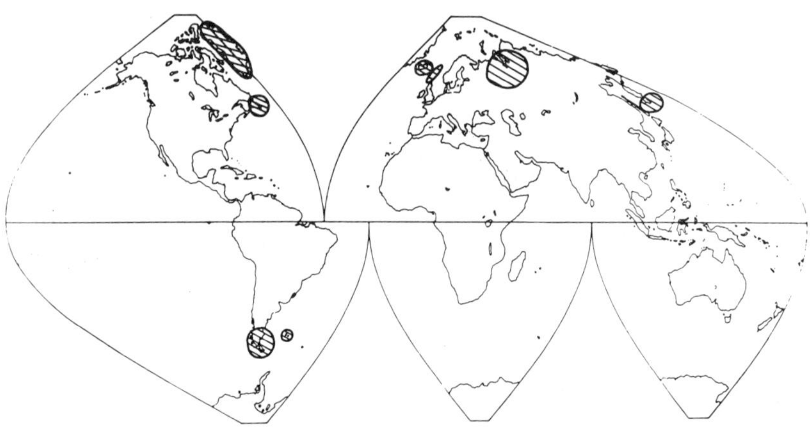

Fig. 6.1 Global occurrence of cholera. Cross-hatched areas indicate regions in the world where cholera has not been recorded.

very beginning of its history, so much so that this country has been considered as the disease's 'homeland'.

However, until now it is not clear why such a situation exists. Why should cholera be endemic in Bengal, and what are the factors responsible for providing a suitable habitat for the disease there?

6.3 SEASONALITY OF CHOLERA

The lower part of the Ganges Delta – being the 'homeland' of cholera – has maintained a clear seasonal pattern of epidemics. For example, in Bangladesh, the Dhaka (urban) and Matlab (rural) areas have been studied as endemic foci of cholera. Systematic studies of cholera epidemiology were started in Bangladesh (then East Pakistan) after the establishment of the Pakistan SEATO Cholera Research Laboratory in 1961. Martin *et al.* (1969) showed in an epidemiological study of endemic cholera in Dhaka from 1964 to 1966 that the outbreaks of cholera every year reached their peaks during November, December and January. They also observed a second peak in April and May in the 1965–1966 epidemic.

McCormack *et al.* (1969) examined the seasonality of cholera in a study carried out from November 1963 to June 1966 in Matlab, and found that each year the cholera epidemic peaked during November to January. In the 1966 epidemic they also observed a small second peak during April and June.

Merson *et al.* (1980), in a study in Matlab in Bangladesh during 1968–1977, showed that the peak incidence of the Classical biotype occurred during November, December and January, while in the case of the El Tor biotype the major peak was during September, October and November, with a second, small peak during March and April.

Glass *et al.* (1982) compiled 15 years' data from Matlab in Bangladesh and examined the seasonality of both El Tor and Classical biotypes. The pattern of seasonality observed for both El Tor and Classical biotypes was similar to that reported by Merson *et al.* (1980).

Samadi *et al.* (1983) showed a significant seasonality of both El Tor and Classical biotypes in urban Bangladesh over a 17-year period between 1964 and 1980, the estimated peak of El Tor being October and for Classical, December. These studies are summarized in Table 6.1.

If the seasonality of cholera in Bengal is evaluated from the available sources of information from 1963 to 1980, it is evident that the occurrence of cholera cases in endemic areas is highly seasonal. It is also clear that the cholera epidemics in Bangladesh always peak during the cooler months of the year, with a second smaller epidemic during the

Table 6.1 Studies on seasonality of cholera in Bangladesh

Date	Place of study	Biotype(s)	Epidemics		Reference
			Winter peak	Summer peak	
1963–1966	Matlab	Classical	November–January	April–June	McCormack *et al.* (1969)
1964–1966	Dhaka	Classical	November–January	April–May	Martin *et al.* (1969)
1968–1977	Matlab	Classical	November–January	March–April	Merson *et al.* (1980)
		El Tor	September–November	March–April	
1966–1980	Matlab	Classical	November–January	March–April	Glass *et al.* (1982)
		El Tor	September–November	March–April	
1964–1980	Dhaka	Classical	October–December	March–May	Samadi *et al.* (1983)
		El Tor	September–November	March–May	

hot season. So far no satisfactory explanation has been put forward to explain this pattern of cholera seasonality.

6.4 ENDEMICITY OF CHOLERA

Why is the Ganges delta a reservoir of infection? What is the mechanism by which the endemicity of this area is maintained?

From time to time various explanations have been put forward, with climatic factors often having been considered responsible for maintaining endemicity. Swaroop (1951) investigated the annual rates of cholera mortality in individual cholera-endemic districts in India for the period of 1901–1945. In this way he located the endemic foci and then tried to correlate these endemic areas with other topographic conditions. He observed the following factors which were common in all endemic areas.

1. All endemic areas were located around rivers.
2. All these tracts lay in areas of high population density.
3. All the areas lay in low-lying lands.
4. All these tracts lay in areas of high absolute humidity.

Among these four factors all except the second seem to indicate that the topography or environment of these places played an important role in maintaining endemicity. Swaroop finally concluded that all these features may indirectly affect the causative organism or the human hosts and their interrelationships.

It may be that under these environmental conditions *V. cholerae* O1 may be able to persist in the environment or be able to colonize a plant or animal that lives in these environmental conditions in the aquatic systems in these areas. The answer to maintaining endemicity lies in the mechanisms by which *V. cholerae* O1 survives during inter-epidemic periods in the endemic areas.

As far as the endemicity of these areas is concerned, four models are available to explain the maintenance of endemic cholera (Miller *et al.*, 1982, 1985). These are: (i) carrier status in animals; (ii) carrier status in humans; (iii) continuous transmission in humans; and (iv) an environmental reservoir. These will be discussed separately.

6.4.1 CARRIER STATUS IN ANIMALS

No animals have so far been demonstrated to be reservoirs of cholera. Sack (1973) investigated canines as reservoirs of cholera in Calcutta, India. Over 500 dogs were examined bacteriologically but no evidence of a *V. cholerae* O1 reservoir was found, although 14% of dogs harboured *V. cholerae* non-O1 in their intestines. These results indicate that non-human reservoirs exist for *V. cholerae* non-O1 in contrast to *V. cholerae* O1. However, toxigenic *V. cholerae* O1 have been isolated from domestic animals, e.g. cows, goats and chicken, only in the locality of concurrent cholera cases in man (Sanyal *et al.*, 1974). It is likely that these animals ingest the *V. cholerae* O1 from the environment at the time the disease is occurring locally, but they do not act as reservoirs.

6.4.2 CARRIER STATUS IN HUMANS

Only a few chronic gallbladder carriers have been demonstrated to date and this is in spite of some very extensive studies on populations exposed to cholera (Gangarosa *et al.*, 1966; Azurin *et al.*, 1967). Chronic carriers of *V. cholerae* O1 in the gallbladder were also found in experimental dogs convalescent from cholera (Sack *et al.*, 1969). However, it has been observed that rough vibrios can be excreted by convalescent patients for a few months after infection (Wallace *et al.*, 1967; Pierce *et al.*, 1970). These rough vibrios proved non-pathogenic after testing in animal models (Wallace *et al.*, 1967; Sigel *et al.*, 1980). Relatively large numbers of carriers would be required to give the observed epidemiological pattern of the disease. Therefore, chronic carriage of *V. cholerae* O1 by humans is unlikely to contribute to the persistence of cholera during inter-epidemic periods.

6.4.3 CONTINUOUS TRANSMISSION IN HUMANS

The possibility of low levels of infection maintaining cholera endemicity is a theory proposed by Gangarosa and Mosley (1974). Studies in Calcutta during 1968 showed that *V. cholerae* O1 can be isolated from one or more members of 83% (19/23) of households having an index cholera case (Joint ICMR-GWB-WHO Cholera Working Group, 1970). Among a total of 492 household contacts, 96 (20%) became infected. Of the 96 carriers identified, 53 excreted *V. cholerae* O1 on only one day, whereas two of the remainder excreted *V. cholerae* O1 for more than 25 days. However, short-term carriage cannot explain the hypothesis of continuous transmission.

6.4.4 ENVIRONMENT AS RESERVOIR

As this is the alternative mechanism to be examined, it will be helpful to list the reasons favouring this hypothesis.

(a) Historical background

The first speculation about the environment as a reservoir came from Robert Koch (1884) when he isolated the comma bacillus from a tank in Calcutta during 1883 epidemic. He wrote:

> 'I succeeded in finding the comma bacillus with all its characteristic peculiarities in a tank that supplies water for drinking and household purposes for all the people living around the immediate neighbourhood where a number of fatal cases had taken place.'

He then discussed the survival and multiplication of the comma bacillus in the environment:

> 'There remains still the important question to be answered, whether the infectious material can reproduce or multiply itself outside the human body. I believe that it can. As the comma bacillus can grow on a gelatin plate, as it can grow on a piece of linen, or in meat broth or on potatoes, it must also be in a position to grow in the open air, specially as we have seen that a comparatively low temperature enables it to develop. I would not certainly assume that the multiplication of the comma bacilli outside the human body takes place in well water or river water without any assistance, for these fluids do not possess that concentration of nutritious substances necessary for the growth of the bacilli. But I can easily imagine that, although the whole mass of the water in a tank or reservoir is too poor in nutritious substances for bacilli to flourish in it, yet some spots may contain sufficient concentration of nutritive substances, for example,

those spots where a gutter, or an outlet of a cesspool, opens into the stagnant water, where vegetable matter, animal refuse etc., lie and are exposed to putrifaction by bacteria. At such points a very active form of life can develop.'

After Koch's isolation of *Vibrio* and postulation of the environment as the possible reservoir of comma bacilli, scientists have carried out numerous investigations both in the laboratory and in the field to explore the possibility of the environment as a reservoir.

(b) Evidence against the environmental hypothesis

The evidence against an environmental hypothesis is that some of the laboratory studies show that the survival of *V. cholerae* O1 in river and tank water is short. Some of the studies are discussed below.

Lahiri *et al.* (1939) studied the viability of *V. cholerae* O1 in natural waters collected from several sources in Calcutta, where the disease was most prevalent. The water sources were spring, tap, river and tank. They found that the survival of the organism in raw water samples varied from 1 hour in samples of spring water to a maximum of 72 hours in certain tank waters.

Survival experiments of *V. cholerae* O1 in peptone water, human faeces, food and water were carried out in Taiwan by Cheng (1963). The water examined had been collected from tap, well, drain, river and sea. He found that *V. cholerae* O1 survived less than 24 hours in well water, 2 days in river water, 1 day in drain water, 5 days in sea water and less than 6 hours in pond water. He concluded that the survival capacity of *V. cholerae* O1 in water varied according to temperature, pH, salt, organic matter, degree of pollution, sunlight and presence of chlorine.

Konchady *et al.* (1969) studied the survival of *V. cholerae* O1 in five different sources of water in Calcutta, namely from tap, open well, river, canal and pond, and noted the duration of survival in all water samples to be about 5 days.

Mukerjee *et al.* (1961) observed the viability of *V. cholerae* O1 in natural water sources in and around Calcutta, using freshly collected water from the Ganges river, tanks and sewage. In total, 97 water samples were tested, the survival varying from 1 to 6 days.

All these studies showed that *V. cholerae* could not survive long in water collected from different natural sources. On the basis of such findings, it was considered that the natural bodies of water do not act as reservoirs because of the short survival time of *V. cholerae* O1 in the water (Felsenfeld, 1974). However, this interpretation may be inadequate because any water sample from a river, tank, well, spring, etc.

represents only a tiny fraction of the total number of habitats in the aquatic environment. All natural bodies of water are composed of hundreds of different kinds of macro and micro flora and fauna. So, this kind of negative conclusion based on survival studies of *V. cholerae* O1 in surface water (excluding its different kinds of plants and animals) in the laboratory is of limited application.

(c) Evidence consistent with environmental hypothesis

Some studies which support an environmental hypothesis are discussed below. For convenience of discussion, the data are grouped into different categories, such as: (i) taxonomical evidence; (ii) epidemiological evidence; (iii) laboratory-based survival studies; and (iv) environmental isolation of *V. cholerae*. These will be discussed separately.

Taxonomical evidence

Taxonomical studies of 142 so-called non-agglutinable (NAG) vibrios were carried out by Sakazaki *et al.* (1967) in 1967. A total of 80 physiological, morphological and biochemical characters were compared with five reference strains of *V. cholerae* El Tor. They also compared the DNA base composition. They found that the different characteristics of five reference strains coincided highly with the majority of the NAG strains (over 90% S value). So, on the basis of numerical taxonomy and DNA base composition, the authors suggested that the NAG vibrios are very closely related to *V. cholerae* O1 biotype El Tor.

Colwell (1970) studied the polyphasic taxonomy of the genus *Vibrio*. A total of 86 strains were included in the study of which 30 were *V. cholerae*, 35 *V. parahaemolyticus*, and 21 representative strains of *Pseudomonas*, *Spirillum*, *Achromobacter*, *Arthrobacter* and other *Vibrio* species. She examined 200 different morphological, physiological and biochemical characteristics of each strain (including DNA base composition), and ultrastructure by electron microscopy. The taxonomic data were analysed by computer using numerical taxonomy programs, the data showing that *V. cholerae* and non-agglutinable vibrios fell into a single relatively homogeneous *V. cholerae* species cluster.

Polynucleotide similarities among bacteria help to assess their genetic and phylogenic relationships. To examine this relationship, Citarella and Colwell (1970) studied selected *Vibrio* species, such as El Tor and Classical biotype of *V. cholerae*, *V. parahaemolyticus*, *V. alginolyticus*, and marine vibrios. They examined the polynucleotide sequence relationships among different species by means of DNA reassociation reactions and chromatography on hydroxyapatite. On the basis of the degree of reassociation and thermal stability, they concluded that there is little evolutionary divergence between *V. cholerae* and non-cholera vibrios.

Such studies have shown that *V. cholerae* O1 and non-O1 are very closely related. *V. cholerae* non-O1 are found throughout the year in the aquatic environment in cholera-endemic areas like Bangladesh (Khan *et al.*, 1984). So, if non-O1 can survive for a long time in the environment it may well be that *V. cholerae* O1 will be able to survive in the same manner. As the number of *V. cholerae* O1 are always less than *V. cholerae* non-O1 in environmental waters, and as there is no medium to differentiate one organism from the other, there always remains a chance of overlooking *V. cholerae* O1 in environmental samples.

Epidemiological evidence

Pollitzer (1959) described simultaneous outbreaks of cholera in different places in Bengal which, geographically, are quite far from one another. Under favourable climatic conditions, the disease then spread in a wave-like form to areas generally free from cholera.

McCormack *et al.* (1969) in an epidemiological study in rural endemic areas in Bangladesh observed the location of cases by village during the early weeks of the 1964 and 1966 epidemics. They observed simultaneous outbreaks of the disease in several villages in widely scattered areas, so they postulated multiple simultaneous introductions of *V. cholerae* into different villages.

Glass *et al.* (1982) in a study of rural endemic cholera in Bangladesh in 1966–1980 observed that the first recognized cases at the start of each yearly cholera season occurred at distant locations in the Matlab area. They also observed that in 1979, the first recognized cases were of different phage types.

If the epidemiological pattern of cholera in an endemic area like Bangladesh is analysed, it can be seen that simultaneous outbreaks of cholera occurred in distant villages without any known interconnections. This epidemiological feature of cholera is consistent with the environmental hypothesis.

Laboratory-based survival studies

Baker *et al.* (1983) studied the survival of *V. cholerae* O1 in artificial seawater and natural seawater microcosms. Two strains of *V. cholerae* O1, one of clinical origin and another isolated from the estuarine environment, were used in the experiments. It was observed that the initial numbers of cells inoculated into microcosms increased 2.5 $\log_{10}$ CFU within 3 days. After 75 days the number of viable cells was still 1–2 $\log_{10}$ CFU higher than the initial inoculum size.

Miller *et al.* (1984), using six Bangladeshi isolates of *V. cholerae* O1, studied the effect of different physicochemical conditions on the survival of toxigenic *V. cholerae* O1 in water in a series of laboratory experiments. Three of the strains used were isolated from cholera patients and three were from polluted water. Miller and colleagues found that toxigenic *V. cholerae* O1 can survive for a long period (months) in water having a salinity of 0.25–3% and a pH of around 8.0 at 25°C; hence, in an appropriate environment, *V. cholerae* O1 can survive for months.

Recent laboratory microcosm studies have also demonstrated that toxigenic *V. cholerae* O1 can survive longer in association with aquatic macrophytes (Islam *et al.*, 1990b) and algae (Islam *et al.*, 1988, 1990a) than in the surrounding water (see later).

Environmental isolations of V. cholerae

V. cholerae O1 and non-O1 have been isolated from the environmental sources from different parts of the world, which are discussed below (see also Table 6.2):

1. **India.** Occasionally, *V. cholerae* O1 have been isolated from surface water of places in India which were free from cholera for a number of years. Read and Pandit (1941) studied the distribution of *V. cholerae* O1 in certain endemic rural areas of Bengal and Bihar in India between 1939–1940. As a negative control, they selected Sindh province which had been free of cholera for the past 10 years. They collected water samples from tanks, wells, lakes, marshes, streams and rivers from Sindh province, and isolated *V. cholerae* O1 from one of the tanks and jhils for 41 days, suggesting that *V. cholerae* O1 can survive long-term in surface water.

   Venkatraman *et al.* (1941) studied the occurrence of *V. cholerae* O1 in natural sources of water in the absence of cholera in the Cauvery delta of India during 1940. This region had been free from cholera from May of the preceding year. The authors examined 1827 stool samples from the inhabitants of the region, all of which were negative. In total, 878 specimens of water were examined from 237 different sources which included rivers, tanks, ponds and wells. *V. cholerae* O1 biotype El Tor were isolated from 21 water samples collected from 17 different tanks. All the *V. cholerae* O1-positive tanks were situated in two small areas at the tail of the delta. It was found that *V. cholerae* O1 were isolated at one time or another during the year from 17 out of 61 tanks in two areas.

   Panja and Ghosh (1947) examined 524 water samples for *V. cholerae* collected from the Hoogly river in Calcutta, India;

Table 6.2 Isolation of *V. cholerae* from surface water sources in various countries

Types of water source	Date	Country	Reference
Tanks and jhils in Bengal and Bihar	1939–1940	India	Read and Pandit (1941)
Tanks in Madras	1940	India	Venkatraman *et al.* (1941)
River	1942	India	Panja and Ghosh (1947)
River and canals	1958–1959	India	Abou-Gareeb (1960)
Ponds, ditches, canals, etc. in Kent	1976–1979	UK	Lee *et al.* (1982)
Streams and ditches in Kent	1978–1980	UK	West and Lee (1982)
Chesapeake Bay	1976–1978	USA	Kaper *et al.* (1979)
Canals near Pecan Island and White Lake	1978	USA	Blake *et al.* (1980)
Chesapeake Bay, sewers and bayous in Louisiana	1977–1980	USA	Colwell *et al.* (1981)
Lagoon in Cagliari	1979	Italy	Salmaso *et al.* (1980)
Two rivers in Queensland	1977	Australia	Rogers *et al.* (1980)
Four rivers in Queensland	1977–1985	Australia	Bourke *et al.* (1986)
River, creaks, canals, ditches, etc.	1985	USA	Rhodes *et al.* (1986)
Rivers	1981		Bockemuhl *et al.* (1986)
Bay, rivers and canal	1984	USA	Kaysner *et al.* (1987)
River	1988	Japan	Venkateswaran *et al.* (1989)
Watershed	ND	USA	Perez-Rosas and Hazen (1989)
River	1990	Malaysia	Isa *et al.* (1990)
River	1989	India	Rai *et al.* (1991)
River	1991	Peru	Tamplin and Carrillo (1991)
Ponds	1988–1989	Bangladesh	Islam *et al.* (1991)
Ponds	1988–1989	Bangladesh	Islam *et al.* (1992)

V. cholerae O1 and non-O1 were isolated from 52 water samples, indicating that Hoogly river water may be responsible for cholera infection if used for domestic purposes.

Abou Gareeb (1960) carried out a bacteriological survey of the waters of the Hoogly river and its associated canals from December 1958 to August 1959. A total of 89 water samples were examined, with *V. cholerae* O1 being isolated in 12 instances. The isolation of *V. cholerae* O1 was spread fairly evenly over the whole study period, which extended fairly evenly over the whole study period, which extended over both epidemic and non-epidemic periods, including the monsoon. This study demonstrated the continuous potential infectivity of the river and canal waters of Calcutta during epidemic and non-epidemic cholera periods.

Nair *et al.* (1988) studied the ecology of *V. cholerae* non-O1 in freshwater environments of Calcutta, by collecting water samples from a lake, canal and a pond once monthly from July 1984 to June 1985. *V. cholerae* non-O1 was found in all sites during the study period, suggesting that the organism is common in freshwater environs of this region.

Rai *et al.* (1991) investigated the persistence of *V. cholerae* during interepidemic periods in Varanasi, India. In 1989, they collected 37 water samples from five ghats of the River Ganges in 1989, finding 18.9% of the samples to be positive for *V. cholerae* non-O1.

2. **United Kingdom.** Lee *et al.* (1982) carried out several surveys in various water sources in Kent to establish the incidence of *V. cholerae* O1 and non-O1 in the aquatic environment, which included various surface water sources such as ponds, lakes, ditches and canals. They observed that, during summer, *V. cholerae* O1 and non-O1 occurred sporadically in very low numbers in water containing >5 mmol Na^+/l. The highest numbers of up to 700 CFU/ml appeared regularly during the summer in static brackish water containing 25–200 mmol Na^+/l.

   West and Lee (1982) investigated the distribution of *V. cholerae* non-O1 in a stream and a ditch in Kent, from 1978 to 1980, finding the organism to occur in both sites during the summer. They also noted that *V. cholerae* non-O1 occurrence was not related to human faecal contamination.

3. **USA.** Kaper *et al.* (1979) carried out an investigation in the Chesapeake Bay, from October 1976 to January 1978, examining water, sediments and shellfish samples from 21 locations. Different physical and chemical parameters of water such as temperature, dissolved oxygen, pH, salinity, etc. were also examined. In total, 65 *V. cholerae* non-O1 were isolated throughout the year's study. No correlation was found between *V. cholerae* isolates and faecal coliform counts; in contrast, *Salmonella* sp. isolates correlated strongly with faecal coliform counts. No physical, chemical or microbiological parameters were observed to have any marked association with the incidence of *V. cholerae* non-O1. The only striking pattern observed was the range of salinity, all strains being isolated from sampling points where salinity was 0.4–1.7%. The authors concluded that the natural habitat of *V. cholerae* non-O1 appears to be natural bodies of water.

   The first case of cholera reported in the USA since 1911 was seen in 1973, in an old man living in Port Lavaca, Texas, *V. cholerae* El Tor Inaba being isolated from the patient's stool (Weissman *et al.*, 1974). Another case, in a 44-year-old man, was detected 5 years later in 1978 in Louisiana, after which 10 more cases were detected in four additional clusters (Blake *et al.*, 1980) in the same area surrounding the Gulf of Mexico. It was found that the only meal which was common to all the infected persons was crab (Blake *et al.*, 1980). After performing a matched triplet case control study, the only significant difference in exposure between cases and controls was seen to be the consumption of crabs.

An attempt was made to isolate *V. cholerae* O1 from seafood and the environment at the sites from which the crabs came. *V. cholerae* O1 was isolated from one of the three boiled crabs, from a shrimp caught in a canal and from two Moore swabs from two canals. All the specimens were cultured after enrichment in alkaline peptone water from 6–12 hours. All isolates were toxigenic *V. cholerae* O1, biotype El Tor serotype Inaba.

It was hypothesized that *V. cholerae* O1 had been able to survive along the Gulf Coast for years in the absence of human disease. The strongest evidence was the fact that the isolates from Texas and all the isolates from Louisiana five years later were of a single phage type unique to the US. Moreover, the Gulf Coast isolates were haemolytic, had vibriophage VcA-3, and had a unique cholera toxin gene pattern on Southern-blot analysis (Wachsmuth *et al.*, 1991). These endemic toxigenic *V. cholerae* O1 strains were clearly different from the seventh pandemic El Tor strains isolated worldwide in the past 30 years.

Finally it has been suggested that toxigenic *V. cholerae* O1 can survive and multiply in the environment and persist indefinitely without human faecal contamination. Colwell *et al.* (1981) isolated toxigenic *V. cholerae* O1, serotype Inaba, from Chesapeake Bay and from sewers and bayous in Louisiana. These organisms were isolated from some areas shown free from faecal contamination (using the presence of *E. coli* as an index); hence, the authors concluded that *V. cholerae* is a component of the autochthonous flora of brackish water, estuaries and salt marshes of coastal areas in temperate zones.

V. cholerae non-O1 were isolated from rivers, creeks, waters, irrigation, canals and ditches in Western Colorado during the summer of 1985. The organism occurred in fresh water with a Na^+ content $\geqslant 5\,mmol/l$, as well as in water of higher salinity, ca. $17\,mmol\ Na^+/l$ (Rhodes *et al.*, 1986).

The incidence of *V. cholerae* O1 and *V. cholerae* non-O1 were studied in shellfish, sediment and waters of California, Oregon and

Washington during the summer of 1984. *V. cholerae* non-O1 was found in 23 estuaries and in 44.6% of the 529 samples examined. *V. cholerae* O1 Inaba was isolated from water samples in Morrow Bay, California (Kaysner *et al.*, 1987).

Perez-Rosas and Hazen (1989) monitored densities of *V. cholerae* non-O1 and faecal coliforms at 12 sites in a rain forest watershed in Puerto Rico for 12 months. Densities of *V. cholerae* non-O1 and faecal coliforms did not correlate. High densities of *V. cholerae* non-O1 were also found at pristine sites at the highest point in the watershed, suggesting that *V. cholerae* non-O1 strains are indigenous to tropical freshwaters.

4. **Italy.** In 1973, outbreaks of cholera were reported simultaneously from three foci in Italy, a coastal town on the Gulf of Naples in the region of Campania, the Adriatic port city of Bari and from Cagliari in Sardinia. Consumption of seafood, particularly mussels, was probably the vehicle of *V. cholerae* infection (Baine *et al.*, 1974). After 6 years, cholera outbreaks again occurred at Cagliari in Sardinia. All but one of the 12 people with *V. cholerae* O1 infection gave a history of recent consumption of marine bivalves known locally as **arselle** (pelecypods). *V. cholerae* O1 was also isolated from samples of water and bivalves obtained from a lagoon on the outskirts of the city of Cagliari. *V. cholerae* O1 biotype El Tor, serotype Ogawa, phage type 4 were identified both from environmental and clinical isolates. All the strains isolated in 1979 outbreaks from different sources were of the same phage type as those from the first outbreak in 1973. As no evidence of cholera was seen in Cagliari between 1973 and 1979, Salmaso *et al.* (1980) suggested that the organism may have been maintained in the marine environment.
5. **Australia.** In Australia the first cholera case was reported in 1977 from Queensland (Rao and Stockwell, 1980; Rogers *et al.*, 1980). Two persons who shared a caravan in a caravan park excreted toxigenic *V. cholerae* O1, biotype El Tor, serotype Inaba and phage type 2. There was no history of foreign travel by the infected persons, and the sanitary facilities were adequate; neither were there detected any cross-connections between the septic tank, surface water disposal and reticulated water systems. Although *V. cholerae* O1 was isolated from the water supply of the caravan park, no *E. coli* was isolated, indicating that the supply was not contaminated by sewerage. Investigation revealed that a river, which was used as a supplementary source of water supply at times of increased demand, contained *V. cholerae* O1. The same strain of *V. cholerae* O1 was isolated intermittently from two rivers for 22 months, and persistently from one of the rivers for 2 months. It was suggested finally that *V. cholerae* O1 not only survived in the river water, but also multiplied there.

Studies showed that *V. cholerae* O1 were isolated at least once a year from one river in Queensland, Australia between 1977 and 1983. There was no indication of human, animal or sewage contamination, neither was any source of importation observed. The continuous isolation from this river over a 7-year period clearly indicated the long-term persistence of *V. cholerae* O1 in this aquatic environment (Bourke *et al.*, 1986).

V. cholerae O1 has also been isolated from water, sediments and plants from eight riverine sites in South-East Queensland, the organism being found to be tolerant of low salt levels (Myatt and Davis, 1989).

6. **Other countries.** Bockemuhl *et al.* (1986) studied the incidence of *V. cholerae* non-O1 at two sites on the Elbe River at Hamburg between June 1981 and December 1982, a total of 107 *V. cholerae* non-O1 were isolated from 147 water samples. The vibrio incidence was not related to the faecal coliform counts of the water and it was concluded that *V. cholerae* non-O1 are indigenous organisms of the Elbe River.

Venkateswaran *et al.* (1989) studied the occurrence of toxigenic vibrios in the freshwater environment of the Ohta river in Japan between August and October, 1988; toxigenic *V. cholerae* non-O1 were isolated from the surface water.

Isa *et al.* (1990), in investigating an outbreak of cholera in Tumpat, Kelantan, Malaysia, found that the Kelantan river water was the reservoir and cholera was transmitted from the river by river clams. The effort of public health education, especially against the drinking of river water and the consumption of raw clams, prevented spread of the disease.

Tamplin and Carrillo (1991) studied the occurrence of *V. cholerae* O1 in various environmental water samples collected from March to August, 1991 in Peru. The water samples were from the Pacific coast (Lima), mountain region (Lago Titicaca) and tropical rain forests (Amazon River). *V. cholerae* O1 were isolated by culture and by fluorescent antibody methods. Frequent isolation of *V. cholerae* O1 was seen in seawaters and river water, *V. cholerae* O1 being isolated in particularly high numbers from the river water (30°C) in the city of Iquitos, in the upper Amazon basin. In contrast, *V. cholerae* O1 was found in low numbers and isolated only infrequently from the cold (12°C) waters of Lago Titicaca near Puno, a high-altitude city.

Islam *et al.* (1995) monitored the occurrence of *V. cholerae* O1 and non-O1 in pond ecosystems in an endemic area of Bangladesh during May 1988 to April 1989. They collected a total of 600 samples every 15 days, which included surface water, plants, phytoplankton, zooplankton and sediment. *V. cholerae* biotype El Tor and serotype Inaba were isolated in February, March, April, September, October and December. Isolation of *V. cholerae* O1 during much of the year suggested that freshwater ponds can act as a reservoir of this organism; *V. cholerae* non-O1 were also isolated throughout the year (Islam *et al.*, 1992).

Other epidemiological and ecological studies have also implicated the aquatic environment as sources of *V. cholerae* in sporadic and epidemic disease outbreaks (Feachem *et al.*, 1981; Glass *et al.*, 1983; Johnston *et al.*, 1983; Hunt *et al.*, 1988; West, 1989; Draser, 1992).

Environmental isolations of *V. cholerae* O1 and non-O1 from diverse geographic areas suggest that *V. cholerae* O1 may be a long-term inhabitant of some aquatic environments. Aquatic environments consist of different kinds of flora and fauna which may form an infinite number of microenvironments; it would be impossible to examine all these microenvironments individually for *V. cholerae*. Microbial ecologists are now seriously considering a possible ecological niche for *V. cholerae* O1 in the aquatic environment, the existence of which could explain the organism's inter-epidemic survival.

6.5 AQUATIC FAUNA AND FLORA AS RESERVOIRS

The question remains, however, of which particular micro or macro environment or niche in the aquatic ecosystem is the reservoir? Do the *V. cholerae* O1 survive in the aquatic environment as free-living organisms or do they prefer an epibiotic host as a suitable ecological niche?

Hood *et al.* (1984) found that the survival of *V. cholerae* is dependent on suspended particulate matter in the water. They observed that the survival time was reduced in filtered and centrifuged sea water, the viable number of cells decreasing in proportion to the filtration and the speed of the centrifugation. Since filtering and centrifugation reduced the particulate load, the viability of the organism may have been related to the presence of particulates. So they suggested that *V. cholerae* prefer an epibiotic form of habitat. As the organisms can be recovered from filtered estuarine and sea water as well as particulate-free water, they also suggested that *V. cholerae* can survive as a planktonic form for a certain period of time.

Lee *et al.* (1982) in a 3-year survey from 1979–1981 of the incidence of *V. cholerae* O1 and non-O1 in water, animals and birds in Kent, UK, observed that about 6% of all the gulls sampled contained *V. cholerae* non-O1. They also collected the water samples from the same ditches where the birds were caught but could not isolate any *V. cholerae* non-O1.

Moreover, *V. cholerae* non-O1 have been isolated from the cavum nasi and pharynx of ducks by Bisgaard *et al.* (1978). They could not, however, isolate *V. cholerae* non-O1 from ducks which had never been outside the houses. The prevalence of *V. cholerae* non-O1 in the cavum nasi was very high when ducks were admitted to open fields; however, the organism could not be isolated from ducks kept in cages but provided with the same drinking water source as in the field. Similar observations were also made by Ogg *et al.* (1989), who isolated *V. cholerae* O1 from cloacal swabs and freshly voided faeces collected from 20 species of aquatic bird in Colorado and Utah during 1986 and 1987. They could not, however, detect *V. cholerae* O1 in water samples collected from the habitat of the birds. These studies might indicate that it was not the water but the plants or animals (e.g. duckweeds, algae, crustaceans, etc.) present in the water that had been eaten by the birds which contained *V. cholerae*. Thus, although the means by which *V. cholerae* survive during inter-epidemic periods remains to be clarified, plant or animal reservoirs seem quite likely media at present.

6.5.1 AQUATIC FAUNA AS RESERVOIRS (TABLE 6.3)

Different kinds of aquatic fauna such as oysters, zooplankton, crabs, etc. have been considered as potential habitats of *V. cholerae* in the aquatic environment.

Dastidar and Narayanaswami (1968) studied the chitinase activity of seven strains of *V. cholerae* Classical biotype, 15 strains of El Tor biotype and four NAG vibrios, detectable amounts of chitinase activity being observed in most strains studied.

Nalin (1976) suggested that *V. cholerae* can survive during the inter-epidemic period attached to copepods in the Ganges delta, because of the vibrio's chitinase production and ability to use chitin as a source of nutrients.

Kaneko and Colwell (1973) examined copepods collected from Chesapeake Bay and isolated *Vibrio parahaemolyticus* from more than 80% of the copepod samples, the number of organisms increasing with the rise in water temperature.

Sochard *et al.* (1979) investigated microorganisms associated with the surface and the gastrointestinal tract of different genera of copepods collected from pelagic, estuarine and fresh water environments. *Vibrio* spp. were isolated from copepods collected in all three environments, the highest isolation (94.4%) being made from those collected from estuarine regions. Five genera of copepods were collected: *Acartia tonsa, Pontellopsis regalis, Pleuromamma* sp., *Labidocera aestiva* and *Centropages furcatus*; among these the highest number of *Vibrio* spp. were isolated from *A. tonsa*. This may be due to the feeding habits of *A. tonsa*, which

Table 6.3 Isolation of vibrios from aquatic fauna

Fauna	Vibrios	Date	Region/country	Reference
Copepods	*V. parahaemolyticus*	1970–1971	Chesapeake Bay, USA	Kaneko and Colwell (1973)
Oysters	*V. cholerae* non-O1	1976–1978	Chesapeake Bay, USA	Kaper *et al.* (1979)
Copepods	*Vibrio* spp.	ND	Gulf of Mexico	Sochard *et al.* (1979)
Crabs, shrimp	*V. cholerae* O1	1978	Area between Vermilion Bay and Mud Lake, USA	Blake *et al.* (1980)
Oysters	*V. cholerae* O1 *V. cholerae* non-O1	1979–1980	Florida, USA	Twedt *et al.* (1981)
Oysters	*V. cholerae* O1	1980	Apalachicola Bay, Florida, USA	Hood *et al.* (1981)
Sea mullet	*V. cholerae* O1	1977–1983	River in Queensland, Australia	Bourke *et al.* (1986)
Zooplankton	*V. cholerae* non-O1	1988	Fukuyama coastal water, Japan	Venkateswaran *et al.* (1989)
Oysters	*V. cholerae* O1	1991	Mobile Bay on the Gulf Coast	De Paola *et al.* (1992)

ND, No data.

is herbivorous while the others are carnivorous. It is possible that the *Vibrio* spp. were attached to phytoplankton eaten by *A. tonsa*; hence this copepod was heavily colonized.

Twedt *et al.* (1981) examined freshly harvested oysters from estuarine waters in Florida, USA, from June 1979 to May 1980. They isolated *V. cholerae* non-O1 from 111 out of 790 oysters. *V. cholerae* O1 serotype Inaba was also isolated from seven samples during the months of May to July. All the *V. cholerae* O1 isolates were non-toxigenic. *V. cholerae* non-O1 was also isolated from oysters collected from Chesapeake Bay (Kaper *et al.*, 1979).

Hood *et al.* (1981) isolated two strains of *V. cholerae* O1 from oysters, *Crassostrea virginica*, collected during April 1980 in estuarine waters in the Apalachicola Bay, Florida. None of these strains was toxigenic.

Huq *et al.* (1983) studied the role of copepods on survival and multiplication of *V. cholerae* O1 in laboratory microcosms, finding that the organism survived longer and multiplied in the presence of copepods; *V. cholerae* was isolated for 13 days from the copepod samples. Patuxent River water (Maryland) was used in the experiments, with salinity ranging from 0.2–2.2%, somewhat higher than fresh water levels; the study outcome had lower-salinity water been used is not clear. In addition, 13 days is considered insufficient time to cover the inter-epidemic period.

As previously discussed, Bourke *et al.* (1986) isolated *V. cholerae* O1 El Tor Inaba from a river in Australia at least once each year from 1977 to 1983, and investigated possible reservoirs and natural ecological niches of the organism. They analysed tissues from 127 rock oysters (*Crassostrea commercialis*), 15 mud crabs (*Scylla serpata*) and five unidentified fish from the river mouth, but they could not isolate *V. cholerae* from any sample. About 120 each of sea mullet (*Mugli cephalus*) and freshwater mullet (*Trachystoma petardi*) (both herbivores) were netted and tested, though *V. cholerae* O1 was isolated from only two sea mullet from two rivers. *V. cholerae* O1 was also isolated from four of 456 batches of river plants collected.

Amako *et al.* (1987) examined the role of chitin in survival of *V. cholerae* O1 in laboratory microcosms; the organisms were found to survive more than 1 week in the presence of chitin at low temperature (0°C), compared with <1 day in its absence.

Venkateswaran *et al.* (1989), in studying the seasonal variation of *V. cholerae* non-O1 and the effect of zooplankton on its distribution in Fukuyama Coastal waters of Japan, showed that zooplankton samples did not harbour more *V. cholerae* non-O1 than the water column.

DePaola *et al.* (1992) investigated the ecology of vibrios in Mobile Bay on the US Gulf Coast. They implemented an intensive sampling programme and isolated *V. cholerae* O1 serotype Inaba, biotype El Tor from the intestinal contents of fish and four Mobile Bay oyster samples; all isolates were toxigenic.

6.5.2 AQUATIC FLORA AS RESERVOIRS

Dr Robert Koch in 1884 suggested that aquatic flora and fauna might be reservoirs of cholera in cholera endemic areas. He examined the maintenance of cholera in the lower part of the Ganges delta by pointing out the geographical location and condition of habitation. He explained,

> 'The lower part of the delta seems entirely uninhabited. This uninhabited district called Sunderbunds [beautiful forest], comprises an area of 7500 English square miles. A luxuriant vegetation and an abundant variety of animal life have developed in this uninhabited district. One can easily imagine what quantities of vegetable and animal matter are exposed to putrefaction in the boggy districts of Sunderbunds and that an opportunity scarcely to be found in any other place in the world is offered here for the development of microorganisms.'

He then postulated,

> 'Under peculiar circumstances a thoroughly special fauna and flora of microorganisms must develop there to which in all probability the comma bacillus belongs.'

To identify the role of vegetative matter as nutrients for the comma bacillus he made some observations. He stated,

> 'I have often formerly made such experiments and it has often happened that a water contains almost no bacteria at all whilst remains of plants especially roots or fruits swimming in it, teemed with bacteria, especially kinds of bacilli and spirilli even in the immediate neighbourhood of these objects, the water was rendered turbid by swarms of bacteria which clearly received their nourishment from the nutritive matter scattered by diffusion at a very small distance.'

Thus, there was an indication from the time of the discovery of *V. cholerae* that aquatic flora may be possible reservoirs of cholera in endemic areas. The aquatic flora, in turn, may be separated into two groups, namely macrophytes and microphytes.

(a) Macrophytes as reservoirs

Macrophytes are further divisible into two groups for convenience of discussion, *viz.* marine and freshwater macrophytes.

Marine macrophytes (Table 6.4)

Chan and McManus (1969) investigated the bacterial population on two marine algae, *Polysiphonia lanosa* and *Ascophyllum nodosum*, and the sea water of their environment. The bacteria were isolated and identified to genus level; bacterial numbers associated with *P. lanosa* and *A. nodosum* were 100- to 10 000-fold higher than from the surrounding waters, vibrios being the predominant organism on *A. nodosum*. Nutrition requirement experiments showed that most of the vibrios isolated had absolute requirements for amino acids. It is known that marine algae produce a number of extracellular products which contain peptides, amides and free nitrogen (Fogg and Westlake, 1955; Fogg and Boalch, 1958); thus, this may explain why the amino acid-requiring bacteria were predominant on algal surfaces.

Laycock (1974) investigated the bacterial population associated with the fronds of the sublittoral brown alga *Laminaria longicruris* collected at monthly intervals from Shag Bay, Nova Scotia, in Canada. He isolated two major groups of bacteria from the seaweed: vibrios and pseudomonads and observed a distinct variation in bacterial count between groups with the time of the year and the region of the frond from which they were isolated. Data indicated that a high proportion of vibrios was maintained by *L. longicruris* throughout the winter. The isolated vibrios could hydrolyse laminaran (an excretion product of *Laminaria* spp.) which may explain the existence of the organisms on this seaweed.

Shiba and Taga (1980) investigated the bacterial flora on different seaweed and of the environmental seawater from December, 1973 to June, 1974 in the Nabeta inlet and Otsuchi Bay, Japan. Two green algae, *Monostroma nitidum* and *Enteromorpha linza*, one red alga *Porphyra suborbiculata* and a brown alga *Eisenia bicyclis* were collected. The viable counts of bacteria attached to the green algae, *M. nitidum* and *E. linza* ranged from 10^4–10^6/cm^2; those on the red alga *P. suborbiculata* ranged from 10^3–10^4/cm^2; counts in seawater samples were 10^3/ml. Bacterial counts on the brown alga *E. bicyclis* were generally lower than on red and green algae, but fluctuated considerably, from 2.2×10^4/cm^2 in March to only 290/cm^2 in January. Such variation was not influenced by changes in the bacterial population in seawater, which was relatively constant during the study period, but thought to depend on the physiological condition of the seaweed, as bacterial levels were lowest when the largest number of germinating leaves was observed.

It was also noted that the counts of heterotrophic bacteria differed among algal species. The bacteria belonging to the vibrios and *Flavobacterium–Cytophaga* group were present on the green algae collected at both stations, but were not dominant in environmental seawater. It was

Table 6.4 Association between marine macrophytes and bacteria

Plants	Bacteria	Date	Country	Source
Polysiphonia lanosa, *Ascophyllum nodosum*	*Vibrio*, *Flavobacterium*, *Pseudomonas*, *Achromobacter*, *Escherichia*, *Sarcina*, *Staphylococcus*	1964	Canada	Chan and McManus (1969)
Laminaria longicruris	*Vibrio* and *Pseudomonas*	1972–1973	Canada	Laycock (1974)
Enteromorpha linza *Porphyra suborbiculata*, *Monostroma bicyclis*	*Vibrio*, *Flavobacterium* and *Cytophaga*	1973–1974	Japan	Shiba and Taga (1980)
Ulva lactuca, *Enteromorpha intestinalis*, *Ceramium rubrum*, *Polysiphonia lanosa*	*Vibrio cholerae* O1	1986	UK	Islam *et al.* (1988)

suggested that a beneficial relationship existed between the green algae and their epiphytic bacteria.

Islam *et al.* (1988) studied the survival and attachment of toxigenic *V. cholerae* O1 to four marine algae, *Ulva lactuca, Enteromorpha intestinalis, Ceramium rubrum* and *Polysiphonia lanosa,* in artificial aquatic ecosystems. Of all the plants tested, *V. cholerae* O1 survived longest (27 days) in association with *U. lactuca.*

Freshwater macrophytes (Table 6.5)

Plant surfaces can influence the distribution of bacteria in different ways, e.g. topography of the leaf surfaces, excretion of metabolites which can act as nutrients for bacteria and by the presence of stimulatory or inhibitory substances (Hossell and Baker, 1976, 1979a,b,c).

The higher numbers of bacteria in mature leaves may be due to multiplication of bacteria or to attachment of bacteria over time from the water column. So, the growth rates of epiphytic bacteria were studied on a common duckweed, *Lemna minor* (Hossell and Baker, 1979c). In detachment experiments carried out with 20 plants, a slight increase in population in 1 hour indicated the bacteria can multiply on *Lemna* plant surfaces.

Baker and Farr (1982) investigated the amount of dissolved organic carbon (DOC) produced by *L. minor* and its utilization by epiphytic bacteria. DOC production by *L. minor* was measured in the presence and absence of epiphytic bacteria; ca. 2% of the carbon fixed by *L. minor* was shown to be secreted as DOC. Production of DOC by axenic cultures of *L. minor* was greater than that by *L. minor* colonized by bacterial epiphytes, suggesting a role of epiphytes as a carbon 'sink' and supporting the idea that carbon fixed by macrophytes is transferred to heterotrophic epiphytic bacteria.

Islam *et al.* (1984) carried out a survival study of toxigenic *V. cholerae* O1 with *L. minor* in artificial aquatic ecosystems. Survival of both environmental and clinical strains of *V. cholerae* O1 was assessed by viable bacterial counts on thiosulphate-citrate bile salt (TCBS) sucrose agar. It was observed that both strains survived longer on *L. minor* than in water on which *L. minor* was floating or in control water (Figure 6.2). *V. cholerae* secretes an enzyme, mucinase (Schneider and Parker, 1982) which may play a role in the environment by degrading naturally present mucin and mucin-like substances present in the plant cells and thus further help to form an association with the plants. It was suggested that plants may serve as an effective environmental reservoir for *V. cholerae* either through a non-specific association, or by interaction with *V. cholerae* in a commensal relationship (Islam *et al.*, 1990b).

The association of vibrios with freshwater macrophytes was first

Table 6.5 Association between freshwater macrophytes and vibrios

Macrophytes	Vibrios	Date	Country	Source
Water hyacinth (*Eichhornia crassipes*), *Monochoria hastata*, *Marsilea quadrifolia*, *Ludwigia repens*	*V. cholerae* non-O1	1977	Bangladesh	Islam and Aziz (1978)
Water hyacinth (*Eichhornia crassipes*)	*V. cholerae* O1 El Tor	1976–1977	Bangladesh	Spira *et al.* (1981)
Water hyacinth (*Eichhornia crassipes*), *Telanthera philoxeroides*, *Panicum* sp.	*V. cholerae* non-O1	1980–1981	Bangladesh	Islam *et al.* (1983)
Duckweed (*Lemna minor*)	*V. cholerae* O1	1986	UK	Islam *et al.* (1990b)

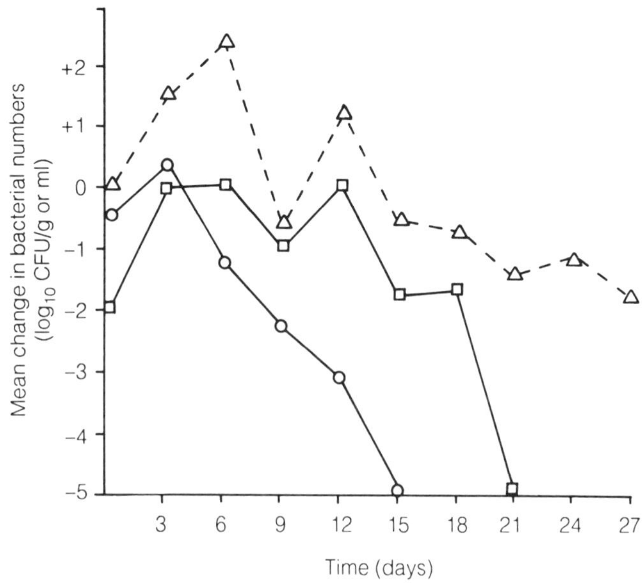

Fig. 6.2 Survival of environmental strain of *Vibrio cholerae* on *Lemna minor* (△) in water on which *L. minor* was floating (○) and in control water without *L. minor*, at 1.0% salinity on TCBS agar medium (□).

reported in the late 1970s when Islam and Aziz (1978) studied freshwater macrophytes in cholera endemic areas in Bangladesh. As part of a study to investigate the association of vibrios with hydrophytic plants (Islam and Aziz, 1981), plants (11 types) and water samples were collected at 15-day intervals from different parts of Dhaka, Bangladesh from July to December, 1977. The association of vibrios with four macrophytes was observed, including *Eichhornia crassipes* (water hyacinth), *Monochoria hastata, Marsilea quadrifolia* and *Ludwigia repens*. Vibrios were isolated from the roots of these plants.

Spira *et al.* (1981) studied the association of *V. cholerae* El Tor with water hyacinth in endemic areas in Bangladesh, as well as in laboratory microcosms. Water and intact water hyacinth plants floating within the boundaries of the same sampling point were collected. In one-third of the samples, *V. cholerae* biotype El Tor were present both in plants and water; the rest of the sample showed 52% isolation from plants alone and 16% from water alone. The difference in percentage of isolation from plants alone and water alone was statistically significant. A significant difference was also observed at all times in the number of

V. cholerae per gram of roots as compared with stems and leaves of water hyacinths, possibly due to the much greater surface area exposed to water compared with the rest of the plant. This study concluded that virulent *V. cholerae* biotype El Tor can be associated with water hyacinth. Vibrios present in water contaminated by stools from cholera victims concentrated on the surface of these plants and increased the length of time the vibrios remained viable in the aquatic environment.

Islam *et al.* (1983) carried out a preliminary survey of isolation of vibrios from water plants, water and soil sediments in three ponds of Dhaka city in Bangladesh. Over a 1-year period from September 1980 to August 1981, *E. crassipes* (water hyacinth), *Telanthera philoxeroides* and *Panicum* sp. were collected every month, though vibrios were found only in the roots of the water hyacinth.

Islam *et al.* (1992) later examined the association of *V. cholerae* non-O1 in various components of pond ecosystems in Bangladesh. Water, plants, phytoplankton and sediment samples were collected from five ponds in and around Dhaka city, Bangladesh at 15-day intervals, between May 1988 and April 1989. The plants included *E. crassipes*, *Nymphoides* sp. and *T. philoxeroides*. Isolation of *V. cholerae* non-O1 was 28% and 47% from plants and phytoplankton respectively. This study demonstrated that *V. cholerae* non-O1 prefer phytoplankton as a habitat rather than higher plants in the aquatic environment.

(b) Microphytes as reservoirs

As with macrophytes, microphytes are sub-divisible into two groups for convenience of discussion; marine and freshwater microphytes.

Marine microphytes (Table 6.6)

Most bacteria in seawater are attached to marine plankton and small particles in seawater. Simidu *et al.* (1971) investigated the generic composition of bacterial flora of marine phyto- and zooplankton by collecting plankton samples from the Nishiura Bay, on the Pacific Coast of the Bozo Peninsula 200 km south-east of Tokyo. They also enumerated and identified the bacterial flora from the seawater. The viable bacterial number ranged from 1.2×10^5–1.1×10^7/ml of plankton sample, whereas the total bacterial number in seawater ranged from 7×10^1–1.2×10^2/ml. The main constituents of the bacterial flora from the plankton samples were the *Vibrio–Aeromonas* group with *Pseudomonas* the next most frequent.

Algal extracellular products, which are released as products of cell metabolism, a process sometimes known as 'excretion', are the sources of microbial nutrients. The excretion products are the source of extracel-

Table 6.6 Association between microphytes and bacteria

Microphytes	Vibrios	Date	Country	Reference
Marine microphytes				
Phytoplankton	*Vibrio* spp. *Aeromonas* sp. *Pseudomonas* sp.	1970	Japan	Simidu *et al.* (1971)
Skeletonema costatum and *Chaetoceros* sp.	*Vibrio* sp. *Pseudomonas* sp. *Spirillum* sp. *Flavobacterium* sp.	1977	France	Martin and Bianchi (1980)
Freshwater microphytes				
Rhizoclonium fontanum	*V. cholerae* O1	1976	UK	Islam *et al.* (1989)
Anabaena variabilis	*V. cholerae* O1	1976	UK	Islam *et al.* (1990a)
Phytoplankton	*V. cholerae* O1	1987	Bangladesh	Tamplin *et al.* (1990)
Phytoplankton	*V. cholerae non*-O1	1988–1989	Bangladesh	Islam *et al.* (1992)
Blue-green alga, *Anabaena* sp.	*V. cholerae* O1	1988–1989	Bangladesh	Islam *et al.* (1994c)

lular organic materials under natural conditions on which a bacterial population can grow without any additional carbon source.

Martin and Bianchi (1980) studied bacterial population dynamics in continuous cultures of marine planktonic algae, maintained by continuous nutrient salt enrichment of natural oligotrophic seawater, in a large-volume tank exposed to external climatic conditions. They followed the development of the population during two experimental phytoplankton blooms, in spring and autumn, 1977. They observed a difference in generic identities of bacterial strains at different times of the study period. During the spring experiments which started on the first week of April, the percentage of *Pseudomonas* was 54% and *Vibrio* sp. was ca. 15%. At the end of the first week of May, the *Pseudomonas* percentage had decreased to 22% and the vibrios increased to 74%.

During the autumn experiment which started on the third week of October, *Vibrio, Pseudomonas, Spirillum* and *Flavobacterium* were 8%, 28%, 5% and 1%, respectively. On the first week of November, the percentages of *Pseudomonas, Spirillum* and *Flavobacterium* increased, whereas *Vibrios* decreased to 1%. After 1 week, i.e. the second week of November, the *Vibrio* percentage started increasing gradually, whereas all other bacteria showed a gradual decrease. By the fourth week of November, the percentage of *Vibrio* had become the highest (32%) of all bacteria isolated.

In both spring and autumn experiments, the phytoplankton blooms were of *Skeletonema costatum* and *Chaetoceros* spp. So there were complex competitive population changes between different bacterial flora in a natural body of water, and *Vibrio* proportions varied from being at low level to becoming the dominant genus present.

Hellebust (1965) estimated the fraction of photo-assimilated carbon excreted during log-phase growth by different classes of marine microphytes. He also investigated the effects of light intensity on excretion of extracellular products by cultures of marine algae. Nine species of algae, i.e. *Cricosphaera carterae, Exuviella* sp., *Chaetoceros simplex, S. costatum, Thalassiosira fluviatilis, Phaeodactylum tricornutum, Chlorella* sp., *Pyramimonas* sp. and *Tetraselmis* sp. were included. All the algae were in log-phase growth during the experiment. The amount of carbon excreted by the marine species ranged from 1.5–24% of that assimilated through photosynthesis. It was also noted that the percentage of excretion varied considerably from species to species, as did the composition of excretion products. Most of the excretion products were of low molecular weight, for example amino acids and sugar alcohols.

Freshwater microphytes (Table 6.6)

The freshwater microphytes comprise two types, the green algae (Chlorophyceae) and the blue-green algae (Cyanophyceae).

1. **Green algae.** Studies have shown that bacteria can utilize algal extracellular products as nutrients (Fogg and Westlake, 1955; Fogg and Boalch, 1958; Ward *et al.*, 1964; Belly *et al.*, 1973) and can maintain a commensal relationship with algae. These studies also showed that the relationship between bacteria and algae is not a matter of chance, but that *V. cholerae* may coexist with some types of algae in the environment and may use the extracellular materials secreted by those algae as nutrients.

   Islam *et al.* (1989) in a laboratory-based study, reported that toxigenic *V. cholerae* O1 gain survival advantages in association with a filamentous green alga, *Rhizoclonium fontanum* in an artificial aquatic environment. *V. cholerae* O1 survived longer with *R. fontanum* ($P < 0.05$) than in water on which *R. fontanum* was floating, or in control water (without *R. fontanum*) (Figure 6.3). Despite this difference in length of survival of *V. cholerae* O1, *R. fontanum* could not be considered as a reservoir. Only those flora with which *V. cholerae* O1 would be able to survive at least the 4–5 months necessary to span the inter-epidemic period should be considered as reservoirs.

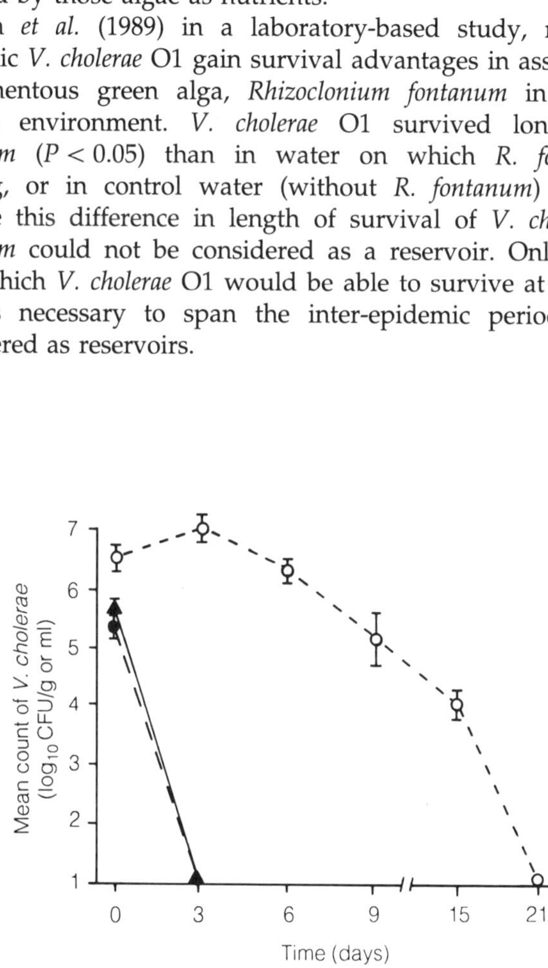

Fig. 6.3 Survival of strain 250 of *Vibrio cholerae* on *Rhizoclonium fontanum*, in water on which *R. fontanum* was floating and in control water at 0.05% salinity on TCBS agar medium. Flask A: ○, *R. fontanum*, ▲, water on which *R. fontanum* was floating. Flask B: ●, control water. Each line represents the arithmetic mean of *V. cholerae* count ($\log_{10}$ CFU/g or ml) $\pm$ 2 SEM of four experiments with strain 250 of *V. cholerae*.

2. **Blue-green algae.** The association between blue-green algae and bacteria during a bloom is well known (Paerl, 1976, 1978a,b; Paerl and Keller, 1978) but the functional aspects of the algae–bacteria relationship in natural communities have not been properly defined.

   V. cholerae have been detected using indirect fluorescent antibody (FA) technique from both clinical and environmental samples (Islam *et al.*, 1993a,b, 1994a,b,c). Sack and Barua (1964) detected *V. cholerae* O1 from the rice-water stools of cholera patients using a FA technique in the early 1960s. Recently, Colwell *et al.* (1985, 1990) used the same technique to detect *V. cholerae* O1 in laboratory microcosms and environmental water samples.

   Brayton *et al.* (1987) carried out a field study in a rural cholera endemic area in Bangladesh. They detected and enumerated the *V. cholerae* O1 from water samples collected from village tubewells, ponds and the adjacent river using either FA direct viable counts or most probable number indices by culture. All water samples yielded significantly higher counts of *V. cholerae* O1 by the former technique than by the latter, the difference providing additional evidence for the existence of viable but non-culturable forms of *V. cholerae* O1 in the natural environment in an area in which cholera is endemic. It is evident from these results that the FA procedure is a more sensitive method of detection than standard culture procedures, as it permits enumeration of both culturable and non-culturable cells.

   Huq *et al.* (1990) detected *V. cholerae* O1 from 876 plankton samples collected from ponds and rivers between February 1987 and January 1990 in Matlab, Bangladesh, using both conventional cultural and FA techniques. *V. cholerae* O1 was detected in 563 samples (64.3%) by the FA method but in only three samples (0.34%) by cultural technique. Thus, a FA technique is clearly more sensitive than a cultural approach for detecting *V. cholerae* O1 in such environmental samples.

   Tamplin *et al.* (1990) performed both laboratory and field-based studies to investigate the attachment of *V. cholerae* O1 to zooplankton and phytoplankton using a FA technique with a monoclonal antibody specific for the A antigen of O1 lipopolysaccharide of *V. cholerae* O1. Zooplankton and phytoplankton were collected from a river and two ponds in Matlab, Bangladesh in April, 1987. Among the zooplankton, five species of copepods, five species of cladocerans and one species of rotifers were present. Among phytoplankton, two species of green algae (*Volvox* sp. and *Pediastrum simplex*) and two of blue-green algae (*Spirulina* sp. and an unicellular Cyanobacteria) were studied. None of the zooplankton showed any binding of *V. cholerae* O1 with whole specimens, whereas three of four phytoplanktons did so. *V. cholerae*

O1 strains attached preferentially to zooplankton molts (exuviae) rather than to whole specimens, supporting the suggestion that phytoplankton provide a better habitat than zooplankton for *V. cholerae* O1, probably because the bacteria have a constant supply of nutrients from the living phytoplankton – a situation not possible with dead zooplankton. The authors did not observe the attachment of *V. cholerae* O1 to any natural zooplankton or phytoplankton (uninoculated) specimens. In a similar study, Islam *et al.* (1994c, see later) also found that *V. cholerae* attached to a blue-green alga, *Anabaena* sp. but not with other algae present in the same sample collected from the natural waters of Bangladesh. Such results indicate that *V. cholerae* may have some specificity for attachment.

Islam *et al.* (1990a) studied the persistence of *V. cholerae* O1 in the mucilaginous sheath of a blue-green alga, *A. variabilis* in an artificial aquatic environment. Data indicated that *V. cholerae* O1 can survive in cultural form in association with *A. variabilis* for up to 120 hours, but survived in water on which the alga was floating and in control water without alga for more than 144 hours (Figure 6.4). Examination of the alga by phase contrast microscopy after 10 days of survival showed that *V. cholerae* O1 entered the mucilaginous sheath of *A. variabilis*. *V. cholerae* inside the mucilaginous sheath of *A. variabilis* became non-culturable but were found dividing by binary fission and clustering around the heterocysts which are the known sites for atmospheric nitrogen fixation. *V. cholerae* O1 was detected inside the mucilaginous sheath for up to 15 months using the FA technique, a time considered long enough for the organism to span the inter-epidemic period. Overall, this study suggested that *A. variabilis* may act as a reservoir of *V. cholerae* O1 in an aquatic environment, and that *V. cholerae* O1 does not lose toxin properties during survival when in association with algae (Islam, 1990a,b, 1991).

The production of mucinase by vibrios (Schneider and Parker, 1982) may be a process by which plant and planktonic mucin is degraded in nature, and one factor which allows *V. cholerae* O1 to associate with mucilaginous blue-green alga, *A. variabilis*.

One important functional aspect of *A. variabilis* and *V. cholerae* O1 is the exchange of carbon dioxide and oxygen. In the association of *A. variabilis* and *V. cholerae* O1, the algal host provides oxygen during photosynthesis that can be utilized in aerobic respiration of *V. cholerae* O1, which produce carbon dioxide, which may be available for algal photosynthesis. Thus, *A. variabilis* and *V. cholerae* O1 may maintain a symbiotic relationship (Islam, 1987; Islam *et al.*, 1994c).

Paerl and Gallucci (1985) observed microscopically that motile bacteria can easily discriminate heterocysts from vegetative cells.

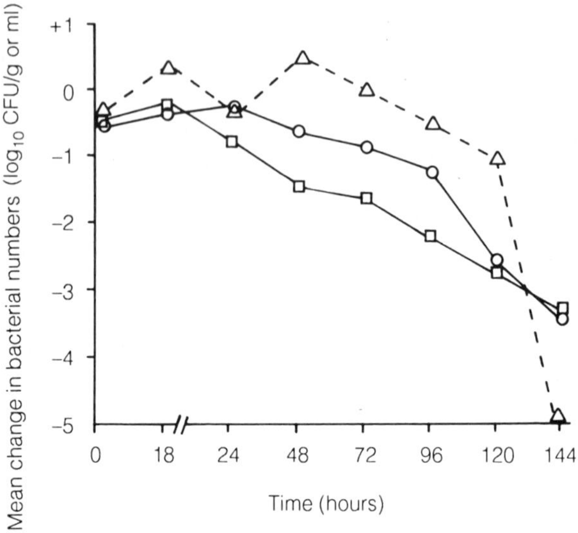

Fig. 6.4 Recovery of *Vibrio cholerae* from *Anabaena variabilis*, water on which *A. variabilis* was floating and control water of 0.05% salinity on TSA media. Flask A: △, *A. variabilis*; ○, water on which *A. variabilis* was floating; Flask B: □, control water.

Bacteria interacting with *A. oscillarioides* filaments were observed to 'bump' onto both vegetative and heterocyst cells, but did not adhere to either. When bacteria encountered heterocysts, flagellar rotation was often increased for 5–10 minutes. The bacteria then attached to the heterocyst–vegetative cell junction, after which flagellar motion stopped. Once attachment took place both hosts and epiphytes started growing. It was observed that bacteria rarely penetrated cyanobacterial cell walls. Islam *et al.* (1990a) also observed that *V. cholerae* O1 aggregated near the heterocyst–vegetative cell junction.

Chemotaxis played a key role in the establishment and maintenance of cyanobacterial and bacterial association (Bell and Mitchell, 1972; Paerl and Gallucci, 1985). Products of nitrogen fixation, which were excreted at heterocyst–vegetative cell junctions are responsible for attracting diverse, naturally occurring heterotrophic bacteria. Amino acids are probably chemotactic agents since they are excreted by nitrogen-fixing *Anabaena*.

These studies demonstrated that heterotrophic bacterial and filamentous blue-green algal association may involve two processes. First, the bacteria attach to the filament and then they utilize the algal metabolites as nutrients; second, the attachment is initiated by chemotaxis and shows site and species specificity.

On the basis of all these findings, field studies were carried out by Islam *et al.* (1994c) to detect *V. cholerae* O1 in blue-green alga from the aquatic environment of Bangladesh. Phytoplankton samples were collected every 15 days between May 1988 and April 1988 from a pond in Dhaka city which is used for bathing, washing, swimming and occasionally drinking purposes. The phytoplanktons were mainly *Anabaena* sp., *Euglena* sp. and *Phaeus* sp. *V. cholerae* O1 was detected by immunofluorescence in the mucilaginous sheath of *Anabaena* sp. in 16 out of 24 plankton samples. *V. cholerae* O1 could be detected only in association with *Anabaena* sp. and not with other algae collected from the pond, e.g. *Euglena* sp. and *Phaecus* sp.

The study clearly demonstrated, therefore, an association between *V. cholerae* O1 and a blue-green alga, *Anabaena* sp. in the natural aquatic environment in Bangladesh. These findings suggest that *V. cholerae* O1 may have a preference for association with blue-green algae, particularly *Anabaena* sp. or related species which have mucilaginous sheaths around them. It is evident, therefore, from these studies that blue-green algae may act as a reservoir of *V. cholerae* O1 in such an aquatic environment.

6.6 CONCLUSION

It now appears that blue-green algae may act as a reservoir of *V. cholerae*. This finding is a major contribution to our understanding of cholera epidemiology. The conventional view of *V. cholerae* O1 as an organism found in the environment only in close association with human infections, and surviving for only a few days in the environment is again challenged by this finding. This observation may provide answers to the long-standing question about the interepidemic survival of *V. cholerae* and may help to explain how the endemicity and seasonality of cholera in endemic areas are maintained. It has been observed that the peak incidence of cholera in endemic areas of Bangladesh occurs together with the bloom of blue-green algae in their natural aquatic environment (Cockburn and Cassanos, 1960; Oppenheimer *et al.*, 1978; Islam *et al.*, 1990a, 1991; Islam, 1994; Islam and Bateman, 1994). However, the implication of this discovery for the control of cholera in humans remains to be elucidated.

ACKNOWLEDGEMENTS

This research was supported by the International Centre for Diarrhoeal Disease Research, Bangladesh (ICDDR,B). The ICDDR,B is supported by countries and agencies which share its concern for the health problems of developing countries. Current donors include: the aid agencies of the Governments of Australia, Bangladesh, Belgium, Canada, China, Denmark, Germany, Japan, the Netherlands, Norway, Republic of Korea, Saudi Arabia, Sweden, Switzerland, the United Kingdom and the United States; international organizations including the Arab Gulf Fund, the United Nations Children's Fund (UNICEF), the United Nations Development Programme (UNDP), the United Nations Population Fund (UNFPA) and the World Health Organization; and private foundations including Ford Foundation, Population Council, Rockefeller Foundation and the Sasakawa Foundation; and private organizations including American Express Bank, Bayer A.G. and CARE.

We are grateful to Professor D.J. Bradley for advice, Dr C.J. Miller for his helpful comments on an earlier draft of this paper and Mr Manzurul Haque for secretarial assistance.

6.7 REFERENCES

Abou-Gareeb, A.H. (1960) The detection of cholera vibrios in Calcutta waters: the river Hooghly and canals. *Journal of Hygiene*, **58**, 21–33.

Amako, K., Shimodori, S., Imoto, T., Miake, S. and Umeda, A. (1987) Effects of chitin and its soluble derivatives on survival of *Vibrio cholera* O1 at low temperature. *Applied and Environmental Microbiology*, **53**, 603–5.

Azurin, J.C., Kobari, K., Barua, D., Alvero, M., Gomez, C.Z., Dkzon, J.J., Nakano, E., Suplido, E. and Ledesma, L. (1967) A long term carrier of cholera: Cholera Dolores. *Bulletin of the World Health Organization*, **37**, 745–9.

Baine, W.B., Mazzotti, M., Greco, D., Izzo, E., Zampieri, A., Angioni, G., Giota, M.D., Gangarosa, E.J. and Pocchiari, F. (1974) Epidemiology of cholera in Italy in 1973. *Lancet*, **ii**, 1370–4.

Baker, J.H. and Farr, I.S. (1982) Organic secretions by duckweed and their relationship with epiphytic bacteria, in *Bacteria and Plants*, Society of Applied Bacteriology Symposium Series, No. 10, (eds M. Rhodes-Roberts and F.A. Skinner), Academic Press, London, p. 250 (abstract).

Baker, R.M., Singleton, F.L. and Hood, M.A. (1983) Effects of nutrient deprivation on *Vibrio cholerae*. *Applied and Environmental Microbiology*, **46**, 930–40.

Barua, D. (1992) History of cholera, in *Cholera*, (eds D. Barua and W.B. Greenough III), Plenum Publishing Corporation, New York, pp. 1–35.

Bell, W. and Mitchell, R. (1972) Chemotactic and growth responses of marine bacteria to algal extracellular products. *Biological Bulletin*, **143**, 265–77.

Belly, R.T., Tansey, M.R. and Brock, T.D. (1973) Algal excretion ^{14}C-labelled compounds and microbial interactions in *Cyanidium caldarium* mats. *Journal of Phycology*, **9**, 123–7.

Bisgaard, M., Sakazaki, R. and Shimada, T. (1978) Prevalence of non-cholera vibrios in cavum nasi and pharynx of ducks. *Acta Pathologica et Microbiologica Scandinavica*, **86**, 261–6.

Blake, P.A., Allegra, D.T., Snyder, J.D. *et al.* (1980) Cholera – a possible endemic focus in the United States. *New England Journal of Medicine*, **302**, 305–9.

Bockemuhl, J., Roch, K., Wohlers, B., Aleksic, V., Aleksic, S. and Wokatsch, R. (1986) Seasonal distribution of facultatively enteropathogenic vibrios (*Vibrio cholerae, Vibrio mimicus, Vibrio parahaemolyticus*) in the freshwater of the Elbe River at Hamburg. *Journal of Applied Bacteriology*, **60**, 435–42.

Bourke, A.T.C., Cossins, Y.N., Gray, B.R.W., Lunney, T.J., Rostron, N.A., Holmes, R.V., Griggs, E.R., Larsen, D.J. and Kelk, V.R. (1986) Investigation of cholera acquired from the riverine environment in Queensland. *Medical Journal of Australia*, **144**, 229–34.

Brayton, P.R., Tamplin, M.L., Huq, A. and Colwell, R.R. (1987) Enumeration of *Vibrio cholerae* O1 in Bangladesh waters by fluorescent-antibody direct viable count. *Applied and Environmental Microbiology*, **53**, 2862–3.

Chan, E.C.S. and McManus, E.A. (1969) Distribution, characterization and nutrition of marine microorganisms from the algae *Polysiphonia lanosa* and *Ascophyllum nodosum*. *Canadian Journal of Microbiology*, **15**, 409–20.

Cheng, C.T. (1963) Survival of El Tor vibrios outside the human population. *Nagasaki Medical Journal*, **38**, 870–6.

Citarella, R.V. and Colwell, R.R. (1970) Polyphasic taxonomy of the genus *Vibrio*: polynucleotide sequence relationships among selected *Vibrio* species. *Journal of Bacteriology*, **104**, 434–42.

Cockburn, T.A. and Cassanos, J.G. (1960) Epidemiology of endemic cholera. *Public Health Reports*, **75**, 791–803.

Colwell, R.R. (1970) Polyphasic taxonomy of the genus *Vibrio*: Numerical taxonomy of *V. cholerae, V. parahaemolyticus* and related *Vibrio* species. *Journal of Bacteriology*, **104**, 410–13.

Colwell, R.R., Seidler, R.J., Kaper, J. *et al.* (1981) Occurrence of *Vibrio cholerae* serotype O1 in Maryland and Lousiana Estuaries. *Applied Environmental Microbiology*, **41**, 555–8.

Colwell, R.R., Brayton, P.R., Grimes, D.J., Roszak, D.B., Huq, S.A. and Palmer, L.M. (1985) Viable but non-culturable *Vibrio cholerae* and related pathogens in environment: implications for release of genetically engineered microorganisms. *BioTechnology*, **3**, 817–20.

Colwell, R.R., Tamplin, M.L., Brayton, P.R. *et al.* (1990) Environmental aspects of *Vibrio cholerae* in transmission of cholera, in *Advances in Research on Cholera and Related Areas*, 7th edn, (eds R.B. Sack and R. Zinnaki), KTK Scientific Publishers, Tokyo, pp. 327–43.

Dastidar, S.G. and Narayanaswami, A. (1968) The occurrence of chitanase in vibrios. *Indian Journal of Medical Research*, **56**, 654–8.

DePaola, A., Ccpers, G.M., Motes, M.L. *et al.* (1992) Isolation of Latin American epidemic strain of *Vibrio cholerae* O1 from US gulf coast. *Lancet*, **339**, 624 (letter).

Drasar, B.S. (1992) Pathogenesis and ecology: the case of cholera. *Journal of Tropical Medicine and Hygiene*, **95**, 365–72.

Feachem, R.G., Miller, C.J. and Drasar, B.S. (1981) Environmental aspects of cholera epidemiology II. Occurrence and survival of *Vibrio cholerae* in the environment. *Tropical Diseases Bulletin*, **78**, 865–80.

Felsenfeld, O. (1974) The survival of cholera vibrios, in *Cholera*, (eds D. Barua and W. Burrows), W.B. Saunders, Philadelphia, pp. 359–66.

Fogg, G.E. and Boalch, G.T. (1958) Extracellular products in pure cultures of a brown alga. *Nature*, **181**, 789–90.

Fogg, G.E. and Westlake, D.F. (1955) The importance of extracellular products of algae in freshwater. *Verhandlungen der Internationalen Vereinigung fur theoretische und angewandte Limnologie*, **12**, 219–32.

Gangarosa, E.J. and Mosley, W.H. (1974) Epidemiology and surveillance of cholera, in *Cholera*, (eds D. Barua and W. Burrows), W.B. Saunders, Philadelphia, pp. 391–403.

Gangarosa, E.J., Ssaghari, H., Emile, J. and Siadat, H. (1966) Detection of *Vibrio cholerae* biotype El Tor by purging. *Bulletin of the World Health Organization*, **34**, 363–9.

Glass, R.I., Becker, S., Huq, M.I., Stoll, B.J., Khan, M.U., Merson, M.H., Lee, J.V. and Black, R.E. (1982) Endemic cholera in rural Bangladesh, 1966–1980. *American Journal of Epidemiology*, **116**, 959–70.

Glass, R.I., Huq, M.I., Lee, J.V., Therlfall, E.J., Khan, M.R., Alim, A.R.M.A., Rowe, B. and Gross, R.J. (1983) Plasmid-borne multiple drug resistance in *Vibrio cholerae* serogroup O1, biotype El Tor: evidence for a point source outbreak in Bangladesh. *Journal of Infectious Diseases*, **147**, 204–9.

Hellebust, J.A. (1965) Excretion of some organic compounds by marine phytoplankton. *Limnology and Oceanography*, **10**, 192–206.

Hood, M.A., Ness, G.E. and Rodrick, G.E. (1981) Isolation of *Vibrio cholerae* serotype O1 from the eastern oyster, *Crassostrea virginica*. *Applied and Environmental Microbiology*, **41**, 559–60.

Hood, M.A., Ness, G.E., Rodrick, G.E. and Blake, N.J. (1984) The ecology of *Vibrio cholerae* in two Florida estuaries, in *Vibrios in the Environment*, (ed. R.R. Colwell), John Wiley & Sons Inc., New York, pp. 399–409.

Hossell, J.C. and Baker, J.H. (1976) The distribution and characterization of bacteria on the surfaces of some river macrophytes. *Journal of Applied Bacteriology*, **41**, 24–5.

Hossell, J.C. and Baker, J.H. (1979a) A note on the enumeration of epiphytic bacteria by microscopic method with particular reference to two freshwater plants. *Journal of Applied Bacteriology*, **46**, 87–92.

Hossell, J.C. and Baker, J.H. (1979b) Epiphytic bacteria of the fresh water plant *Ranunculus penicillatus*; enumeration, distribution and identification. *Archiv für Hydrobiologie*, **86**, 322–37.

Hossell, J.C. and Baker, J.H. (1979c) Estimation of the growth rates of epiphytic bacteria and *Lemna minor* in a river. *Freshwater Biology*, **9**, 319–27.

Hunt, M.D., Woodward, W.E., Keswick, B.H. and DuPont, H.L. (1988) Seroepidemiology of cholera in Gulf Coastal Texas. *Applied and Environmental Microbiology*, **54**, 1673–7.

Huq, A., Small, E.B., West, P.A., Huq, M.I., Rahman, R. and Colwell, R.R. (1983) Ecological relationships between *Vibrio cholerae* and planktonic crustacean copepods. *Applied and Environmental Microbiology*, **45**, 275–83.

Huq, A., Colwell, R.R., Rahman, R. *et al.* (1990) Detection of *Vibrio cholerae* O1 in the aquatic environment by fluorescent-monoclonal antibody and culture methods. *Applied and Environmental Microbiology*, **56**, 2370–3.

Isa, A.R., Othman, W.M. and Ishak, A. (1990) Cholera outbreak in Tumpat, Kelantan. *Medical Journal of Malaysia*, **45**, 187–93.

Islam, M.S. (1987) Studies of aquatic flora as possible reservoirs of toxigenic *Vibrio cholerae* O1. PhD Thesis, Faculty of Medicine, University of London.

Islam, M.S. (1990a) Effect of various biophysicochemical conditions on toxigenicity of *Vibrio cholerae* O1 during survival with a green alga, *Rhizoclonium fontanum*, in an artificial aquatic environment. *Canadian Journal of Microbiology*, **36**, 464–8.

Islam, M.S. (1990b) Increased toxin production by *Vibrio cholerae* O1 during survival with green alga, *Rhizoclonium fontanum*, in an artificial aquatic environment. *Microbiology and Immunology*, **34**, 557–63.

Islam, M.S. (1991) Toxigenicity and toxin genes of *Vibrio cholerae* O1 isolated from an artificial aquatic environment. *World Journal of Microbiology and Biotechnology*, **7**, 269–71.

Islam, M.S. (1994) Reservoirs, seasonality and endemicity of cholera in Bangladesh: role of environment, in *Proceedings of the Third Annual Scientific Conference of ICDDR,B*, Dhaka, 15–16 January, pp. 9–10.

Islam, M.S. and Aziz, K.M.S. (1978) Association of vibrios with some hydrophytic plants, in *Proceedings of the 3rd Annual Science Conference of Bangladesh Association for the Advancement of Science*, Chittagong, January 8–12, p. 52, abstract No. 121.

Islam, M.S. and Aziz, K.M.S. (1981) Association of vibrios with some hydrophytic plants. *Bangladesh Journal of Microbiology*, **1**, 70–2.

Islam, M.S. and Bateman, O.M. (1994) Cholera control in developing countries. *Waterlines*, **12**, 20–3.

Islam, M.S., Rashid, M.H., Aziz, K.M.S., Huq, M.I., Chowdhury, A.A. and Rahim, M.Z. (1983) A preliminary survey of isolation of vibrios from water plants, water and soil sediments in ponds of Dhaka, in *Proceedings of the 2nd Asian Conference on Diarrhoeal Disease*, Calcutta, India, February 21–24, p. 23.

Islam, M.S., Drasar, B.S. and Bradley, D.J. (1984) Survival of *Vibrio cholerae* O1 in artificial aquatic ecosystems. *Journal of Medical Microbiology*, **18**, vii.

Islam, M.S., Drasar, B.S. and Bradley, D.J. (1988) Survival and attachment of toxigenic *Vibrio cholerae* O1 in association with four marine algae. *Bangladesh Journal of Microbiology*, **5**, 41–8.

Islam, M.S., Drasar, B.S. and Bradley, D.J. (1989) Attachment of toxigenic *Vibrio cholerae* O1 to various freshwater plants and survival with a filamentous green alga, *Rhizoclonium fontanum*. *Journal of Tropical Medicine and Hygiene*, **92**, 396–401.

Islam, M.S., Drasar, B.S. and Bradley, D.J. (1990a) Long-term persistence of toxigenic *Vibrio cholerae* O1 in the mucilaginous sheath of a blue green alga, *Anabaena variabilis*. *Journal of Tropical Medicine and Hygiene*, **93**, 133–9.

Islam, M.S., Drasar, B.S. and Bradley, D.J. (1990b) Survival of toxigenic *Vibrio cholerae* O1 with a common duckweed, *Lemna minor*, in artificial aquatic ecosystems. *Transactions of the Royal Society of Tropical Medicine and Hygiene*, **84**, 422–4.

Islam, M.S., Alam, M.J. and Khan, S.I. (1995) Occurence and distribution of culturable *Vibrio cholerae* O1 in aquatic environments of Bangladesh. *International Journal of Environmental Studies*, **47**, 217–23.

Islam, M.S., Alam, M.J. and Neogi, P.K.B. (1992) Seasonality and toxigenicity of *Vibrio cholerae* non-O1 isolated from different components of pond ecosystems of Dhaka city, Bangladesh. *World Journal of Microbiology and Biotechnology*, **8**, 160–3.

Islam, M.S., Hasan, M.K., Miah, M.A., Qadri, F., Yunus, M., Sack, R.B. and Albert, M.J. (1993a) Isolation of *Vibrio cholerae* O139 Bengal from water in Bangladesh. *Lancet*, **342**, 430 (letter).

Islam, M.S., Drasar, B.S. and Sack, R.B. (1993b) The aquatic environment as a reservoir of *Vibrio cholerae*: a review. *Journal of Diarrhoeal Disease Research*, **11**, 197–206.

Islam, M.S., Hasan, M.K., Miah, M.A., Huq, A., Bardha, P.K., Sack, R.B. and Albert, M.J. (1994a) Specificity of Cholera Screen test during an epidemic of cholera-like disease due to *Vibrio cholerae* O139 synonym Bengal. *Transactions of the Royal Society of Tropical Medicine and Hygiene*, **88**, 424–5.

Islam, M.S., Hasan, M.K., Miah, M.A., Yunus, M., Zaman, K. and Albert, M.J. (1994b) Isolation of *Vibrio cholerae* O139 Bengal from the aquatic environment of Bangladesh: implications for disease transmission. *Applied and Environmental Microbiology*, 60, 1684–6.

Islam, M.S., Miah, M.A., Hasan, M.K., Sack, R.B. and Albert, M.J. (1994c) Detection of non-culturable *Vibrio cholerae* O1 in a blue-green alga from aquatic environment of Bangladesh. *Transactions of the Royal Society of Tropical Medicine and Hygiene*, **88**, 298–9.

Johnston, J.M., Martin, D.L., Perdue, J., McFarland, L.M., Caraway, C.T., Lippy, E.C. and Blake, P.A. (1983) Cholera on a Gulf Coast oil rig. *New England Journal of Medicine*, **309**, 523–6.

Joint ICMR–GWB–WHO Cholera Study Group. (1970) Cholera carrier studies in Calcutta, 1968. *Bulletin of the World Health Organization*, **43**, 379–87.

Kamal, A.K. (1974) The seventh pandemic of cholera, in *Cholera*, (eds D. Barua and W. Burrows), W.B. Saunders, Philadelphia, pp. 1–14.

Kaneko, T. and Colwell, R.R. (1973) Ecology of *Vibrio parahaemolyticus* in Chesapeake Bay. *Journal of Bacteriology*, **113**, 24–32.

Kaper, J., Lockman, H., Colwell, R.R. and Joseph, S.W. (1979) Ecology, serology and enterotoxin production of *Vibrio cholerae* in Chesapeake Bay. *Applied and Environmental Microbiology*, **37**, 91–103.

Kaysner, C.A., Abeyta, C. Jr, Wekell, M.M., Depaola, A. Jr, Stoll, R.F. and Leitch, J.M. (1987) Incidence of *Vibrio cholerae* from estuaries of the United States west coast. *Applied and Environmental Microbiology*, **53**, 1344–8.

Khan, M.U., Shahidullah, M., Haque, M.S. and Ahmed, W.U. (1984) Presence of vibrios in surface water and their relation with cholera in a community. *Tropical Geography and Medicine*, **36**, 335–40.

Khan, M.U., Mosley, W.H., Chakraborty, J., Sarder, A.M. and Khan, M.R. (1981) The relationship of cholera to water source and use in rural Bangladesh. *International Journal of Epidemiology*, **10**, 23–5.

Koch, R. (1884) An address on cholera and its bacillus. *British Medical Journal*, **2**, 403–7, 453–9.

Konchady, D., Prescott, L.M., Datta, A. and De, S.P. (1969) Survival of *V. cholerae* in some of the waters of Calcutta. *Indian Journal of Medical Research*, **57**, 60–5.

Lahiri, M.N., Das, B.C. and Malik, K.S. (1939) The viability of *V. cholerae* in natural waters. *Indian Medical Gazette*, **74**, 742–4.

Laycock, R.A. (1974) The detrital food chain based on seaweeds. 1. Bacteria associated with the surface of *Laminaria* fronds. *Marine Biology*, **25**, 223–31.

Lee, J.V., Bashford, D.J., Donovan, T.J., Furniss, A.L. and West, P.A. (1982) The incidence of *Vibrio cholerae* in water, animals and birds in Kent, England. *Journal of Applied Bacteriology*, **52**, 281–91.

Martin, A.R., Mosley, W.H., Sau, B.B., Ahmed, S. and Huq, I. (1969) Epidemiological analysis of endemic cholera in urban East Pakistan, 1964–1966. *American Journal of Epidemiology*, **89**, 572–82.

Martin, Y.P. and Bianchi, M.A. (1980) Structure, diversity and catabolic potentialities of aerobic heterotrophic bacterial populations associated with continuous culture of natural marine phytoplankton. *Microbial Ecology*, **5**, 265–79.

McCormack, W.M., Mosley, W.H., Fahimuddin, M. and Benenson, A.S. (1969) Endemic cholera in rural East Pakistan. *American Journal of Epidemiology*, **89**, 393–404.

Merson, M.H., Black, R.E., Khan, M.U. and Huq, M.I. (1980) Epidemiology of cholera and enterotoxigenic *Escherichia coli* diarrhoea, in *Cholera and Related Diarrhoeas*, 43rd Nobel Symposium, Stockholm, (eds O. Ouchterlony and J. Holmgren), S. Karger, Basel, pp. 34–45.

Miller, C.J., Drasar, B.S. and Feachem, R.G. (1982) Cholera and estuarine salinity in Calcutta and London. *Lancet*, **i**, 1216–18.

Miller, C.J., Drasar, B.S. and Feachem, R.G. (1984) Response of toxigenic *Vibrio cholerae* O1 to physico-chemical stresses in aquatic environments. *Journal of Hygiene (Cambridge)*, **93**, 475–95.

Miller, C.J., Feachem, R.G. and Drasar, B.S. (1985) Cholera epidemiology in developed and developing countries: new thoughts on transmission, seasonality and control. *Lancet*, **ii**, 261–3.

Mukerjee, S., Rudra, B.C. and Roy, U.K.G. (1961) Observations on cholera endemicity in Calcutta and survival of *V. cholerae* in the water sources. *Annals of Biochemistry and Experimental Medicine*, **21**, 31–40.

Myatt, D.C. and Davis, G.H. (1989) Isolation of medically significant *Vibrio* species from reverine sources in south east Queensland. *Microbios*, **60**, 111–23.

Nair, G.B., Sarker, B.K., De, S.P., Chakrabarti, M.K., Bhadra, R.K. and Pal, S.C. (1988) Ecology of *Vibrio cholerae* in the freshwater environments of Calcutta, India. *Microbial Ecology*, **15**, 203–15.

Nalin, D.R. (1976) Cholera, copepods and chitinase. *Lancet*, **ii**, 958.

Ogg, J.E., Ryder, R.A. and Smith, H.L. Jr. (1989) Isolation of *Vibrio cholerae* from aquatic birds in Colorado and Utah. *Applied and Environmental Microbiology*, **55**, 95–9.

Oppenheimer, J.R., Ahmed, M.G., Huq, A., Haque, K.A., Alam, A.K.M.A., Aziz, K.M.S., Ali, S. and Haque, A.S.M.M. (1978) Limnological studies of three ponds in Dacca, Bangladesh. *Bangladesh Journal of Fisheries*, **1**, 1–28.

Paerl, H.W. (1976) Specific associations of the blue green algae *Anabaena* and *Aphanizomenon* with bacteria in freshwater blooms. *Journal of Phycology*, **12**, 431–5.

Paerl, H.W. (1978a) Microbial organic carbon recovery in aquatic ecosystems. *Limnology and Oceanography*, **23**, 927–35.

Paerl, H.W. (1978b) Role of heterotrophic bacteria in promoting N_2 fixation by *Anabaena* in aquatic habitats. *Microbial Ecology*, **4**, 215–31.

Paerl, H.W. and Gallucci, K.K. (1985) Role of chemotaxis in establishing a specific nitrogen-fixing cyanobacterial–bacterial association. *Science*, **227**, 647–9.

Paerl, H.W. and Keller, P.E. (1978) Significance of bacterial-*Anabaena* (Cyanophyceae) associations with respect to N_2 fixation in fresh water. *Journal of Phycology*, **14**, 254–60.

Panja, G. and Ghosh, S.K. (1947) Isolation of cholera vibrios from Hooghly river water at Calcutta. *Indian Journal of Medical Research*, **35**, 1–2.

Perez-Rosas, N. and Hazen, T.C. (1989) In situ survival of *Vibrio cholerae* and *Escherichia coli* in a tropical rain forest water shed. *Applied and Environmental Microbiology*, **55**, 495–9.

Pierce, N.E., Banwell, J.G., Gorbach, S.L., Mitra, R.C. and Mondal, A. (1970) Convalescent carriers of *Vibrio cholerae*: detection and detailed investigation. *Annals of Internal Medicine*, **72**, 357–64.

Pollitzer, R. (1959) *Cholera*, Monograph No. 43, World Health Organization, Geneva.

Rai, R.N., Tripathy, V.C. and Joshi, R.D. (1991) Persistance of *Vibrio cholerae* in interepidemic period – preliminary observations on analysis of water. *Journal of Communicable Diseases*, **23**, 44–5.

Rao, A. and Stockwell, B.A. (1980) The Queensland cholera incidence of 1977. 1. The index case. *Bulletin of the World Health Organization*, **58**, 663–4.

Read, W. and Pandit, S. (1941) Distribution of *V. cholerae* and El Tor type strains in certain rural areas in India. *Indian Journal of Medical Research*, **29**, 403–5.

Rhodes, J.B., Smith, H.L. Jr and Ogg, J.E. (1986) Isolation of non-O1 *Vibrio cholerae* serovars from surface waters in western Colorado. *Applied and Environmental Microbiology*, **51**, 1216–19.

Rogers, R.C., Cuffe, R.G.C.J., Cossins, Y.M., Murphy, D.M. and Bourke, A.T.C. (1980) The Queensland cholera incidence of 1980. 2. The epidemiological investigation. *Bulletin of the World Health Organization*, **58**, 665–9.

Sack, R.B. (1973) A search for canine carriers of vibrios. *Journal of Infectious Diseases*, **127**, 709–12.

Sack, R.B. and Barua, D. (1964) The fluorescent antibody technique in the diagnosis of cholera. *Indian Journal of Medical Research*, **52**, 848–54.

Sack, R.B., Carpenter, C.C.J., Yardley, J.H. and Subang, A.E. (1969) Experimental canine cholera. IV. The chronic carrier state for *Vibrio cholerae*. *Journal of Infectious Diseases*, **119**, 165–71.

Sakazaki, R., Gomez, C.Z. and Sebald, M. (1967) Taxonomical studies on the so-called NAG vibrios. *Japanese Journal of Medical Science and Biology*, **20**, 265–80.

Salmaso, S., Greco, D., Bonfiglio, B. *et al.* (1980) Recurrence of pelecypod-associated cholera in Sardinia. *Lancet*, **ii**, 1124–7.

Samadi, A.R., Chowdhury, M.K., Huq, M.I. and Khan, M.U. (1983) Seasonality of classical and El Tor for cholera in Dhaka, Bangladesh: 17-year trends. *Transactions of the Royal Society of Tropical Medicine and Hygiene*, **77**, 853–6.

Sanyal, S.C., Singh, S.J., Tiwari, I.C., Sen, P.C., Marwah, S.M., Hazarika, U.R., Singh, H., Shimada, T. and Sakazaki, R. (1974) Role of household animals in maintenance of cholera infection in a community. *Journal of Infectious Diseases*, **130**, 575–9.

Schneider, D.R. and Parker, C.D. (1982) Purification and characterization of the mucinase of *Vibrio cholerae*. *Journal of Infectious Diseases*, **145**, 474–82.

Shiba, T. and Taga, N. (1980) Heterotrophic bacteria attached to seaweeds. *Journal of Experimental Medicine Biology and Ecology*, **47**, 251–8.

Sigel, S.P., Lanier, S., Baselski, V.S. and Parker, C.D. (1980) *In vivo* evaluation of pathogenicity of clinical and environmental isolates of *Vibrio cholerae*. *Infection and Immunity*, **28**, 681–7.

Simidu, U., Ashino, K. and Kaneko, E. (1971) Bacterial flora of phyto- and zoo-plankton in the inshore water of Japan. *Canadian Journal of Microbiology*, **17**, 1157–60.

Sochard, M.R., Wilson, D.F., Austin, B. and Colwell, R.R. (1979) Bacteria associated with the surface and gut of marine copepods. *Applied and Environmental Microbiology*, **37**, 750–9.

Spira, W.M., Huq, A., Ahmed, Q.S. and Saeed, Y.A. (1981) Uptake of *Vibrio cholerae* biotype El Tor from contaminated water by water hyacinth (*Eichhornia carssipes*). *Applied and Environmental Microbiology*, **42**, 550–3.

Swaroop, S. (1951) Endemicity of cholera in India. *Indian Journal of Medical Research*, **39**, 141–83.

Tamplin, M.L. and Carrillo, P.C. (1991) Environmental spread of *Vibrio cholerae* in Peru. *Lancet*, **338**, 1216–17 (letter).

Tamplin, M.L., Gauzens, A.L., Huq, A., Sack, D.A. and Colwell, R.R. (1990) Attachment of *Vibrio cholerae* serogroup O1 to zooplankton and phyto-plankton of Bangladesh waters. *Applied and Environmental Microbiology*, **56**, 1977–80.

Twedt, R.M., Madden, J.M., Hunt, J.M. *et al.* (1981) Characterization of *Vibrio cholerae* isolated from oysters. *Applied Environmental Microbiology*, **41**, 1475–8.

Venkateswaran, K., Kiiyukia, C., Takaki, M., Nakano, H., Matsuida, H., Kawakami, H. and Hashimoto, H. (1989) Characterization of toxigenic vibrios isolated from the freshwater environment of Hiroshima, Japan. *Applied and Environmental Microbiology*, **55**, 2613–18.

Venkatraman, K.V., Krishnaswami, A.K. and Ramakrishnan, C.S. (1941) Occurrence of *Vibrio* El Tor in natural sources of water in the absence of cholera. *Indian Journal of Medical Research*, **29**, 419–24.

Wachsmuth, K., Bopp, C.A. and Fields, P.I. (1991) Difference between toxigenic *Vibrio cholerae* O1 from South America and US Gulf Coast. *Lancet*, **337**, 1097–8.

Wallace, C.K., Anderson, P.N., Lewis, G.W., Serge, G.V., Pierce, N.F., Brown, T.C., Sanyal, S.C. and Waldman, R.H. (1967) Probable gall bladder infection in convalescent cholera patients. *Lancet*, **i**, 865–8.

Ward, C.H., Moyer, J.E. and Vela, G.R. (1964) Studies on bacteria associated with *Chlorella pyrenoidosa* TX71105 in mass culture. *Developments in Industrial Microbiology*, **6**, 213–22.

Weissman, J.B., Dewitt, W.E., Thompson, J., Muchnick, C.N., Portnoy, B.L., Feeley, J.C. and Gangarosa, E.J. (1974) A case of cholera in Texas in 1973. *American Journal of Epidemiology*, **100**, 487–98.

West, P.A. (1989) The human pathogenic vibrios – a public health update with environmental perspectives. *Epidemiology and Infection*, **103**, 1–34.

West, P.A. and Lee, J.V. (1982) Ecology of *Vibrio* species including *Vibrio cholerae* in natural waters of Kent, England. *Journal of Applied Bacteriology*, **52**, 435–48.

Zoological microhabitats of *Vibrio cholerae* 7

S.M. Lewin

7.1 INTRODUCTION

Despite considerable research, the routes of transmission and the basis for the seasonal and pandemic cycles of cholera have not been explained. Those areas of the world where clinical cholera cases appear without apparent importation on a regular seasonal basis are endemic foci for cholera. While cholera transmission can occur in many parts of the world, not all such places permit the inter-epidemic maintenance of *Vibrio cholerae*. Two types of cholera endemic focus exist. Primary endemic foci are present independent of the occurrence of a pandemic, and are found mainly in Bangladesh and India. Secondary endemic foci are only present during a pandemic and are currently found in parts of Africa, the USA, Australia and Europe (Miller, 1984). Common factors between the different foci may give clues as to how *V. cholerae* is maintained in these areas during the non-epidemic part of the year. Low-lying riverine areas are strongly associated with epidemic areas and frequently outbreaks are initiated in fishing communities. The mechanisms for the maintenance of endemicity may also be linked with seasonal variation in the distribution of cholera cases.

In the past 20 years increasing evidence has led to a change in the majority opinion that human cholera is spread by contamination of food and water to a belief that *V. cholerae* O1 is an autochthonous member of the aquatic environment causing disease in humans who have become accidentally exposed. There have been many indications that such an aquatic environmental reservoir exists; however, its nature has not been elucidated.

Cholera and the Ecology of Vibrio cholerae. Edited by B.S. Drasar and B.D. Forrest.
Published in 1996 by Chapman & Hall, London. ISBN 0 412 61220 8.

Evidence for an aquatic reservoir of *V. cholerae* has been derived from laboratory based studies of physicochemical stresses on the survival of *V. cholerae* (Miller *et al.*, 1984) and field studies that have isolated *V. cholerae* from the aquatic environment in the absence of any evident association with clinical cholera (Bashford *et al.*, 1979; Colwell *et al.*, 1977, 1981). *V. cholerae* O1 has been isolated from surface water and sea water (Hughes *et al.*, 1981; Gyobu *et al.*, 1984); plankton (Colwell *et al.*, 1980); shellfish (Dutt *et al.*, 1971; Blake *et al.*, 1977); fish (Pandit and Hora, 1951; Pollitzer, 1959; Merson *et al.*, 1977; Gyobu *et al.*, 1984) and plants (Spira *et al.*, 1981). Furthermore, *V. cholerae* has been shown to survive long periods of low nutrient stress as coccoid ultramicrobacteria, a free-living planktonic form which it is postulated is a resting phase (Xu *et al.*, 1982). However, none of these potential habitats has yet been shown to be an important reservoir of cholera.

This chapter focuses on the ecology of *V. cholerae* O1 and specifically on the role of zoological microhabitats. The various potential aquatic zoological microhabitats will be reviewed and only passing mention will be made of terrestrial animals and birds since, as already mentioned, it is generally accepted that *V. cholerae* is an autochthonous aquatic resident. Also, the evidence collected to date does not support animal-to-animal or human-to-human transmission as a long-term inter-epidemic means of maintenance. While the major concentration will be on *V. cholerae* O1, non-O1 *V. cholerae* will also be discussed as appropriate, since information on the non-O1s may well be relevant to *V. cholerae* O1. Furthermore, the emergence of a new serogroup, *V. cholerae* O139, apparently with pandemic potential, necessitates discussion.

7.2 *VIBRIO CHOLERAE*: O1 AND NON-O1

V. cholerae O1 and *V. cholerae* non-O1 comprise a single taxonomic species as evidenced by numerical taxonomy, DNA/DNA hybridization and nucleic acid sequence data (Colwell and Spira, 1992). Non-O1s have also been shown to produce a similar cholera toxin (CT) to O1 *V. cholerae*; however, most environmental non-O1 isolates produce none or very minor amounts of toxin. They differ also by the expression of different O antigens and *V. cholerae* O1 has been given precedence owing to its historical and epidemiological significance in human disease. Before establishing the close relatedness of *V. cholerae* O1 and non-O1, the importance of knowledge gathered in respect of non-O1 *V. cholerae* for the ecology and epidemiology of *V. cholerae* O1 was not appreciated and it seems likely that there may be similarities in the microhabitat of each. Non-O1 *V. cholerae* are much easier to isolate than O1 from the environment and, indeed, are more frequently isolated. This may possibly be due to some competitive advantage. However, until recently non-O1s were not commonly isolated in the clinical situation, being only a sporadic cause of gastroenteritis. The emergence of a new serotype of *V. cholerae*, O139 with pandemic potential has resulted in a dramatic change in the predominant serotypes causing cholera in parts of Asia and will be discussed at the end of the chapter.

7.3 NON-AQUATIC ANIMALS AS RESERVOIRS FOR *V. CHOLERAE*

The only evidence for domestic animals harbouring *V. cholerae* O1 has been during times when cholera has been in the local community (Sanyal *et al.*, 1974). A total of 195 animals including cows, buffaloes, donkeys, horses, goats and chickens were tested over a year in Varanasi, India. Of the 54 isolates of *V. cholerae* from 21 animals only eight were *V. cholerae* O1. These came from three cows, two dogs and two chickens. In only one cow was *V. cholerae* O1 detected on a second occasion, therefore indicating that there was no long-term carriage. The remaining 46 strains were non-O1, were isolated from cows, goats, dogs and chickens and were found to be excreted over much longer periods than the O1 strains. A separate study of 500 dogs in Calcutta, India, did not isolate *V. cholerae* O1 at all and found only one dog with an antibody titre indicative of recent O1 infection or cross-reactive antibody, although 14% of the dogs tested did carry non-O1 *V. cholerae* (Sack, 1973).

A study in Kent, UK, did not isolate any *V. cholerae* from sheep, cattle, pigs or poultry but did culture *V. cholerae* non-O1 from 6% of gulls sampled (a number of species) and from two mute swans (*Cygnus olor*) (Lee *et al.*, 1982). Non-O1 *V. cholerae* were also isolated from farmed ducks in Denmark (Bisgaard and Kristensen, 1975). *V. cholerae* O1 and non-O1 were isolated from the cloacal swabs and faeces of 20 species of aquatic bird in Colorado and Utah in the USA; two of the species, the Great Blue Heron (*Ardea herodius*) and the Ring-billed gull (*Larus delawarensis*), harboured O1 vibrios while three others yielded isolates that initially typed with group O1 antiserum but subsequently became non-typable (Ogg *et al.*, 1989). It does appear that some aquatic birds can harbour *V. cholerae*, though more frequently of non-O1 serotypes. Such carriage of *V. cholerae* most likely does not represent a long-term reservoir, but is a reflection of diet which is predominantly seafood.

7.4 PLANKTON

Plankton have been studied extensively as a potential aquatic zoological microhabitat for *V. cholerae*. Sochard and colleagues (1979) investigated the surface and gut flora of laboratory-reared *Acartia tonsa* (an estuarine, neritic copepod) and wild collected estuarine and marine copepods (*A. tonsa, Pontellopsis regalis, Pleuromamma* sp., *Labidocera aestive* and *Centropages furcatus*). *Vibrio* sp. predominated in marine and estuarine copepods, while *Pseudomonas* sp. predominated in healthy laboratory specimens. However, when aseptic dissection techniques were used, the laboratory specimens yielded solely *Vibrio* sp.; *V. cholerae*, however, was not among those isolated (Sochard *et al.*, 1979). Subsequently, *V. cholerae* was isolated from plankton samples taken from Bangladesh waters and Chespeake Bay (Colwell *et al.*, 1980). Huq and colleagues (1983) then demonstrated the attachment of *V. cholerae* O1 and non-O1 serovars to the surface, oral region and the egg sac of live chitin-containing copepods in natural water samples. Furthermore, scanning electron microscopy revealed multiplication of the vibrios on the surface, although the highest colonization of vibrios was in the oral region and the egg sac. Attachment did not occur in dead copepods as was shown by the decrease in *V. cholerae* plate counts at about the same time at which the copepods began to die (Huq *et al.*, 1984a). Attachment prolonged survival and was optimal at pH 8.5, 1.5% salinity and a temperature of about 30°C (Huq *et al.*, 1984b). These observations, coupled with the appearance of epidemic cholera in Bangladesh during September or October following an increase in the zooplankton population in the months of August and September, led to the proposal that plankton could be an important factor in the ecology of *V. cholerae* and in the epidemiology of cholera facilitating the inter-epidemic maintenance of the organism (Huq *et al.*, 1984a). Subsequent studies on the attachment of *V. cholerae* O1 to a variety of zooplankton (copepods, cladocerans and rotifers) and phytoplankton demonstrated preferential attachment to molted zooplankton exoskeletons (exuviae), with high numbers of organisms on the body and appendage parts, over whole specimens and equal attachment to whole specimens and exuviae of phytoplankton (Tamplin *et al.*, 1990). Staining of uninoculated zooplankton revealed that endogenous bacteria were also primarily attached to exuviae. *V. cholerae* O1 was, however, not observed in uninoculated specimens.

It is interesting to note that Huq and colleagues (1983, 1984a) observed attachment and multiplication of *V. cholerae* O1 and non-O1 with live copepods, but not with dead ones, while Tamplin and colleagues (1990) found greater attachment to moulted exoskeletons over whole zooplankton. It might have been expected that attachment would be greater to the whole specimens based on the earlier data (Huq *et al.*, 1983, 1984a). It would appear therefore that the association of *V. cholerae* with plankton involves more host factors than simply chitin, although attachment to chitin does appear to aid survival under unfavourable conditions (Nalin *et al.*, 1979).

A subsequent study used the fluorescent-monoclonal antibody staining procedure (FA) for viable but non-culturable *V. cholerae* O1 concomitantly with conventional culture methods to detect *V. cholerae* O1 in environmental samples of water and plankton in Bangladesh (Huq *et al.*, 1990). Using the FA technique, 63.6% of plankton samples taken from several fixed sites over the 3-year duration of the study were positive for *V. cholerae* O1 while none yielded the organism by conventional culture methods. *V. cholerae* O1 was present throughout the year, with seasonal variations, in the water and on plankton as determined by the FA technique. Even at sites where cases of cholera were reported, only 3.2% of plankton samples and 4.3% of water samples were positive by conventional methods while 78.49% of plankton were positive by the FA staining. *V. cholerae* O1 was detected on the surfaces of zooplankton and phytoplankton, though preferentially on copepods, and on intact and fragmented plankton. The results are in agreement with the work of Tamplin and colleagues (1990) and may not disagree with Venkateswaran and colleagues' (1989) investigation of the role of zooplankton in the seasonality of *V. cholerae* non-O1 in Japanese coastal waters. This latter study used only conventional methods of isolation and the results did not support an association with zooplankton in unfavourable conditions to aid survival of *V. cholerae*. The degree of nutrient availability appeared to be more important in the distribution of *V. cholerae*. Since a decline in nutrient concentration may lead to the viable but non-culturable form of *V. cholerae* it would no longer be detected.

7.5 SEAFOOD

7.5.1 VIBRIOS AS PART OF THE INDIGENOUS MICROFLORA OF FISH

(a) Freshwater

Vibrios are not often isolated from freshwater fish from either temperate or tropical climes. In only one study of the intestinal flora of freshwater fish, non-O1 vibrios were commonly cultured from 10 species of fish sampled from the Danube delta (Ciufecu *et al.*, 1979).

(b) Marine water

Members of the Vibrionaceae form part of the indigenous microflora of marine fish in both temperate and tropical climes. *Vibrio parahaemolyticus* is widely distributed in tropical coastal waters and is commonly isolated from healthy fish (Bockemühl and Triemer, 1974) while *Vibrio alginolyticus*, *Vibrio vulnificus* and *V. cholerae* non-O1 have also been associated with fish disease (Austin and Allen-Austin, 1985). Non-O1 cholera vibrios are regularly isolated from seafood and are recognized as important causative agents of gastrointestinal illness (Kay *et al.*, 1984; Piergentili *et al.*, 1984). In 1977, *V. cholerae* non-O1 was first isolated from diseased wild and cultured ayu (*Plecoglossus altivelis*) in Japan and subsequently shown to be pathogenic for ayu (Muroga *et al.*, 1979). Since then, non-O1 cholera vibrios have been shown to be a common and widespread cause of vibriosis in Japan (Kiiyukia *et al.*, 1992). A 3-year study of the ecology of *V. cholerae* non-O1 in Toyama prefecture in Japan gave an isolation rate in excess of 50% from fish during the summer months (July–September), independent of the fish species tested (Kodama *et al.*, 1983). Since non-O1 serovars were isolated year-round from the seawater the researchers did not believe the isolates from the fish simply resulted from contamination, but were part of the normal resident flora.

Fish surface microflora was at one time believed to simply reflect the flora of the surroundings (Colwell, 1962) but Austin (1982) demonstrated a considerable diversity of bacterial types on the surfaces of healthy turbot (*Scophthalus maximus*) with the major number of bacterial taxa isolated from the system found on the fish, rather than in the surrounding water. Horsely (1973) isolated *Vibrio* sp. from the gills of marine-caught Atlantic salmon (*Salmo salar*) but not from the water. Vibrios were shown to be the predominant flora of the digestive tract of the plaice (*Kareius bicoloratus*; Simidu and Hasuo, 1968) and the jackmackerel (*Trachurus japonicus*; Aiso *et al.*, 1968). Separate analysis of the bacterial composition of the stomach and the intestine revealed that *Vibrio, Pseudomonas, Achromobacter, Flavobacterium, Sarcina, Corynebacterium,* and *Bacillus* were all isolated from the stomach, while only *Vibrio* sp. were isolated from the intestine. Since the surrounding sea water contained large numbers of *Pseudomonas* and *Achromobacter*, in addition to vibrios, the fish flora was not just a reflection. MacDonald *et al.* (1986) recorded that about 45% of the gastrointestinal flora of the Dover sole (*Solea solea* L.) comprised *Vibrio* sp. They further demonstrated a preponderance of chitinolytic bacteria and proposed that such bacteria may have a role in the nutrition of the fish host. This is particularly interesting in light of the fact that *V. cholerae* is chitinolytic, and of the research by Huq and colleagues (1983, 1984a) on the predilection of *V. cholerae* for chitin-containing copepods.

7.5.2 *V. CHOLERAE* O1 AND SEAFOOD

The epidemiological importance of fish in cholera was first recognized in Japan by Donitz (1886; in Pollitzer, 1959) who observed the 1877 Tokyo outbreaks. In Haneda, a coastal village through which the major fishing traffic passed on to Tokyo, there was a particularly high incidence of cholera cases, especially among fishermen involved in transporting the catches to the capital. The 1937 report of the Eastern Bureau of the League of Nations Health Organization recorded a similar history (Pollitzer, 1959). In the Philippines the disease was more prevalent among Japanese fishermen and the consumption of poor quality dried fish appeared to be causally associated with the early victims of the 1907 epidemic in Northern Samar (Heiser, 1908).

More recent work on seafood and cholera has comprised reports of isolated food-borne outbreaks in which fish or shellfish have been implicated as the vehicle of transmission (Table 7.1). A variety of seafood has been incriminated in cholera outbreaks, but *V. cholerae* has rarely been isolated from the food item consumed (Merson *et al.*, 1977)

Table 7.1 Cholera outbreaks in which seafood has been implicated

Year	Country	Source of infection	Reference
1961	Philippines	Raw shrimp	Joseph *et al.* (1965)
1969	Malaysia	Shellfish	Dutt *et al.* (1971)
1973	Italy	Raw mussels, seafood	De Lorenzo *et al.* (1974) Baine *et al.* (1974)
1974	Portugal	Shellfish	Blake *et al.* (1977)
1974	Guam	Local raw fish	Haddock (1987)
1977	Guam	Raw oysters	Haddock (1987)
1977	Gilbert Islands	Partially dried raw salt fish, raw seafood	McIntyre *et al.* (1979)
1978	USA, Louisiana	Poorly cooked crabmeat	Blake *et al.* (1980)
1979	Sardinia	Shellfish	Salmaso *et al.* (1980)
1981	USA, Louisiana	Turtle/fish	Shandera *et al.* (1983)
1981	USA, Louisiana	Shrimp	Shandera *et al.* (1983)
1982	Singapore	Squid	Goh *et al.* (1984)
1982	Truk	Mixed raw seafood	Holmberg *et al.* (1984)
1982	Guam	Local raw fish	Haddock (1987)
1984	USA, Maryland	Crab	Lin *et al.* (1986)
1985	Guam	Local raw fish	Haddock (1987)
1985	Guam	Imported raw crab	Haddock (1987)
1986	Guam	Raw oysters	Haddock (1987)
1986	USA, Florida	Raw oysters	Klontz *et al.* (1987)
1986	USA, Atlanta	Raw oysters	Pavia *et al.* (1987)
1986	USA, Louisiana	Crab/shrimp	Lowry *et al.* (1989)
1988	USA, Louisiana	Raw oysters	Anonymous (1989)
1991	USA, New Jersey	Imported crab, Ecuador	Finelli *et al.* (1992)
1991	USA, NEw York	Imported crab, Ecuador	Anonymous (1991)
1991	USA, Florida	Raw oysters and ceviche eaten in Ecuador	Anonymous (1991)

and few studies have investigated the potential association of *V. cholerae* with seafood under controlled conditions.

(a) Cholera outbreaks involving seafood

Seafood incriminated in cholera outbreaks may have been contaminated in the environment before death, or, as a result of handling procedures following death. Unfortunately the majority of studies in which food handling is held responsible for an outbreak of cholera fail to identify the specific food items concerned (Goh *et al.*, 1987). However, in 1978 in Louisiana a cluster of cholera cases were traced to inadequately cooked crabmeat from which *V. cholerae* O1 was cultured (Blake *et al.*, 1980), an outbreak in Singapore in 1982 was attributed to squid prepared by infected food handlers (Goh *et al.*, 1984) and the epidemic of 1982 in Truk resulted from a combination of infected raw seafood and contamination of foods inside the home (Holmberg *et al.*, 1984).

A number of studies from around the world have demonstrated an association between the consumption of raw seafood and the onset of cholera (Joseph *et al.*, 1965; De Lorenzo *et al.*, 1974; McIntyre *et al.*, 1979; Klontz *et al.*, 1987).

Asia

A retrospective investigation of an outbreak in the Philippines in 1961 found that ingestion of a small shrimp, called lipon or alamang, was associated with cholera cases. Cooked shrimps were not implicated and this was further substantiated by the greater numbers of cases among adults as compared with children (who are usually given cooked seafood) during the early stages of the epidemic. The explosive nature of the outbreak with rapid and extensive dissemination followed by a dramatic decline after 1 week also implied a common source or vehicle of infection. Other specific observations were that the initial outbreaks arose in fishing villages or in the coastal areas of large cities and relatively few cases were recorded in inland areas (Joseph *et al.*, 1965).

In the 1969 Kelantan outbreak of cholera in Malaysia, the consumption of raw or partially cooked shellfish was again incriminated in the transmission of *V. cholerae* (Dutt *et al.*, 1971). A considerable number of people were infected with *V. cholerae* O1 and with non-O1s through the agency of shellfish. The spread of cholera among the first recognized cases in the newly infected Tarawa Atoll of the Gilbert Islands was apparently due to the consumption of raw, partially dried, salted lagoon fish (McIntyre *et al.*, 1979). No other raw or cooked seafood, drinking water source, or lagoon-related activity was implicated at that stage. Later in the epidemic the following raw seafoods were all associated with cholera: sardines, clams, lagoon fish, and salt fish. The individual association of these fish, however, could not be determined since people most often ate more than one variety of seafood. The frequent consumption of raw salt fish as compared with infrequent consumption was also related to the onset of illness. There was no correlation between eating raw ocean fish, or any of the seafoods after cooking or the source of drinking water and the occurrence of cholera.

On Guam, a retrospective examination of *V. cholerae* infections over the years 1974–1986 revealed four outbreaks associated with the consumption of local raw fish and two associated with the consumption of raw seafood (crab or fish) imported from the Philippines (Haddock, 1987). All of the isolates were *V. cholerae* biotype El Tor, serotype Ogawa. Only in one instance was the vibrio isolated from the suspect food, a local fish, *Mulloidichthys auriflama* (goatfish), as well as the patient. An environmental survey of Agana Bay from where the fish was caught failed to isolate any further *V. cholerae* O1 from fish, seaweed or water. Four other outbreaks were the result of infection contracted abroad; two from the Philippines by ingestion of raw oysters and one each from Thailand and Truk from unknown sources.

Europe

Anecdotal evidence suggested contaminated shellfish caused the 1974 cholera outbreak in Portugal (Blake *et al.*, 1977). Further investigation led to the isolation of *V. cholerae* from 42% of 154 shellfish samples (comprising 44/114 clams, 11/22 cockles, 7/15 oysters and 2/3 mussels) and 24% of seawater samples taken from the Ria, an area of mud flats and islands. In Tavira, shellfish could not be demonstrated as a cause of transmission, although it seemed probable that the first case was due to eating infected cockles since the patient consumed semi-cooked cockles 3 days before becoming ill and cockles sampled from the same area a month later were shown to be contaminated with *V. cholerae*. In Faro City, cockles were epidemiologically implicated and bottled spring water from a local spring, the Fonte de Bispo, was also likely to be involved in transmission.

Mussels were implicated by De Lorenzo and colleagues (1974) in cholera disseminated in Naples during the 1973 August to October outbreak. Some 31.5% of gastroenteritis patients, admitted during the outbreak who had eaten raw mussels in the 5 days before admission were affected by cholera, while 10.9% of similar patients who had not consumed raw mussels were affected. Furthermore, there was a striking decline in the number of cholera cases once the consumption of mussels was forbidden. While mussels were probably the major source of infection there were three endemic foci; Campania, Puglia and Cagliari

(Sardinia) in which different seafoods likely played a role (Baine *et al.*, 1984). In Campania, the cases followed a holiday mussel feast and the food histories of patients were in agreement with mussels as the source. In Puglia, a variety of raw seafood including fish, shellfish and baby squid was implicated by the food histories and further substantiated by the high concentration of cases near the fishermens' wharves in Bari. In Cagliari, the best evidence for the role of seafood was obtained since *V. cholerae* O1 was isolated from arselle in the Santa Gilla beds. The route of introduction of cholera into the Naples area was unclear but may have been due to infected mussels imported from Tunisia. Bivalves were also indicated in the 1979 epidemic in Sardinia (Salmaso *et al.*, 1980). Of 10 bacteriologically confirmed cases and two asymptomatic excretors, 11 individuals had recently consumed arselle (the term incorporates a number of species of marine bivalves belonging to the class Pelecypoda). *V. cholerae* O1 was isolated from water samples and bivalves in the Santa Gilla lagoon. Since arselle had been incriminated in the 1973 outbreak (Baine *et al.*, 1974) the recurrence of cholera may well have been due to persistence of *V. cholerae* O1 in the environment.

USA

Better evidence for the environmental persistence of *V. cholerae* O1 has been obtained in the United States. Since 1911 cholera was absent from the US but in 1973 the first domestically acquired case occurred in Port Lavaca, Texas. The source of the infection could not be identified and seafood was not incriminated on this occasion (Weissman *et al.*, 1974). Since then there have been a number of outbreaks of cholera of endemic origin in which seafood has been the implicated vehicle of transmission.

The best-characterized outbreak occurred in 1978 when five separate clusters of cholera on the Southwestern Louisiana coast of the Gulf of Mexico were shown to be significantly related to the consumption of crab (Blake *et al.*, 1980). All the infected individuals had eaten crab in the 5 days before falling ill, while matched controls had not. *V. cholerae* O1 was isolated from the refrigerated remains of one crab left over by the patients in one of the clusters. In all instances the crabs had been either boiled or steamed and therefore an investigation of the effect of cooking on the survival of *V. cholerae* O1 in artificially infected blue crabs was carried out. Crabs had to be boiled for at least 10 minutes or steamed for at least 30 minutes so that no vibrios could be isolated. Crabs steamed for a lesser period could appear fully cooked, yet culture *V. cholerae* O1. Environmental sampling isolated *V. cholerae* O1 from shrimp and from two Moore swabs, but not from crabs or oysters.

In 1981, two independent cases of cholera occurred in southeastern

Texas, close to the Gulf of Mexico (Shandera *et al.*, 1983). In both instances it was not established how the infection was contracted. In the first case *V. cholerae* O1 was isolated from mud and shallow standing water of unknown origin beside the house; however, the patient had also eaten partially cooked turtle and fish caught in the bayous near his home a few days before falling ill. While the turtle and fish were unavailable for testing, that they were partially cooked does not eliminate them from suspicion, as evidenced by the 1978 Louisiana outbreak when inadequately cooked crab was shown to be the source of cholera infection (Blake *et al.*, 1980). With the second individual, frozen shrimp were suggested but culturing of other shrimp from the same lot was negative. In 1984 a single case of cholera occurred in Maryland, again in which no definitive source of infection was traced (Lin *et al.*, 1986). However, the individual had consumed crab cakes made from Texas Gulf Coast crabs and cooked shrimp imported from French Guiana and packed in Tampa, Florida, 2 days before developing cholera. The crab was deemed the most probable source, partly because of the various unique characters of the *V. cholerae* O1 isolate shared with other United States isolates (see Table 7.1).

In 1986 an outbreak involving 18 people in 12 clusters occurred in Louisiana (Lowry *et al.*, 1989). A further two cases were exported to Florida and Georgia via raw oysters. In the Florida case, oysters were implicated as the vehicle of transmission although none of them was available for testing. The oysters were consumed on two separate occasions and were traced to Louisiana and possibly Texas (Klontz *et al.*, 1987). In Atlanta a single individual was infected with toxigenic *V. cholerae* O1. However, five eating companions who also consumed oysters were not affected, no other cases were identified and the organism could not be isolated from oyster meats obtained from the implicated oyster beds in Texas (Pavia *et al.*, 1987). During August to October of 1986, 18 cases of cholera occurred in disparate locations in southern Louisiana. None of the individuals had travelled abroad in the past year. Stool cultures from seven individuals yielded toxigenic *V. cholerae* O1 biotype El Tor, serotype Inaba, the remaining 11 did not have cultures done but gave high vibriocidal antibody titres, although five of these were asymptomatic. While no common source was conclusively identified, 17/18 ate crabs or shrimp within 3 days of the onset of symptoms and inadequate cooking or improper handling appeared to play a significant role. The seafood was harvested from a wide variety of sites along the Louisiana coast of the Gulf of Mexico, but *V. cholerae* O1 could not be isolated from raw, boiled processed crabs, raw or cooked shrimp, raw shucked or unshucked or cooked oysters, or water samples taken from multiple sites (Lowry *et al.*, 1989). *V. cholerae* non-O1 and other *Vibrio* sp. were, however, routinely recovered.

In August 1988 a further case of cholera associated with the consumption of raw oysters from the Louisiana coast of the Gulf of Mexico was identified in Colorado. Of nine people who ate oysters from the same batch, only one individual fell sick and cultured *V. cholerae* O1. Phage typing showed the organism to be identical to all the others associated with the Gulf of Mexico endemic focus (Anonymous, 1989). A further five cases of cholera associated with the consumption of oysters from the Gulf of Mexico were identified during August to October of 1988 in various US states (Anonymous, 1989).

In all of the aforementioned cases the outbreaks were caused by haemolytic, toxigenic *V. cholerae* O1 biotype El Tor serotype Inaba isolates. Nearly all of the pandemic strains being isolated elsewhere in the world were poorly or non-haemolytic by that period, indicating that the US strains were of endemic origin. Further indications of their common source and endemic origin were the possession of identical DNA restriction enzyme patterns, phage sensitivities and CT gene sequences with the exception of the 1986 oyster isolate from Florida/Louisiana (Klontz *et al.*, 1987) (Table 7.2). All the other isolates hybridized with two CT gene restriction fragments (6 and 7 kb) while the Florida/Louisiana isolate only hybridized to the 6 kb fragment. Whether this isolate is an evolutionarily early form of the Gulf coast strains, a mutant or a recent introduction is open to speculation. The strong basis of similarity between the other isolates does indicate that the Gulf Coast provides a long-term reservoir for this strain.

South America

In January 1991 an outbreak of cholera started in Chancay and Chimbote, just north of Lima in Peru, which marked the start of epidemic cholera in South America (Anonymous, 1991a). The outbreaks spread rapidly along the Peruvian coast and in 1992 many other Latin American countries were quickly affected including, Ecuador, Columbia, Brazil, Mexico and Chile. How or why epidemic cholera suddenly appeared in South America in the 1990s is not known. Clinical and environmental surveys conducted before 1991, and in particular in the 1970s, failed to isolate toxigenic *V. cholerae* O1 at all, although a few non-toxigenic biotype El Tor *V. cholerae* O1 were detected, indicating that the isolation methods used were sufficiently sensitive (Gangarosa and Tauxe, 1992). The indications have been that seafood may be a significant vehicle for cholera transmission. The consumption of raw or nearly raw seafood, frequently as ceviche (which uses varying amounts of lemon juice), is highly popular in Latin America (Tauxe and Blake, 1992). Public health education in

Table 7.2 Unique characters shared by USA Gulf Coast isolates of *Vibrio cholerae* O1

Year	Location	Source	Common characters unique to Gulf Coast isolates*	Reference
1973	Port Lavaca	Texas	tox^+/Inaba/El Tor/ Haemolytic CT gene/ phage sensitivity/ RE pattern	Weissman *et al.* (1974)
1978	S. Louisiana	Louisiana	tox^+/Inaba/El Tor/ Haemolytic CT gene/ phage sensitivity/ RE pattern	Blake *et al.* (1980)
1981	Jefferson County	Texas	tox^+/Inaba/El Tor/ Haemolytic CT gene/ phage sensitivity/ RE pattern	Shandera *et al.* (1983)
1984	Maryland	Texas	tox^+/Inaba/El Tor/ Haemolytic CT gene/ RE pattern	Lin *et al.* (1986)
1986	Atlanta	Louisiana	tox^+/Inaba/El Tor/ Haemolytic CT gene	Pavia *et al.* (1987)
1986	Florida‡	Louisiana	tox^+/Inaba/El Tor/ Haemolytic RE pattern/phage sensitivity	Klontz *et al.* (1987)

*CT gene unique *Hind*III restriction site.
‡ oyster isolate – hybridized only with single 6 kb fragment of CT.
CT, cholera toxin; RE, restriction endonuclease pattern.
tox^+ = toxigenic.

Peru stressed with success that raw seafood should be avoided, including ceviche. As a consequence few infected patients or matched controls reported consumption of raw seafood. By contrast, in Ecuador, no such education was undertaken and as a result investigations have shown cholera to be associated with the consumption of crab, raw fish and raw seafood in general (Blake, 1993). There is, however, a lack of studies to demonstrate the nature of the association, whether certain varieties of seafood provide a long-term environmental reservoir for *V. cholerae* O1 or there is simply contamination. A survey conducted over 6 months (March to August) in Peru isolated *V. cholerae* O1 from the skin and intestines of finfish from polluted coastal waters around Lima and also in plankton samples and molluscs (Tamplin and Parodi, 1991). Both culture methods and fluorescent monoclonal antibody tests were used for detection. Whether fish, plankton or molluscs were sampled at the other sites tested is not stated.

Since 1991 a number of cases of cholera have arisen in the US resulting from the consumption of seafood imported from South America. In 1991, crab meat brought in by a traveller from Ecuador was implicated in an outbreak of cholera among a group of New Jersey residents. Crab salad was the only food consumed in common by the infected individuals and no-one who did not eat the crab exhibited any symptoms. Unfortunately none of the salad remained for testing. The crab had been boiled and shelled before being taken to the US, but the boiling was for an unspecified period of time and the crab was not refrigerated during transportation. Therefore it is unclear whether the crab became infected before harvest, at purchase or after cooking (Finelli *et al.*, 1992). In an unrelated outbreak four further cases of cholera associated with the consumption of crab meat imported from Ecuador occurred in New York in 1991. The crab had been purchased in Ecuador, boiled and shelled and the meat and claws frozen, then carried back in a suitcase and on arrival in New York refrozen, not having thawed. A salad was made after thawing the crab in a double boiler for 15–20 minutes without further cooking (Anonymous, 1991c). *V. cholerae* O1 could not be cultured from the remaining crab, but CT was detected by the use of polymerase chain reaction (PCR). Infection of the crab by the index case was not deemed likely as it was not indicated by the circumstances. Another imported case of cholera occurred in Florida in a woman who returned from Ecuador where she had consumed raw oysters and ceviche (Anonymous, 1991b).

(b) Ecological studies of cholera in seafood

Despite numerous reports implicating shellfish or fish in the transmission of cholera, controlled studies to determine the nature and significance of the association have been scarce and results tend to be inconclusive.

A study of the incidence of *Vibrio* sp. isolated from blue crabs (*Callinectes sapidus*) collected over 1 year from Galveston Bay, Texas determined a significant difference in the internal and external flora of the crabs in respect of *Vibrio* sp. with vibrios the predominant flora in the haemolymph of most crabs (Davis and Sizemore, 1982). Since the flora most closely resembled that of the digestive tract the researchers postulated either that it might be the source, or that the haemolymph is selective for vibrios. The predominant isolates were *V. parahaemolyticus* and *V. vulnificus*, occurring in the highest numbers during the summer and autumn. An organism biochemically identical to *V. cholerae* but not typeable by either O1 or non-O1 antisera was detected in 3.5% of crabs sampled in the winter and spring months.

The preferential colonization of the hindgut of the blue crab by *V. cholerae* has been demonstrated in a separate study (Huq *et al.*, 1986). The hindgut of crustacea possesses a chitin lining derived from the ectodermal epithelium, whereas the midgut is endodermal in origin and is not chitinaceous (K.H. Chu, PhD thesis, 1984, cited in Huq *et al.*, 1986). These results correlate with the findings of Huq and colleagues (1983) that *V. cholerae* attached to chitinaceous plankton, thus prolonging survival of the vibrio and the findings of Tamplin and colleagues (1990) that *V. cholerae* attaches preferentially to exuviae of zooplankton over whole zooplankton. However, another investigation of attachment to live versus dead copepods demonstrated a preference for live copepods, indicating the importance of other factors in attachment (Colwell *et al.*, 1985) (see pp. 202–206).

The affinity of *V. cholerae* for chitin might also facilitate its survival within the intestines of fish. In an investigation of the intestinal flora of the Dover sole (*Solea solea* L.) it was suggested that the gut flora might be involved in the breakdown of substrates, including chitin, and hence in the nutrition of the fish host (MacDonald *et al.*, 1986). *Vibrio* sp. comprised 45% of the intestinal flora and many of the organisms isolated were chitinolytic. Certainly, in recent years *V. cholerae* non-O1 have been isolated with increasing frequency as a common and widely distributed cause of fish disease in Japan (Kiiyukia *et al.*, 1992).

Early studies of the role of fish in the environmental survival of *V. cholerae* were conducted primarily in *Hilsa ilisha* and were inconclusive. Extensive surveys of *Hilsa* fish in the River Hooghly did not isolate *V. cholerae* O1 from any specimens, although non-O1 vibrios were commonly isolated (Krishnan, 1953; Lahiri, 1954). However, artificially infected climbing perch (*Anabas testudineus*) and murrel (*Opicephalus punctatus*) excreted vibrios (O1 or non-O1) for up to 30 days after infection, though after that period only non-O1 organisms could be cultured from dissected fish (Pillay *et al.*, 1954).

In a more recent study Wang *et al.* (1987) investigated experimental infection of loach (*Misgurnus anguillicaudatus*), finless eel (*Monopterus albus*), Crucian carp (*Carassius auratus*) and snakehead fish (*Ophicephalus argus*) with a strain of *V. cholerae* O1 biotype El Tor. *V. cholerae* could be isolated from pond water containing loaches or finless eels for up to 6–7 months, despite weekly water changes and from the gallbladder and intestine of these two fish species. *Vibrio* was not isolated from water in vats containing carp or snakehead fish beyond 2 weeks, or from these species of fish at all. Although the salinity of the water is not given it seems likely that it is freshwater since the species of fish they investigated were all freshwater species.

Studies conducted in this laboratory on the survival of *V. cholerae* O1 in association with a species of Tilapia (*Oreochromis spilurus* ssp. *spilurus*) in low-salinity microcosms (0.05% and 0.1% sea salt) at pH 8.5 and 25°C demonstrated only a marginal improvement in survival. The presence of fish or fish products (microcosms in which fish had previously been maintained) extended the period for which *V. cholerae* could be cultured from low-salinity waters by only days. However, the removal of fish from microcosms resulted in an increased rate of decline of *V. cholerae* cultured from the water and transfer of contaminated fish led to inoculation of the fresh tank with similar survival periods to that observed for single tank experiments. The results indicated that an association with Tilapia does not entirely overcome the adverse effects of low salinity on *V. cholerae* and that it is unlikely that a long-term environmental association would occur.

The results of such studies indicate that the association of *V. cholerae* O1 with fish would appear to be species-specific. Although the Tilapia study did show some improvement in survival, culture of *V. cholerae* from the fish invariably required enrichment procedures, despite the use of a high inoculum (5×10^6 organisms/ml). Wang and colleagues (1987) used a much lower inoculum size (10^2–10^3 organisms/ml) yet were able to isolate *V. cholerae* from two of the species they investigated after 6 months by direct plate culture methods. Neither study used fluorescent monoclonal antibody techniques, so persistence in a viable but non-culturable form was never determined.

Oysters have also been investigated as a source of *V. cholerae*, since they are consumed raw and therefore would seem likely to impose an increased risk of infection. In a year-long study of 790 freshly harvested oyster shellstocks 111 yielded 611 strains of *V. cholerae* non-O1 and 7 yielded 11 strains of *V. cholerae* O1. Over half of the non-O1-positive shellstocks were harvested in the months of June to August with 50% of the non-O1 isolates occurring in July. All seven O1-positive shellstocks were harvested in May to August (Twedt *et al.*, 1981). However, none of the isolates was toxigenic, although two non-O1 *V. cholerae* did induce fluid accumulation in infant rabbits. Another investigation tested the effect of the cell-free haemolymph, or serum, from the Eastern oyster (*Crassostrea virginica*) on a variety of bacterial species including *Vibrio* sp., *Pleisiomonas* sp., *Pseudomonas* sp., *Aeromonas* sp. and *E. coli*. Only *V. cholerae* was agglutinated by the serum. Agglutination was observed with all biovars and serovars of *V. cholerae* O1, but not with all *V. cholerae* non-O1. The significance of these data for the ecology of *V. cholerae* is uncertain, but it may indicate a lesser ability to colonize or invade oyster tissues by *V. cholerae* O1 as compared with non-agglutinable *Vibrio* sp. and non-O1 *V. cholerae* (Tamplin and Fisher, 1989).

7.6 AMOEBAE

7.6.1 BACTERIAL SYMBIONTS

It is established that amoebae use bacteria as a food source; however, symbiotic relationships between infecting bacteria and their amoebic hosts have been demonstrated on a number of occasions although the identity of the bacterial species has not always been established (Jeon, 1972; Hall and Voelz, 1985; Mirelman, 1987). In such systems the bacteria can have a profound effect on the activity of the amoebae. For example, the initial infection of *Amoeba discoides* with a rod-shaped bacterium was harmful to the amoeba, but after a number of years the amoeba became dependent on the infecting organism for its normal functions. The bacterium appeared to be an obligate intracellular parasite since it could not be cultured extracellularly under the conditions tried (Jeon, 1972). The virulence of *Entamoeba histolytica*, the causative agent of amoebic dysentery, is increased by association with certain species of bacteria, including *Escherichia coli*, that are capable of attaching to the amoeba (Mirelman, 1987).

7.6.2 HUMAN PATHOGENS AND AMOEBAE

The pathogen *Legionella pneumophila*, causative agent of Legionnaires' disease is commonly isolated from environmental sites as diverse as hospital showerheads (Cordes *et al.*, 1981), cooling towers, humidifiers (Bartlett *et al.*, 1983), mud, streams (Morris *et al.*, 1979) and lakes (Fliermans *et al.*, 1981). However, its fastidious nature in laboratory culture combined with evidence that legionellae infect macrophages in man led some researchers to postulate that it is not a free-living aquatic organism (Tison *et al.*, 1980) and to investigate the possibility that it might be associated with free-living amoebae (Rowbotham, 1980). Subsequent investigation has supported this contention as *L. pneumophila* has been shown to infect and multiply within amoebae of the genera *Acanthamoeba* and *Naegleria* (Rowbotham, 1980; Anand *et al.*, 1983; Holden *et al.*, 1984) leading to the suggestion that amoebae form the primary reservoir of *L. pneumophila* in the environment. The nature of the infection caused by *L. pneumophila* in amoebae is influenced by the temperature of incubation. Anand and colleagues (1983) demonstrated that co-culture of *L. pneumophila* serogroup 1 and *Acanthamoeba palestinensis* at 35°C resulted in the intracellular multiplication of the bacteria, leading to the destruction of the amoebic host, while at 22°C the acanthamoebae phagocytosed and digested the majority of the legionellae, although some viable bacteria did remain. This might indicate that a long-term association between amoebae and *L. pneumophila* can occur in the natural environment.

King and colleagues (1988) demonstrated the increased resistance of coliforms and certain pathogenic bacteria to free chlorine residuals on ingestion by protozoa. Differences in chlorine resistance were observed between the protozoa tested for the same bacterial species. For example, chlorine resistance was significantly higher for *Citrobacter freundii, Enterobacter agglomerans* and *Klebsiella pneumoniae* when associated with *Acanthamoeba castellani* (amoeba) as compared with *Tetrahymena pyriformis* (ciliate). Undigested bacteria could be observed by electron microscopy for as long as 24 hours after initial chlorine exposure and many of the coliforms actually increased in numbers.

7.6.3 *V. CHOLERAE* AND AMOEBAE

The results of an investigation in this laboratory have shown that *V. cholerae* O1 is capable of survival within amoebae (Thom *et al.*, 1992). The presence of *Acanthamoeba polyphaga* or *Naegleria gruberi* trophozoites increased the survival of *V. cholerae* over that of control microcosms lacking amoebae. Different strains of *V. cholerae* isolated from environmental and clinical sources not only survived, but multiplied, in microcosms containing amoebae trophozoites over a period of 24 hours. Control cultures of *V. cholerae* incubated alone exhibited a decline in viable count over the same period. When investigation of the survival of *V. cholerae* extended over a longer period the vibrio was shown to persist in higher numbers in amoebae co-cultures than in controls. *V. cholerae* could still be cultured from amoebae co-cultures after 8 days when many of the controls were negative. No single set of conditions appeared to favour survival of the vibrio within amoebae. The effect of encystment by amoebae upon the survival of ingested vibrios was also investigated. Amoebae were induced to encyst after ingestion of *V. cholerae* and then treated with 0.5% v/v hydrochloric acid to destroy all remaining trophozoites or free *V. cholerae*. Cysts were then placed in a decystment medium for up to 96 hours and the emergent trophozoites subjected to a single freeze–thaw cycle and viable counts of *V. cholerae* made. *V. cholerae* was shown to survive within cysts of *N. gruberi* though not in *A. polyphaga* while all control cultures were sterile. Again, no specific conditions appeared to favour survival, although incubation at the lower temperatures (25°C and 30°C) did result in more frequent isolations of *V. cholerae* over 37°C.

Free-living amoebae are ubiquitous in the environment and could provide a protective niche under unfavourable conditions and an ideal means of dispersal. Although the pathogenesis of *V. cholerae* is not manifested by intracellular invasion, unlike other enteric pathogens of man, such as shigellae, enteroinvasive *Escherichia coli*, salmonellae and

Yersina enterocolitica (Formal *et al.*, 1983) an invasive ability would not be necessary since the amoebae would effect the internalization of the vibrio. A study by Palmer and Falkow (1986) on the phylogenetic relationship between *Chlamydia*, obligate intracellular parasites, and other bacteria demonstrated nucleotide sequence homology with *Ch. psittaci*, *V. cholerae* and *Ps. aeruginosa*. Viewing the results of both studies in conjunction it seems that *V. cholerae* is indeed capable of an intracellular habitat.

7.7 FREE-LIVING *V. CHOLERAE*

The miniaturization of bacteria in response to starvation has been observed by a number of researchers (Novitsky and Morita, 1976; Kjelleberg *et al.*, 1982). *V. cholerae* too has the ability to survive long periods of low nutrient stress as coccoid ultramicrobacteria, formed by reductive division and size reduction, which can be restored to normal size and shape by nutrient supplementation (Hood *et al.*, 1986). Investigation of these non-growing or starved marine bacteria, including *V. cholerae*, has revealed that they are active and utilize exogenous as well as endogenous substrates in the maintenance and reorganization of cellular processes during periods of non-growth (Kjelleberg *et al.*, 1987). The term 'dormant' would not seem to be applicable and the phrase 'viable but non-culturable' has been coined owing to the difficulty in culturing such organisms. There is evidence with some species of bacteria that dwarfing and starving cells adhere more effectively to surfaces than do normal cells (Dawson *et al.*, 1981). It is possible therefore that such forms may be associated with animate or inanimate surfaces and have not been detected owing to the culture methods used. Using a monoclonal fluorescent antibody technique, Huq and colleagues (1990) were able to detect far higher numbers of *V. cholerae* in Bangladesh waters and in association with plankton than with conventional culture methods (see pp. 202–219). The use of this technique has been limited so far and it would seem to be of value to apply it to the investigation of other potential surface associations and zoological microhabitats.

7.8 *V. CHOLERAE* O139: A NEW PANDEMIC STRAIN?

It is impossible to conclude without at least briefly addressing the recent emergence of a highly pathogenic new serotype of *V. cholerae*.

In October 1992, there was a sudden increase in the number of patients reporting to hospitals in and around Madras with an Asiatic cholera-like diarrhoea followed swiftly by similar large outbreaks in Madurai, Vellore, Amravati, Visakhapatanan, Nagpur and Mysore.

Calcutta was affected by the end of November and Southern Bangladesh reported a large epidemic in January 1993 which soon spread to the areas around Dhaka, possibly as a consequence of the 'Bishwa Istama', a massive religious Islamic gathering, and peaked in March (Albert, 1993; Rabbani and Mahalanabis, 1993). The timing of the epidemic represented a shift in the cholera season, since the peaks for the El Tor biotype occur during September to November with a second minor peak during March to April and for the Classical biotype during November to January with another small peak in March to April. A further departure from the norm was the predominance of cases among adults as opposed to children, which is indicative of the emergence of a new epidemic strain in a population of virgin immunity. Indeed, despite the similarity of the disease to O1 cholera, non-O1 *V. cholerae* were isolated from the majority of cases. When the isolates were further analysed it was discovered that they did not belong to any of the 138 non-O1 serogroups established for *V. cholerae*. Serogroup O139 was assigned, and the Indian and Bangladesh strains were tested against the O139 antiserum and shown to belong to the same grouping (Shimada *et al.*, 1993).

In contrast to other *V. cholerae* non-O1s this new serogroup produces CT in similar high quantities to *V. cholerae* O1. Furthermore, the organism retains many genetic similarities to *V. cholerae* O1 and in particular seems to more closely resemble the El Tor biotype. Hall and colleagues (1993) demonstrated strong homology of the O139 CT to the El Tor CT and showed that the restriction fragment pattern of genomic DNA from O139 closely resembles that of *V. cholerae* O1 biotype El Tor, indicating a close relationship between the two strains. No other non-O1 gave a similar pattern and even the Classical biotype showed marked differences to the El Tor biotype. Experiments to determine environmental regulation of virulence factor expression also indicated a relatedness to *V. cholerae* O1 biotype El Tor, leading Hall and colleagues (1993) to conclude that O139 is an O antigen mutant of an El Tor strain. Nevertheless, immunity to O1 El Tor does not appear to confer any immunity to O139, as shown by the high incidence of disease in the adult population. Others reported a case of bacteraemia caused by O139 (Jesudason *et al.*, 1993), not a characteristic of *V. cholerae* O1 of either biotype, but more a feature of non-O1s. This could be an isolated incident or an indication of non-O1 origins. Further characterization of the organism is definitely required.

What of the epidemiology and ecology of this new pathogen? As yet very little is known. A significantly higher incidence was observed among males as opposed to females, which may provide clues as to the epidemiology. Water appears to play a major role in transmission since *V. cholerae* O139 has been isolated from a variety of water sources

(Islam *et al.*, 1993). The apparent relatedness of the organism to *V. cholerae* O1 biotype El Tor would imply a similar ecology to *V. cholerae* O1, although differences in seasonality may be a contradiction. Higher isolation rates for O139 (12% of water samples were positive for O139 as opposed to 1% for O1) with 100% of isolates being positive for CT as compared with <1% of non-O1 isolates, plus the decrease in cholera caused by O1 would also indicate that O139 is more resilient than O1 (Islam *et al.*, 1993).

In Bangladesh (Islam *et al.*, 1993) and India (Jesudason and John, 1993) the *V. cholerae* O139 serotype has superseded *V. cholerae* O1. Furthermore, cases have been reported in Nepal, Thailand, Malaysia, Pakistan, China and Saudi Arabia (Albert, 1993). Obviously studies of the ecology and epidemiology of *V. cholerae* must now include O139 in order to mitigate the considerable pandemic potential of this new pathogen.

7.9 SUMMARY

The ecology of *V. cholerae* O1 has been under investigation for many years, yet still has not been conclusively elucidated. An environmental habitat is now widely accepted but its nature is not resolved. It is possible that a series of organisms and factors play a role in the inter-epidemic maintenance of *V. cholerae*. Both protozoa and zooplankton form part of the diet of fry of certain fish species, and shellfish and could therefore provide a means of infection of such animals. Alternatively – or in addition – starving forms of *V. cholerae* (the viable but non-culturable form) may adhere to such surfaces and revert to a culturable organism with the availability of nutrients. Plankton have been most extensively studied for an association with *V. cholerae* but there is a lack of controlled studies of the role of fish or shellfish in the transmission of cholera and the interrelation of different potential zoological microhabitats needs to be addressed. The FA staining procedure for viable but non-culturable *V. cholerae* should be incorporated in such investigations.

7.10 REFERENCES

Aiso, K., Simidu, U. and Hasuo, K. (1968) Microflora in the digestive tract of inshore fish in Japan. *Journal of General Microbiology*, **52**, 361–4.

Albert, M.J. (1993) Personal reflections on the discovery of *Vibrio cholerae* O139 synonym Bengal: a tribute to team work and international collaboration. *Journal of Diarrhoeal Diseases Research*, **11**, 207–10.

Anand, C.M., Skinner, A.R., Malic, A. and Kurtz, J.B. (1983) Interaction of *L. pneumophila* and a free-living amoeba (*Acanthamoeba palestinensis*). *Journal of Hygiene (Cambridge)*, **91**, 167–78.

Anonymous (1989) Toxigenic *Vibrio cholera* O1 infection acquired in Colorado. *Morbidity and Mortality Weekly Report*, **38**, 19–20.

Anonymous (1991a) Cholera – Peru 1991. *Morbidity and Mortality Weekly Report*, **40**, 287–9.

Anonymous (1991b) Cholera – New Jersey and Florida. *Morbidity and Mortality Weekly Report*, **40**, 287–9.

Anonymous (1991c) Cholera – New York 1991. *Morbidity and Mortality Weekly Report*, **40**, 516–18.

Austin, B. (1982) Taxonomy of bacteria isolated from a coastal, marine fish-rearing unit. *Journal of Applied Bacteriology*, **53**, 253–68.

Austin, B. and Allen-Austin, D. (1985) A review. Bacterial pathogens of fish. *Journal of Applied Bacteriology*, **58**, 483–506.

Baine, W.B., Zampieri, A., Mazzotti, M., Angioni, G., Greco, D., Di Gioia, M., Izzo, E., Gangarosa, E.J. and Pocchiari, F. (1974) Epidemiology of cholera in Italy in 1973. *Lancet*, **ii**, 1370–4.

Bartlett, C.L.R., Kurtz, J.B., Hutchison, J.G.P., Turner, G.C. and Wright, A.E. (1983) Legionella in hospital and hotel water supplies. *Lancet*, **ii**, 1315.

Bashford, D.J., Donovan, T.J., Furniss, A.L. and Lee, J.V. (1979) *Vibrio cholerae* in Kent. *Lancet*, **ii**, 436–7.

Bisgaard, M. and Kristensen, K.K. (1975) Isolation, characterisation and public health aspects of *Vibrio cholerae* NAG isolated from a Danish duck farm. *Avian Pathology*, **4**, 271–6.

Blake, P.A. (1993) Epidemiology of cholera in the Americas. *Gastroenterology Clinics of North America*, **22**, 639–60.

Blake, P.A., Rosenberg, M.L., Bandiera Costa, J., Soares Ferriera, P., Levy Guimares, C. and Gangarosa, E.J. (1977) Cholera in Portugal, 1974. I. Modes of transmission. *American Journal of Epidemiology*, **105**, 337–43.

Blake, P.A., Allegra, D.T., Snyder, J.D. *et al.* (1980) Cholera – a possible endemic focus in the United States. *New England Journal of Medicine*, **302**, 305–9.

Bockemühl, J. and Trimer, A. (1974) Ecology and epidemiology of *Vibrio parahaemolyticus* on the coast of Togo. *Bulletin of the World Health Organization*, **51**, 353–60.

Ciufecu, C., Nacescu, N., Florian, V., Cvasniuc, T. and Strusievics, B. (1979) Fermentation and non-fermentative gram negative oxidase positive bacteria isolated from freshwater fishes. Note II. *Archives Roumains de Pathologie Experimentale et de Microbiologie*, **38**, 295–8.

Colwell, R.R. (1962) The bacterial flora of Puget Sound fish. *Journal of Applied Bacteriology*, **25**, 147–58.

Colwell, R.R. and Spira, W.M. (1992) The ecology of *Vibrio cholerae*, in *Cholera*, (eds D. Barua and W.B. Greenough III), Plenum Medical Book Company, New York, pp. 107–27.

Colwell, R.R., Kaper, J. and Joseph, S.W. (1977) *Vibrio cholerae*, *Vibrio parahaemolyticus* and other vibrios: occurrence and distribution in Chesapeake Bay. *Science*, **198**, 394–6.

Colwell, R.R., Kaper, J., Seidler, R. *et al.* (1980) Isolation of O1 and non-O1 *Vibrio cholerae* from estuaries and brackish water environments, in *Proceedings of the Fifteenth Joint Conference on Cholera*, The US–Japan Cooperative Medical Science Program, pp. 44–60.

Colwell, R.R., Seidler, R., Kaper, J. *et al.* (1981) Occurrence of *Vibrio cholerae* serotype O1 in Maryland and Louisiana estuaries. *Applied and Environmental Microbiology*, **41**, 555–8.

Colwell, R.R., Singleton, F.L., Huq, A., Xu, H.-S. and Roberts, N. (1985) Ecology of *Vibrio cholerae, Vibrio parahaemolyticus* and related vibrios in the natural environment, in *Bacterial Diarrheal Diseases*, (eds Y. Takeda and T. Miwatani), KTK Scientific Publishers, Tokyo, pp. 273–95.

Cordes, L.G., Wiesenthal, A.M., Gorman, G.W. *et al.* (1981) Isolation of *Legionella pneumophila* from hospital shower heads. *Annals of Internal Medicine*, **94**, 107–7.

Davis, J.W. and Sizemore, R.K. (1982) Incidence of *Vibrio* species associated with blue crabs (*Callinectes sapidus*) collected from Galveston Bay, Texas. *Applied and Environmental Microbiology*, **43**, 1092–7.

Dawson, M.P., Humphrey, B.A. and Marshall, K.C. (1981) Adhesion: a tactic in the survival strategy of a marine vibrio during starvation. *Current Microbiology*, **6**, 195–9.

DeLorenzo, F., Scoscia, M., Manzillo, G. and Balestrieri, G.G. (1974) Epidemic of cholera El Tor in Naples, 1973. *Lancet*, **ii**, 669.

Dutt, A.K., Aiwi, S. and Velauthan, T. (1971) A shellfish borne cholera outbreak in Malaysia. *Transactions of the Royal Society of Tropical Medicine and Hygiene*, **65**, 815–18.

Finelli, L., Swerdlow, D., Mertz, K., Ragazzoni, H. and Spitalny, K. (1992) Outbreak of cholera associated with crab brought from an area with epidemic disease. *Journal of Infectious Diseases*, **166**, 1433–5.

Fliermans, C.B., Cherry, W.B., Orrison, L.H., Smith, S.J., Tison, D.L. and Pope, D.H. (1981) Ecological distribution of *Legionella pneumophila. Applied and Environmental Microbiology*, **41**, 9–16.

Formal, S.B., Hale, T.L. and Sansonetti, P.J. (1983) Invasive enteric pathogens. *Reviews of Infectious Diseases*, **3**, 5702–7.

Giangarosa, E.J. and Tauxe, R.V. (1992) Epilogue. The Latin American cholera epidemic, in *Cholera*, (eds D. Barua and W.B. Greenough III), Plenum Medical Book Company, New York, pp. 351–8.

Goh, K.T., Lam, S., Kumarapathy, S. and Tan, J.L. (1984) A common source foodborne outbreak of cholera in Singapore. *International Journal of Epidemiology*, **13**, 210–15.

Goh, K.T., Lam, S. and Ling, M.K. (1987) Epidemiological characteristics of an institutional outbreak of cholera. *Transactions of the Royal Society of Tropical Medicine and Hygiene*, **81**, 230–2.

Gyobu, Y., Kodama, H., Vetake, H. and Katsuda, S. (1984) Studies on the enteropathogenic mechanism of non-O1 *Vibrio cholerae* isolated from the environment and fish in Toyama prefecture. *Microbiology and Immunology*, **28**, 735–45.

Haddock, R.L. (1987) Cholera in a Pacific island. *Journal of Diarrhoeal Diseases Research*, **5**, 181–3.

Hall, J. and Voelz, H. (1985) Bacterial endosymbionts of *Acanthamoeba* sp. *Journal of Parasitology*, **71**, 89–95.

Hall, R.H., Khambaty, F.M., Kothary, M. and Keasler, S.P. (1993) Non-O1 *Vibrio cholerae. Lancet*, **342**, 430.

Heiser (1908) As quoted in *Cholera Monograph Series*, **43**, World Health Organization, Geneva, pp. 859.

Holden, E.P., Winkler, H.W., Wood, D.O. and Leinbach, E.D. (1984) Intracellular growth of *Legionella pneumophila* within *Acanthamoeba castellanii* Neff. *Infection and Immunity*, **45**, 18–24.

Holmberg, S.D., Harris, J.R., Kay, D.E., Hargett, N.T., Parker, R.D.R., Kansou, N., Rao, N.U. and Blake, P.A. (1984) Foodborne transmission of cholera in Micronesian households. *Lancet*, **i**, 325–7.

Hood, M.A., Guckert, J.B., White, D.C. and Deck, F. (1986) Effect of nutrient deprivation on lipid, carbohydrate, DNA, RNA and protein levels in *Vibrio cholerae*. *Applied and Environmental Microbiology*, **52**, 788–93.

Horsely, R.W. (1973) The bacterial flora of the Atlantic salmon (*Salmo salar*) in relation to its environment. *Journal of Applied Bacteriology*, **36**, 377–86.

Hughes, J.M., Boyce, J.M., Levine, R.J., Khan, M., Aziz, K.M.A., Huq, M.I. and Curlin, G.T. (1981) Epidemiology of El Tor cholera in rural Bangladesh: Dacca: 1981; *International Centre for Diarrhoea Diseases Research*, Scientific Report no. 52.

Huq, A., Small, E.B., West, P.A., Huq, M.I., Rahman, R. and Colwell, R.R. (1983) Ecological relationships between *Vibrio cholerae* and planktonic crustacean copepods. *Applied and Environmental Microbiology*, **45**, 275–83.

Huq, A., Small, E.B., West, P.A. and Colwell, R.R. (1984a) The role of planktonic copepods in the survival and multiplication of *Vibrio cholerae* in the aquatic environment, in *Vibrio's in the Environment*, (ed. R.R. Colwell), John Wiley and Sons Inc., New York, pp. 521–34.

Huq, A., West, P.A., Small, E.B., Huq, M.I. and Colwell, R.R. (1984b) Influence of water temperature, salinity and pH on survival and growth of toxigenic *Vibrio cholerae* serovar O1 associated with live copepods in laboratory microcosms. *Applied and Environmental Microbiology*, **48**, 420–4.

Huq, A., Huq, S.A., Grimes, D.J., O'Brien, M., Chu, K.H., McDowell Capuzzo, J. and Colwell, R.R. (1986) Colonization of the gut of the blue crab (*Callinectes sapidus*) by *Vibrio cholerae*. *Applied and Environmental Microbiology*, **52**, 586–8.

Huq, A., Colwell, R.R., Rahman, R., Ali, A., Chowdhury, M.A.R., Parveen, S., Sack, D.A. and Russek-Cohen, E. (1990) Detection of *Vibrio cholerae* O1 in the aquatic environment by fluorescent monoclonal antibody and culture methods. *Applied and Environmental Microbiology*, 56, 2370–3.

Islam, M.S., Hassan, M.K., Miah, M.A., Quadri, F., Yunus, M., Sack, R.B. and Albert, M.J. (1993) Isolation of *Vibrio cholerae* O139 Bengal from water in Bangladesh. *Lancet*, **342**, 430.

Jeon, K.W. (1972) Development of cellular dependence on infective organisms. *Science*, **176**, 1122–3.

Jesudason, M.V., Cherian, A.M. and John, T.J. (1993) Blood stream invasion by *Vibrio cholerae* O139. *Lancet*, **342**, 431.

Jesudason, M.V. and John, T.J. (1993) A major shift in prevalence of non-O1 and El Tor *Vibrio cholerae*. *Lancet*, **341**, 1090.

Joseph, P.R., Tamayo, J.F., Mosley, W.H., Alvero, M.G., Dizon, J.J. and Henderson, D.A. (1965) Studies of cholera El Tor in the Philippines 2. A retrospective investigation of an explosive outbreak in Bacolod city and Talisay, November, 1961. *Bulletin of the World Health Organization*, **33**, 637–43.

Kay, B.A., Sack, R.B., Spira, W.M. *et al.* (1984) *Vibrio cholerae* non-O1 isolated from five people with diarrhoea in Lima. *Lancet*, **i**, 218.

Kiiyukia, C., Nakajima, A., Nakai, T., Muroga, K., Kawakami, H. and Hashimoto, H. (1992) *Vibrio cholerae* non O1 isolated from Ayu fish (*Plecoglossus altivelis*) in Japan. *Applied and Environmental Microbiology*, **58**, 3078–82.

King, C.H., Shotts, E.B. Jr, Wooley, R.E. and Porter, K.G. (1988) Survival of coliforms and bacterial pathogens within protozoa during chlorination. *Applied and Environmental Microbiology*, **54**, 3023–33.

Kjelleberg, S., Humphrey, B.A. and Marshall, K.C. (1982) Effect of interfaces on small, starved marine bacteria. *Applied and Environmental Microbiology*, **43**, 1166–72.

Kjelleberg, S., Hermansson, M., Marden, P. and Jones, G.W. (1987) The transient phase between growth and nongrowth of heterotrophic bacteria, with emphasis on the marine environment. *Annual Reviews of Microbiology*, **41**, 25–49.

Klontz, K.C., Tauxe, R.V., Cooke, W.L., Riley, W.H. and Wachsmuth, I.K. (1987) Cholera after consumption of raw oysters. *Annals of Internal Medicine*, **107**, 846–8.

Kodama, H., Gyobu, Y., Tokuman, N., Okada, I., Uetake, H., Shimada, T. and Sakazaki, R. (1983) Ecology of non-O1 *Vibrio cholerae* in Toyama prefecture. *Microbiology and Immunology*, **28**, 311–25.

Krishnan, K.V. (1953) Investigation on the probable role of hilsa fish in maintaining cholera endemicity in India, in *The Indian Council of Medical Research – Technical Report of the Scientific Advisory Board for the Year 1952*, pp. 4–6.

Lahiri, M.N. (1954) *Annual report of the All-India Institute of Hygiene and Public Health*, Calcutta, p. 57.

Lee, J.V., Bashford, D.J., Donovan, T.J., Furniss, A.L. and West, P.A. (1982) The incidence of *Vibrio cholerae* in water, animals and birds in Kent, England. *Journal of Applied Bacteriology*, **52**, 281–91.

Lin, F.-Y.C., Morris, J.G. Jr, Kaper, J.B., Gross, T., Michalski, J., Morris, C., Libonati, J.P. and Israel, E. (1986) Persistence of cholera in the United States: isolation of *Vibrio cholerae* O1 from a patient with diarrhea in Maryland. *Journal of Clinical Microbiology*, **23**, 624–6.

Lowry, P.W., Pavia, A.T., McFarland, L.M. *et al.* (1989) Cholera in Louisiana. Widening spectrum of seafood vehicles. *Archives of Internal Medicine*, **149**, 2079–84.

MacDonald, N.L., Stark, J.R. and Austin, B. (1986) Bacterial microflora in the gastro-intestinal tract of Dover sole (*Solea solea* L.) with emphasis on the possible role of bacteria in the nutrition of the host. *FEMS Microbiology Letters*, **35**, 107–11.

McIntyre, R.C., Tira, T., Flood, T. and Blake, P.A. (1979) Modes of transmission of cholera in a newly infected population on an atoll: implications for control measures. *Lancet*, **i**, 311–14.

Merson, M.H., Martin, W.T., Craig, J.P., Morris, G.K., Blake, P.A., Craun, G.F., Feeley, J.C., Camacho, J.C. and Gangarosa, E.J. (1977) Cholera on Guam, 1974. *American Journal of Epidemiology*, **105**, 349.

Miller, C.J. (1984) The ecology of toxigenic O1 *Vibrio cholerae* and its implications for the epidemiology of cholera. PhD Thesis, University of London.

Miller, C.J., Drasar, B.S. and Feachem, R.G. (1984) Response of toxigenic *Vibrio cholerae* O1 to physico-chemical stresses in aquatic environments. *Journal of Hygiene (Cambridge)*, **93**, 475–95.

Mirelman, D. (1987) Ameba–bacterium relationship in amebiasis. *Microbiology Reviews*, **51**, 272–84.

Morris, G.K., Patton, C.M., Feeley, J.C. *et al.* (1979) Isolation of the Legionnaires disease bacterium from environmental samples. *Annals of Internal Medicine*, **90**, 664–6.

Muroga, K., Takahashi, S., Yamanoi, H. and Nishibuchi, M. (1979) Non-cholera vibrio isolated from diseased Ayu. *Bulletin of the Japanese Society of Scientific Fisheries*, **45**, 829–34.

Nalin, D.R., Daya, V., Reid, A., Levine, M.M. and Cisneros, L. (1979) Adsorption and growth of *Vibrio cholerae* on chitin. *Infection and Immunity*, **25**, 768–70.

Novitsky, J.A. and Morita, R.Y. (1976) Morphological characterization of small cells resulting from nutrient starvation of a psychrophilic vibrio. *Applied and Environmental Microbiology*, **32**, 617–22.

Ogg, J.E., Ryder, R.A. and Smith, H.L. (1989) Isolation of *Vibrio cholerae* from aquatic birds in Colorado and Utah. *Applied and Environmental Microbiology*, **55**, 95–9.

Palmer, L. and Falkow, S. (1986) Characterization of cloned genes from *Chlamydia trachomatis*, in *Microbiology – 1986*, (eds L. Lieve, P.F. Bonventre, J.A. Morello, S.D. Silver and H.C. Wu), ASM, Washington, DC, pp. 91–5.

Pandit, C.G. and Hora, S.L. (1951) The possible role of the hilsa fish, *Hilsa ilisha* (Ham) in maintaining cholera endemicity in India. *Indian Journal of Medical Science*, **5**, 343–56.

Pavia, A.T., Campbell, J.F., Blake, P.A., Smith, J.D., McKinley, T.W. and Martin, D.L. (1987) Cholera from raw oysters shipped interstate. *Journal of the American Medical Association*, **258**, 2374.

Piergentili, P., Castellani-Pastoris, M., Fellini, R.D., Farisano, G., Bonello, C., Rigoli, E. and Zampieri, A. (1984) Transmission of non-O group 1 *Vibrio cholerae* by raw oyster consumption. *International Journal of Epidemiology*, **13**, 340–3.

Pollitzer, R. (1959) *Cholera*, WHO Monograph, no. 43.

Rabbani, G.H. and Mahalanabis, D. (1993) New strains of *Vibrio cholerae* O139 in India and Bangladesh: lessons from the recent epidemics. *Journal of Diarrhoeal Diseases Research*, **11**, 63–6.

Rowbotham, T.J. (1980) Preliminary report on the pathogenicity of *Legionella pneumophila* for freshwater and soil amoebae. *Journal of Clinical Pathology*, **33**, 1179–83.

Sack, R.B. (1973) A search for canine carriers of *Vibrio*. *Journal of Infectious Diseases*, **127**, 709–12.

Salmaso, S., Greco, D., Bonfiglio, B. *et al.* (1980) Recurrence of pelecypod-associated cholera in Sardinia. *Lancet*, **ii**, 1124–7.

Sanyal, S.C., Singh, S.J., Tiwari, I.C., Sen, P.C., Marwah, S.M., Hazarika, V.R., Singh, H., Shimada, T. and Sakazaki, R. (1974) Role of household animals in maintenance of cholera infection in a community. *Journal of Infectious Diseases*, **130**, 575–9.

Shandera, W.X., Hafkin, B., Martin, D.L. *et al.* (1983) Persistence of cholera in the United States. *American Journal of Tropical Medicine and Hygiene*, **32**, 812–17.

Shimada, T., Nair, G.B., Deb, B.C., Albert, M.J., Sack, R.B. and Takeda, Y. (1993) Outbreak of *Vibrio cholerae* non-O1 in India and Bangladesh. *Lancet*, **341**, 1347.

Simidu, U. and Hasuo, K. (1968) Salt dependency of the bacterial flora of marine fish. *Journal of General Microbiology*, **52**, 347–54.

Sochard, M.R., Wilson, D.F., Austin, B. and Colwell, R.R. (1979) Bacteria associated with the surface and gut of marine copepods. *Applied and Environmental Microbiology*, **37**, 750–9.

Spira, W.M., Huq, A., Ahmed, Q.S. and Sayeed, Y.A. (1981) Uptake of *Vibrio cholerae* biotype eltor from contaminated water by water hyacinth (*Eichhornia crassipes*). *Applied and Environmental Microbiology*, **42**, 550–3.

Tamplin, M.L. and Fisher, W.S. (1989) Occurrence and characteristics of agglutination of *V. cholerae* by serum from the Eastern Oyster, *Crassostrea virginica*. *Applied and Environmental Microbiology*, **55**, 2882–7.

Tamplin, M.L., Gauzens, A.L., Huq, A., Sack, D.A. and Colwell, R.R. (1990) Attachment of *Vibrio cholerae* serogroup O1 to zooplankton and phytoplankton of Bangladesh waters. *Applied and Environmental Microbiology*, **56**, 1977–80.

Tamplin, M.L. and Parodi, C.C. (1991) Environmental spread of *Vibrio cholerae* in Peru. *Lancet*, **338**, 1216.

Tauxe, R.V. and Blake, P.A. (1992) Epidemic cholera in Latin America. *Journal of the American Medical Association*, **267**, 1388–90.

Thom, S.M., Warhurst, D. and Drasar, B.S. (1992) Association of *Vibrio cholerae* with fresh water amoebae. *Journal of Medical Microbiology*, **36**, 303–6.

Tison, T.J., Hope, D.H., Cherry, W.B. and Fliermans, C.B. (1980) Growth of *Legionella pneumophila* in association with blue-green algae (Cyanobacteria). *Applied and Environmental Microbiology*, **39**, 456–9.

Twedt, R.M., Madden, J.M., Hunt, J.M. *et al.* (1981) Characterization of *Vibrio cholerae* isolated from oysters. *Applied and Environmental Microbiology*, **41**, 1475–8.

Venkateswaran, K., Takai, T., Navarro, I.M., Nakano, H., Hashimoto, H. and Siebeling, R.J. (1989) Ecology of *Vibrio cholerae* non-O1 and *Salmonella* spp. and role of zooplankton in their seasonal distribution in Fukuyama coastal waters, Japan. *Applied and Environmental Microbiology*, **55**, 1591–8.

Wang, J.-B., Cheng, M.-S., Gao, J.-L., Fan, Z.-Y., Hong, Y.-W. and Te, X.-X. (1987) Experimental *Vibrio* El tor infected water fish breeding. Stressing reservoir establishment in loach and finless eels. *Chinese Medical Journal* (*English translation*), **100**, 142–5.

Weissman, J.B., DeWitt, W.E., Thompson, J. Muchnick, C.N., Portnoy, B.L., Feeley, J.C. and Gangarosa, E.J. (1974) A case of cholera in Texas, 1973. *American Journal of Epidemiology*, **100**, 487–98.

Xu, H.-S., Roberts, N., Singleton, F.L., Attwell, R.W., Grimes, D.J. and Colwell, R.R. (1982) Survival and viability of nonculturable *Escherichia coli* and *Vibrio cholerae* in the estuarine environment. *Microbial Ecology*, **8**, 313–23.

The geography of cholera

8

A.E. Collins

8.1 BACKGROUND TO SPATIAL PERSPECTIVES IN DISEASE INCIDENCE

Awareness of spatial differences in disease incidence has existed for thousands of years. In biblical times before the Exodus, the Israelite refugees witnessed how the plagues that afflicted Egypt avoided the land of Goshen but selectively attacked the people living in the other areas (Exodus 9). Work relating to geographical dimensions of disease pre-dates the laboratory orientation of scientific and clinical medicine which was emerging around the time that Robert Koch discovered the cholera vibrio. Distributions of diseases, of which cholera serves as an excellent example, have been utilized to explain disease occurrence by elucidating causal mechanisms, describing its natural history, or providing guidance in the administration of health services.

Studies of disease incidence with an essentially spatial perspective traditionally emphasize both the search for non-uniformity in disease distributions and the search for association with factors responsible for the pattern displayed. These approaches may be accredited with the titles of medical cartography and disease ecology respectively. Good examples of both are the early and well-known studies by John Snow who successfully isolated a contaminated hand pump water supply as the cause of a cholera epidemic in London in 1857. It was Snow in his publication, 'On the mode of communication of cholera' (1855) who illuminated the fact that those administrative areas served by the Southwark and Lambeth water-works were associated with the high rates of death from cholera. He had noted that five years on from the

Cholera and the Ecology of Vibrio cholerae. Edited by B.S. Drasar and B.D. Forrest.
Published in 1996 by Chapman & Hall, London. ISBN 0 412 61220 8.

initial outbreak a change in pattern had come about in that the Southwark and Vauxhall Company water supply was still associated with high cholera mortality, but the Lambeth supply was not and that they had moved their waterworks to a point higher up the Thames, thus obtaining a supply of water free from the sewage of London. A less accurate correlation was made by William Farr in 1852, who made the mistake of postulating a direct relationship between cholera cases and altitude. The error of Farr in attributing cholera mortality directly to residence at low altitude illustrates in a simple way a fundamental problem in ecological associative studies in general, namely that concepts of cause and effect become extremely complex if disease distributions are shaped by an indeterminate range of factors. In this case there was correlation but no clear causative mechanism was identified. Individual chains of causation represent only a fraction of the reality and each component itself is the result of a complex genealogy or spatial history.

However, a theory of the natural tendency of zoonotic diseases to become localized in a specific habitat still carries much weight in modern epidemiology and may in particular help in explaining distributions of cholera. It is a perspective that has been referred to as 'landscape epidemiology' by Pavlosky *et al.* (1955) in Soviet work. This delimits the foci of infectious, zoonotic diseases by analysing the associations of factors such as vegetation, animal and insect life, soil type and acidity, precipitation regime, and other elements of the natural landscape. More well known (in western countries) is the similar perspective adopted by Jaques May (1950) who described the epidemiological constraints of various diseases as requiring the coincidence of two, three or four factors, identifying geographical elements, or 'geogens' as fundamental to determining their existence and distribution. Merson *et al.* (1978), in their epidemiological study of cholera and enterotoxigenic *Escherichia coli* diarrhoea in Bangladesh, asked the question, 'are we dealing with a phenomenon that is related to the biology of the organism or the host or simply to a difference in exposure to contaminated vehicles?' Interestingly, this echoes an early concept captured by Pasteur in the 19th century that 'the germ is nothing, the terrain is everything'. A geographical understanding of 'terrain' – somewhat different to Pasteur's principally biological context – includes factors pertaining to the physical and to the human environment and provides a good framework for understanding distributions of communicable diseases such as cholera.

Also particularly significant in the development of medical geographical thinking was the inclusion by May of the role of culture as a buffer between disease agents and human infection. Behaviour, which may be considered as the observable aspect of culture, often has spatial expression and may create some of the environmental conditions which link to disease and health. Increased awareness of social and behavioural aspects of health therefore means that consideration must also be given to processes determining health, such as the location of health care, rather than simple descriptions of a particular disease and its distribution. It should also be emphasized that behavioural and socioeconomic factors often become confounding variables when carrying out studies in search of a straight environmental link. For example, investigations into links between water quality and incidence of cholera may be seriously flawed if consideration is not given to factors such as water storage, hygiene behaviour, socioeconomic status, and provision and use of health services.

Generally, the broader understanding of 'environment' must be included if our understanding of 'health' is to include the full parameters of the WHO's 1948 definition of 'a state of complete physical, mental and social well being and not merely the absence of disease or infirmity'. Something of the synthesis that is required in viewing the complexity of health outcomes is expressed by Learmonth (1988) who describes 'a holism that extends to consider community health as a whole, ultimately to put health and disease into a community and societal context, always – ideally at least – with respect for and conservation of the ecological balance of people, plants and animals in a particular setting' (Learmonth, 1988). The concept of balance is also echoed by Howe (1982) who has suggested that health equates with ecological equilibrium while ill-health may be considered as a state of maladjustment, disharmony or ecological disequilibrium. The main spatial perspective of this approach is that individual biomes or regions may broadly categorize the conditions under which a wide variety of well-being or ill-health may flourish. Changes in the balance of these ecological systems, whether in a physical environmental or socioeconomic sense, are influential factors in bringing about new patterns of disease. A simplified system approach to analysis of cholera geography is represented in Figure 8.1.

Because of its added emphasis on culture and environment, disease ecology requires contextual understanding, the term 'environment' taken as inclusive of external economic and social phenomena that influence the functioning of the local system. The 1992 World Bank Development Report in a section entitled Environmental Priorities for Development to some extent joins in with this perspective stating that 'too little is known about how risks and diseases are distributed and interact with each other, and uncertainty remains over the extent to which modest changes in infrastructure account for long-run health improvements'.

Disruption of ecosystems can result in new disease distributions and

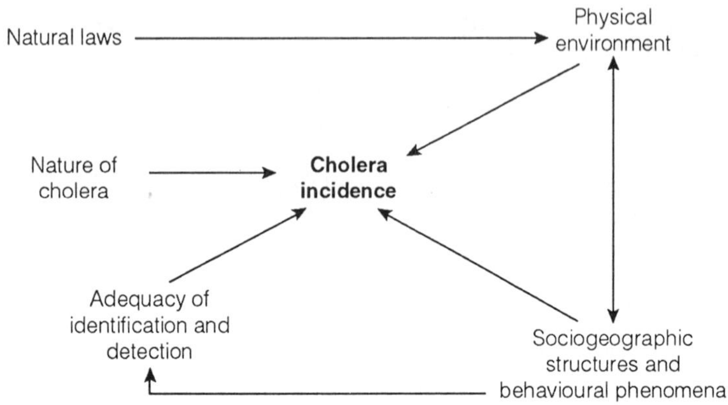

Fig. 8.1 A systems approach to cholera geography. (From Collins, 1993.)

higher rates of incidence for affected areas. Hughes and Hunter (1970) wrote about the 'hidden costs' of ecologically naïve development projects in Africa and how often ecological disruption had negative health outcomes. During the present decade this could be extended to the emergent issues of political and economic change, structural adjustment policies, a changing world order, and their impact on environment and health in many parts of the world. Stock (1986) describes a study strategy that involves looking at 'the political economic nature of ill-health occurring in particular ecological niches. ... In other words, a fully-developed holism, oriented toward understanding the underlying as well as immediate causes of ill-health'.

With these advances in medical geographical perspectives, spatial association remains fundamental. For example, the failed aspects of structural adjustment policies in many Third World countries have been expressed in the form of greater regionalization of resources and an accentuated urban to rural division of wealth (Timberlake, 1985; Reed, 1993; Rich, 1994). Additionally, the effects of changing emphases in development policy is apparent in terms of spatial differentiation of well-being within the urban areas of the same countries (Bergstrom and Ramalingaswami, 1992; Rich, 1994). While confirming the impact of physical and social factors on the health of the urban poor, the WHO (1991) confirm that our understanding of the way in which they interact with the urban development process is incomplete and imprecise, partly due to inadequate information. However, part of the explanation is contained in the 1995 World Health Report (WHO, 1995) which maintains that increasingly the poor pay the price of social inequality with their health.

Variation in exposure to contaminated environments, often degraded through human poverty, forms the theme of many existing commentaries on the spatial distribution of cholera, though opinion as to the most significant cause is varied. Typically this type of associative environmental approach to cholera focuses on secondary transmission, correctly highlighting the importance of factors such as congestion, socioeconomic and behavioural disruption, and poor provision of appropriate health care. For example, Meade *et al.* (1987) comment on how diseases, particularly those found in the tropics, display an occurrence heavily modified by activities that promote economic development for some areas. Certainly, the 1991 spread of cholera through Latin America has cruelly exposed areas with deteriorating social conditions which were the legacy of widespread economic stagnation.

Connection has been made between cholera and 'policies of hard-nose foreign financiers' (Donovan, 1991), the breakdown of government infrastructure resulting in inadequate housing and sanitation (Robinson, 1991), lack of personal hygiene and education (Pan American Health Organization, 1991), overcrowding, lack of adequate and safe water supplies, and insanitary disposal of excreta (WHO, 1991). Other suggested links have been an over-burdened medical service and poor communications (Ferguson, 1977), and peoples' lack of commitment to maintaining a sanitary environment in post-colonial areas, the maintenance of which they consider to be the sole responsibility of the Government (Adesina, 1987). Rapid increase in incidence of gastro-intestinal infections, such as cholera, has also been linked to the importation of pathogens by migrants from rural areas where the disease is endemic (WHO, 1986), or among groups of newly arriving refugees to camps, such as has been described by Mulholland (1985) in an account of a cholera epidemic in a refugee camp in eastern Sudan in 1985.

However, the last few years have thrown up extensive new information on the nature of the disease and its terrain demanding a further extension of geographical perspectives on cholera. This is particularly the case regarding factors relating to the role of primary transmission of the pathogen from environmental reservoirs, now seen as a feature which may influence the distribution of the disease. Recent advancements in microbiological knowledge of *Vibrio cholerae*, a recent surge of incidence of cholera linked to the seventh pandemic during the early 1990s, of which South America and Africa particularly suffered, and the appearance of the new strain, *Vibrio cholerae* O139, in India are recent developments highly relevant to geographical aspects of disease and health and that require further investigation. Interest in the changing biology of cholera, the continued and widespread suffering it causes, and its importance as a global issue has rightfully been reflected by a recent increase of academic papers published in a wide range of journals. With reference to these, and some original field investigation, the remainder of this chapter considers a number of aspects of the new geography of cholera and shows how geographical perspectives, more than ever, have a fundamental role to play in understanding incidence of the disease.

8.2 GEOECOLOGY OF CHOLERA

One of the more significant recent developments in the study of cholera relates to information derived from laboratory research carried out in the 1980s. This has pointed to the existence of aquatic environmental reservoirs in which *V. cholerae* survives for prolonged periods and from which a toxigenic form, under certain conditions of temperature, alkalinity and salinity, may emerge to support epidemic conditions (Miller *et al.*, 1984, 1985, 1986; Barua and Greenough, 1992). The need for attention to be paid to the possibility of these pools existing in nature was suggested after laboratory investigation into the impact of physicochemical stress on *V. cholerae* O1 (Miller *et al.*, 1984, 1986). Work by Miyaki *et al.* (1967), Pandit *et al.* (1967) and Prescott and Bhattacharjee (1969) had earlier suggested survival times dependent on factors such as temperature, pH, osmotic pressure, moisture content, salt or carbohydrate concentration, and the presence of organic matter and bacterial flora as key factors in determining survival times in different foodstuffs. Experimental studies have also shown that gastric acid is one of the body's main lines of resistance against cholera (Cash *et al.*, 1984) and vibrios are known to be unable to survive in carbonated water due to its low pH (WHO, 1986). Other investigations carried out on water suggested the key factors to be temperature, pH, salt, bacterial and organic content (Singleton *et al.*, 1982; Colwell and Spira, 1992). Shellfish have often been suspected of spreading cholera and several recent studies in coastal areas of the USA have served to reawaken the possibility of the non-human animal reservoir (Kaysner *et al.*, 1987; Doran *et al.*, 1989). The aquatic reservoir and the non-human animal reservoir theories are open to convergence, in that the feature of a brackish environment can be considered common to both circumstances. This has recently been emphasized further by the isolation of *V. cholerae* O1 from intestines and skin of fish in polluted coastal waters along the coastline of Peru (Tamplin and Parodi, 1991).

However, survival in aquatic environments with favourable conditions has been only part of the accumulated findings from research on *V. cholerae*. It is now also suggested that, in unfavourable environmental conditions, toxigenic isolates of *V. cholerae* O1 and other vibrios are able to enter a period of dormancy (Xu *et al.*, 1982; Colwell *et al.*, 1985; Colwell and Spira, 1992). If, in addition to nutrient depletion, the cells are subjected to reduction in temperature and/or elevation in salinity, the cells rapidly go non-culturable but remain viable and potentially pathogenic (Colwell *et al.*, 1985). This has meant that in the past potentially toxigenic pathogens capable of causing major epidemics of cholera may have remained hidden in the natural environment. New techniques of isolation which use fluorescent antibody or gene probes help overcome this problem of evasion, meaning that an entirely new appreciation of the space–time distribution of *V. cholerae* and incidence of cholera is possible. There are also possibly new implications for understanding the distribution and behaviour of many other bacterial enteric pathogens.

Linked to these developments, Martins *et al.* (1993) partly explain low numbers in positive samples of sewage water in Brazil between 1974 and 1983 in terms of the ability of *V. cholerae* O1 to undergo environmental adaptation and be non-culturable. Martins *et al.* conclude that *V. cholerae* O1 was therefore likely to be already present in the aquatic environment of São Paulo before the present outbreak of cholera in that country. However, Wachsmuth *et al.* (1993) state that there is still no real evidence of a non-toxigenic precursor for the toxigenic Latin American isolates and that the relatively nearby isolates found along the US Gulf Coast are not directly related to them. The Latin American isolates are considered more likely to be a clonal variant of the seventh pandemic that has been described elsewhere as being introduced by travellers. Wachsmuth draws attention to a shift from Inaba to Ogawa serotype within the Latin American clone. Possibly this helps to explain why the epidemic took off so rapidly, affecting so many people in that region.

The findings of Salazar-Lindo (1993) suggest that the Latin American isolates are clonal and distinct from the other three clones; the seventh pandemic, the US Gulf Coast, and the Australian isolates. He suggests that the toxigenic cholera vibrio organism was introduced into the aquatic environment of the Peruvian coast long before the outbreak flourished and long enough to evolve into a different clone. Instability in the *ctx* genes in the chromosomes of the recently appeared O139 isolates in Bengal and Bangladesh has been reported as unusual (Lida *et al.*, 1993) and 'could be key to understanding the genetic mechanism for the sudden appearance of O139 strain'. Questions of gradual evolution or genetic instability leading to 'spontaneous appearance' of the toxigenic form of the organism are likely to contribute to an ongoing debate on the biogeography of *V. cholerae* into the future.

A further topic of interest is environmental change that impacts on the nature of the organism. Drasar (1992) has indicated that 'survival is only part of the story' and that the implication of an aquatic reservoir is that 'they form an essential component of the ecosystem'. Interaction takes place between the dormant/starved non-toxigenic vibrios and the environmental reservoir, and between hypertoxigenic strains and people who become cholera patients. He also draws attention to possible continuity in the action of cholera toxin (CT) in these reservoirs and the intestine. The ecosystem hypothis opens up much insight to distributions of *V. cholerae* at micro and macro scale and forms a good basis from which to begin an analysis of the possible implications of environmental change.

8.2.1 LOCALIZED ENVIRONMENTAL CHANGE AND EXPOSURE TO *V. CHOLERAE*

There now exists background information to indicate that different biotypes and serotypes of *V. cholerae* inhabit distinct zones determined by a combination of spatially defined environmental preferences and the relative immunity of the population that occupy an area.

Craig (1988) signalled caution in making generalizations in comparing epidemics of different biotypes when analysing time–space clustering of cholera in Matlab, Bangladesh. Spatial differentiation between biotypes has been demonstrated in Bangladesh during the epidemics of 1988–1989 (Siddique *et al.*, 1991) where there was clustering of the classical biotype in the southern region and of the El Tor biotype in all other regions. While the reason for the interplay between *V. cholerae* O1 biotypes in Bangladesh was not entirely clear, the ecological changes in various regions, caused by soil erosion and construction of barrages and dams in the river system, were thought to be possible associating factors. The appearance of the El Tor biotype coincided with increasingly severe flooding in the northeastern and middle-belt regions. Siddique reasons that since El Tor is hardier and more viable in water than the older Classical biotype, it may have found a more suitable habitat in those areas. Also, a drastic reduction of fresh water flow caused by the Farakka barrage in the catchment area changed the dry season hydrodynamics in southern Bangladesh and resulted in an increase in salinity and incursion of brackish water deeper inland. Siddique suggests that the Classical biotypes in the south may have become adapted to this changing environment.

Interestingly, *V. cholerae* O139 may be even hardier with a survival advantage over *V. cholerae* O1. Islam *et al.* (1993) note that whereas *V. cholerae* O1 is normally isolated from less than 1% of water samples during epidemics, 12% of water samples in his study of 92 water samples from ponds, lakes, rivers, and canals in rural Matlab and urban Dhaka, Bangladesh yielded *V. cholerae* O139. Also, attention is drawn here to the much higher toxigeneity of the new serogroup (Islam *et al.*, 1993). A reciprocal seasonal pattern in prevalence of *V. cholerae* O139 with *V. cholerae* O1, identified by Jesudason and Jacob John (1993) also implies a tendency for one strain to fill the niche of the other in response to a controlling environmental factor. However, it should be remembered that *V. cholerae* O139 has been reported as having an overall similarity with *V. cholerae* O1 El Tor (Higa *et al.*, 1993).

Observation of the nature of previous pandemics of cholera should now warn us that the distribution of the more recent serogroup is unlikely to be explained solely in terms of the biology of the organism or the host but indeed also through differences in exposure to its habitat. Two factors stand out as particularly important. First, non-O1 serotypes are known to be widely prevalent in natural aquatic environments (Mandal, 1993) and secondly, previous infection with the El Tor serotypes does not provide immunity to infection from *V. cholerae* O139 (Bhattacharya *et al.*, 1993).

Reports from the cholera epidemic in Angola also present a situation of shifting serotypes (Colombo *et al.*, 1993). In 1988 all strains were Ogawa; by 1991, the prevalent epidemic strain was also Ogawa, but by 1992 this had changed to Inaba, identified both in patients and in the Bengo river which serves the Luanda piped water system. As in the case of Mozambique, which will now be explored in more detail, this pattern might relate to environmental and demographic change caused by continuous instability.

8.2.2 ENVIRONMENT AND CHOLERA IN MOZAMBIQUE

Work carried out by the present author in Quelimane, Mozambique in 1991 is of interest in that the serotype Ogawa of the biotype El Tor was identified, rather than Inaba which was being associated with the main epidemic spreading through Africa at that time. This lent substantial weight to the understanding that the situation in Quelimane was endemic (Collins, 1993). *V. cholerae* O1 El Tor serotype Ogawa, well adapted to the saline and alkaline aquatic reservoirs of Quelimane, may have become part of the indigenous fauna of that area, ready to strike in epidemic proportions should suitable circumstances arise. The larger estuarine city of Beira 300 km to the south also suffered a prolonged cholera epidemic in the early 1990s but serotype Ogawa generally did not occur. Also, environmental tests in that city has found the groundwater to be more acidic and less affected overall by a saline incursion than Quelimane.

The study on the distribution of incidence of cholera in Quelimane suggested that **increased salinity** in the well-water supply of some parts of the city was associated with a higher morbidity rate in those areas during the course of a year-long epidemic. Evidence provided by environmental tests suggested that, along with population density, this may have been a key factor in explaining the pattern observed. It was also noted that degradation of the physical environment through devegetation and compaction of soil within a littoral region, may have led to the salinization and increased alkalinity considered preferable to the longevity and toxigeneity of *V. cholerae* in those areas. Saline water is generally described as having a chloride content in excess of 300 mg/l if relating it to acceptable drinking water levels. It therefore does not take much intrusion of seawater (with ca. 19 g/l chloride ions to cause deterioration. The actual movement of saline water into the freshwater environment is difficult to predict due to the complexity of mixing related to pumping rates and density variations between the two water types (Owen *et al.*, 1991). Freshwater, being less dense than saline will constitute the upper body of water. Provided the density difference can be maintained there will be little upward movement of saline water. However, over-pumping may destroy this balance, causing upconing and a resultant deterioration in quality.

The principal cause of a changed surface environment in those areas of Quelimane supporting unusually high rates of cholera incidence appeared to be the rapid population increase due to an influx of displaced people from war-torn rural areas. Change to natural environments which may impact on health, such as those suggested at Quelimane, represents a potentially fruitful area for research on communicable diseases in the future.

However, another part of the explanation for the observed distributions in Mozambique is likely to have been a lack of resistance of different communities to the different Ogawa and Inaba serotypes which produce only type-specific immunity (Preston, 1993). For example, in Quelimane, the flank of the city worst affected hosted the highest number of displaced persons from rural areas. Many of the areas of origin of these groups were both isolated and not known to have been cholera zones. The incomers would therefore have been especially susceptible to the cholera of Quelimane, while their reduced socioeconomic status from being displaced is likely to have also contributed to infection vulnerability.

The relationship between cholera and **malnutrition** is a further factor that should be considered in this context. Work by Mulholland (1985) in the May–June, 1985 cholera epidemic in a refugee camp in Eastern Sudan and that of Moren *et al.* (1991) investigating a cholera outbreak in a Mozambican refugee camp in Malawi in 1988 both point to its possible significance. Immunologic responses to infection in general are modified by nutritional status (Scrimshaw *et al.*, 1968). Additionally, Rutherford and Mahanjane (1985) suggested that a large portion of diarrhoeal morbidity and mortality in the Mozambican famine of 1983 may have not been infectious but rather due to starvation-induced malabsorption. Hypochlorhydria through loss of gastric acids, which are one of the body's main lines of resistance against cholera, predisposes to higher incidence (Cash *et al.*, 1974). If the evidence of Thomason *et al.* (1981) that acute malnutrition predisposes to hypochlorhydria is correct, then it is reasonable to suspect that this is the mechanism by which cholera selectively attacks the malnourished (Nalin *et al.*, 1978). It can also be noted here that in Quelimane the three zones displaying the lowest expected rate of incidence were densely populated but historically established with a population largely indigenous to Quelimane. In Beira, the displaced were distributed more evenly around the city, or concentrated into special reception areas in more rural areas some distance away. It is reasonable to suggest that aquatic reservoirs and location-specific immunity played less of a role since the cholera there was associated with a more widespread epidemic situation affecting the region as a whole.

It should also be noted here that while both cities have extremely poor sanitation and have suffered the effects of virtual siege conditions during a war lasting years, Beira for the majority of the time was able to maintain some of its key function as an important strategic port at the end of the 'Beira Corridor'. Movement of people into and out of Beira was therefore much greater than in Quelimane which for long periods only received occasional internal flights. Beira's relatively greater contact with the outside world meant that it was more susceptible to the seventh pandemic *V. cholerae* O1 El Tor serotype Inaba which caused 153 367 cases and 13 998 deaths in Africa that year. By the end of 1992 the total for Mozambique was 30 802 cases with 726 deaths, the highest in Africa that year (WHO, 1993). To conclude here, the pattern of incidence between the two cities may show that the relative balance between endemicity and epidemicity for specific areas is expressed by different serotypes with a range determined by predisposing environmental and demographic processes.

Figures 8.2 and 8.3 show two types of environment in the Mozambican cities. Figure 8.2 is of a densely inhabited area where cholera is particularly rife during an epidemic. It is an area which suffers from inundations of salt water and where the soil has become devegetated, compacted and alkalinized. Figure 8.3 is taken in an area slightly further inland which also supports a high population density but where cholera incidence during an epidemic is much less. This area is well vegetated with palms and mango trees, is away from inundations of salt water and has a sandy acidic surface environment.

Fig. 8.2 Mozambican city environment. Caline, denuded and a focus of high cholera incidence. (Photograph courtesy of the author.)

Fig. 8.3 Mozambican city environment. A well site 7 km from that shown in Figure 8.2, with high population density, lower cholera incidence and a vegetated surface environment. (Photograph courtesy of the author.)

8.2.3 GLOBAL ENVIRONMENTAL CHANGE

Environmental change at the global scale may also relate to variation in disease incidence if, for example, organisms such as *V. cholerae* are considered as being sensitive to the physical parameters of temperature, alkalinity, and salinity as indicated by the laboratory research. Thus, if *V. cholerae* is able to survive in coastal waters, as do other salt water organisms such as algae and plankton, and if its successful multiplication is similarly dependent on temperature pH, salinity and available nutrients, then it is reasonable to postulate changes similar to that displayed by some oceanic algal blooms. Epstein (1992) has suggested that the recent pattern of cholera in the Americas may represent the first detectable impact of climatic change on the distribution of water-based and vector-borne diseases. Some evidence for this was that unusually large algae and plankton blooms were reported at sea at the same time as plankton in the harbour near Lima, Peru was found to be contaminated with *V. cholerae* (Tamplin and Parodi, 1991). Epstein points out that the unexpected intensity of the outbreak, which penetrated cities and towns along the Pacific coast in January and February 1991, was consistent with multiple entry points from marine life blooms, with fish, molluscs and crustacea as vectors.

Based on this hypothesis Epstein *et al.* (1993a) go on to suggest that an outbreak of water-borne disease which is a consequence of proliferating plankton and associated bacteria may be considered in part a consequence of sunlight, pH, currents, winds, and river run-offs that govern their location and timing. Major anthropogenic influences on algal blooms – apart from possible global warming – are caused by pollution, through excess nutrient from sewage and fertilizer effluents, soil erosion and acid rain, all of which affect the balance of nitrogen and phosphorus. Additionally attention might be given to over-harvesting of fish and shellfish which reduces algivorous grazing and loss of wetlands which filter nitrogen and phosphorus, store carbon, and support fish and seabirds. The complex nature of the world's climate and ocean circulation and the unpredictability of change caused by human inputs mean that predictions of the full implications of the changing distribution of associated bacteria remains difficult to assess.

Further impacts of climate change on cholera in some countries may eventually relate to sea-level rise, reduction of rainfall, and decreases in crop yields. Rising sea-levels could change the balance between salt and freshwater in the low-lying estuarine environments of cholera-prone regions such as parts of India, Bangladesh, Pakistan, Indonesia, Thailand, Egypt, Nigeria and Mozambique. In the event of further rising sea-levels, pumping stations would have to be located further inland and coastal well-water would become subject to more frequent incursions of brackish ground water. Flooding by seawater would also leave more inland lagoons of water with physical conditions favourable to the longevity of *V. cholerae*. This would be intensified by the spread of unprotected sewage in flood waters.

Reduction of rainfall in the interiors of continents, another possible outcome of change to the global atmosphere–hydrosphere–cryosphere balance, could also increase incidence of inland brackish environments. The processes involved are two-fold. Firstly, through a shift in the freshwater and saltwater margin in coastal zones and secondly, through increased evaporation and capillary rise of saline groundwater. A warning of the effects of change in the margin where fresh and saline water meet was experienced in the estuarine cities of Beira and Quelimane during the drought of 1992 which affected most of Southern Africa. In the case of Beira the urban water source on the Pungwe river, was eventually turned off on the grounds of being too saline to drink. In the case of Quelimane, the artificial lagoons alongside the river Licuar, from which the city's water is pumped, dried up altogether. Though the pumping station is 40 km inland, there are now plans to resite the pumping station further upstream away from the low-lying and saline coastal plain. Desalination plants are generally expensive to maintain and are therefore only usually considered by countries with the combination of extreme freshwater shortages and extensive financial resources, such as Saudi Arabia.

Salinization and alkalinization are increasingly human-induced phenomena. Further examples of impact are provided by situations of high population density without access to an urban water network, such as in Jakarta where large quantities of private wells were sunk. This caused heavy over-pumping of the aquifer and a fall in the water-table to the extent that in northern Jakarta, seawater has contaminated the groundwater in a 5–10 km-wide continuous belt along the coastal plain (Briscoe, 1993). According to the Council on Environmental Quality (1981), about half the world's irrigated lands have been damaged to some extent by salinization or alkalinization. Kovda (1972) estimated that about 60–70% of all irrigated land was gradually being transformed into 'saline deserts'. This occurs particularly where continual addition of water, without adequate drainage, tends to cause a gradual raising of the general or local water-table. As a result of higher evaporation rates, excess compounds such as sodium chloride, magnesium and calcium carbonate, and sulphate are precipitated in the soil pores and on the soil surface and raised to ground level by capillary action. In time this process becomes evident in a whitish surface, characteristic of 'white alkali' soils (Tivy and O'Hare, 1989). Additionally, where there is a high concentration of free sodium chloride in the soil, downward leaching produces compounds of sodium with either carbonates or hydroxides, both of which make the soil solution excessively alkaline. Since organic matter is dispersed and goes into solution during the same process, the soils become highly compacted and impermeable when they dry out. The surface becomes covered with a black surface scum of very alkaline organic matter, known as 'black alkalis', which is practically sterile (Tivy and O'Hare, 1989). A correlation between these zones, reservoirs of higher quantities of *V. cholerae* and incidence of cholera has not yet been fully investigated, but white and black alkalis are often a part of the landscape of urban and semi-urban environments in many of the endemic cholera zones of the world.

A further factor, which also relates to global climate and environmental change, is the link between nutritional well-being and communicable disease incidence in general. Estimates of the effects of climate change on the amount of food produced, world food prices, and people at risk from hunger in developing countries made by the Goddard Institute for Space Studies and the Environmental Change Unit in Oxford (Rosenzweig *et al.*, 1992) indicated that crop yields are likely to decline in low-latitude regions but could increase at mid-high latitudes. Greatest declines in world cereal production were predicted in the developing countries where there could be an average estimated reduction of between 8 and 12%. The number of people at risk from hunger was estimated to increase between 5% and 50% with the greatest absolute increase predicted to be in Africa (Haines and Parry, 1993). Additionally, a reduction in availability of drinking water in some regions equates with using inappropriate water sources. The outcome of these environmental pressures consistently occurring in the same parts of the world is likely to be further displacement of populations from those areas. Congestion of the displaced into fewer viable resettlement locations with inadequate water supply and sanitation further enhances the probability of communicable disease epidemics.

Macro scale environmental and demographic differences are likely to explain the changing distribution of non-O1 *V. cholerae*, which has recently taken on massive proportions in India and Bangladesh, though the full repercussions of this has not unfolded at the time of writing this chapter. This first recorded epidemic caused by a non-O1 *V. cholerae* accounted for over 15 000 cases and 230 deaths in Calcutta in its initial stages alone (Sarkar *et al.*, 1993). In Bangladesh the epidemic began in December, 1992, in the south and spread throughout the country with a total of 107 297 cases and 1473 deaths by the end of March 1992 (ICDDR,B, 1993). The first report of the dispersion of *V. cholerae* O139 outside of the region was to Bangkok (Chongsa-nguan *et al.*, 1993). Chongsa-nguan *et al.* point out that while the seventh pandemic strain of cholera, the El Tor, took 3 years to spread from

Celebes island in Indonesia through Bangkok to India from its first isolation in 1961, the O139 strain of *V. cholerae* has disseminated much faster. The organism is reported as causing local outbreaks in geographically distant sites and of having both the potential of pandemic spread and of virtually replacing *V. cholerae* O1, as happened in India and Bangladesh (Islam *et al.*, 1993; Jesudason and Jacob John, 1993). The spontaneity of occurrences of cholera cases may suggest a possible globalization of conditions required for the succession of new cholera strains. At this point in the disease's history we may be wise to consider the implications of a possible loss of diversity between regions as equating with a loss of protection against location-specific disease. The relative ease with which it was removed from some regions through changes to the human environment is mirrored by the relative ease new strains with a slightly different ecological range can return to the same regions. This is demonstrated by the disappearance of cholera from South America in 1867 and its dramatic reappearance in January 1991.

Also, immunity to the new strain of the disease is already known to be low. This is demonstrated by the fact that the first few years of *V. cholerae* O139 has revealed more adults to be suffering from it than children. Once widespread immunity has developed, most adults are protected and more cases are found in young children (Swerdlow and Ries, 1993). However, specific immunity to the O1 serotype does not provide immunity to the O139 serotype. Greater virulence of the organism during an age of increased travel and large population movements ensures that the disease can be efficiently spread. The phenomenon of forced population displacement, experienced by millions in recent times, provides a lethal combination of socially, economically and biologically vulnerable people, concentrated in zones where they may have little localized immunity to endemic strains of disease. The existence of new serotypes of *V. cholerae*, such as O139, among the predisposing conditions of poor sanitation, continued lack of clean water provision, and a deteriorating physical environment in many parts of the Third World will ensure that huge epidemics of cholera are likely to continue. Where there exists suitable environmental reservoirs for its survival, the long-term results will be the creation of further endemic zones and a new geography of cholera.

8.3 NEW THOUGHTS ON SEASONALITY

Geographical perspectives also contribute to understanding the ecology of cholera through concern with seasonality. Recent knowledge of cholera and *V. cholerae* provide us with some new insights on seasonal patterns of cholera incidence. Whereas the WHO reported in 1970 that not much was known about seasonality (WHO, 1970), we are now more informed on the seasonal role of estuarine salinity, climate, water availability and quality, agricultural output, human mobility and other behavioural factors. Colwell and Spira (1992) suggested that the seasonality of cholera in Bengal may be explained by primary transmission controlled by environmental factors such as temperature, salinity, nutrient concentration, and zooplankton blooms, as well as by seasonal variations in seafood harvesting and consumption, and in direct water contact.

Salinity in estuarine zones is likely to relate to cholera incidence both directly and indirectly. Firstly, as a direct environmental control on survival times and toxigeneity of *V. cholerae*, as has already been described. Secondly, in affecting the availability of potable water supplies, as occurs when a freshwater pumping site becomes saline during a drought, or when wells are made inactive by a change in the balance of fresh and saline groundwater in a zone of wells. Miller *et al.* (1982) have shown how the salinity of estuaries fluctuates in accordance with seasonal variation in river flow and, with reference to Calcutta and London in the 19th-century, suggested this may explain the seasonality of cholera in estuarine cities.

Many studies on cholera incidence include the role of rainfall in seasonal variation. However, these have revealed a variety of patterns between the epidemic curve and monthly rainfall pattern and a variety of explanations are given of principal causal mechanisms. Two patterns of incidence in relation to patterns of rainfall are broadly: (i) the epidemic curve rising steeply during the rainy season and tailing off with the coming of the dry season; and (ii) the epidemic curve taking off in the driest period of the year and being sustained throughout the rains before tailing off. An example of the first pattern has been provided by Mbwette (1987) for Tanzania and of the second by Utsalo *et al.* (1992) in Calabar, Nigeria and Collins (1993) in Quelimane, Mozambique. The pattern has been reported as variable between different locations in Bangladesh by McCormack *et al.* (1969) who noted that in Dacca, the cholera season was during the dry winter months (December and January), but in nearby rural areas, a second epidemic occurred during May–July. Moe *et al.* (1991) explain their result of no significant seasonal difference of diarrhoea during the course of a 1-year study in the Philippines in terms of a lack of seasonal variation in rainfall.

An explanation for a surge in cholera with the onset of rains is that **faecal contamination** is washed into drinking water supplies, such as rivers and open wells. Old and leaky piped water supplies also become more susceptible to contamination at this time as pathogens become more readily diffused in the wet earth around underground pipes. The deterioration of water quality during the wet season in rural Malawi has been quantified by Lindskog and Lindskog (1988) in terms of total coliforms, faecal coliforms and faecal streptococci in wells, rivers and springs. A similar pattern was recorded by Moore *et al.* (1965) in Costa Rica, Feachem *et al.* (1978) in Lesotho, and Barrell and Rowland (1979) in Gambia. However, Wright (1986) records a gradual increase in pollution during the dry season, interpreted as an increasing concentration of bacteria as water volumes decreased. Although Collins (1993) has shown that the cholera epidemic in Quelimane, Mozambique took off at the driest part of the year, just before the onset of rains, further research in Quelimane, Beira and Gorongosa has shown that faecal coliforms generally decrease as water levels lower during the dry season.

These differences may support the view that patterns of faecal contamination which increase during the wet season do not serve as definite indicators of cholera seasonality if *V. cholerae* is controlled by an additional range of environmental control factors. For example, in monitoring non-O1 *V. cholerae* in Lima sewage lagoons, Ventura *et al.* (1992) show that a seasonal variation in incidence was distinct but that there was no seasonal variation of faecal coliforms. Moe *et al.* (1991) conclude that indicator bacteria may mislead in that only beyond a certain threshold will they correlate well with incidence of diarrhoea, thus directing attention to important factors in transmission other than water contamination. It should also be remembered here that sampling of stored household water often reveals that contamination with bacteria continues to take place after collection. This has been shown by Lindskog and Lindskog (1988), Awad el Karim *et al.* (1985) and Swerdlow *et al.* (1992). The seasonal aspect of post-collection contamination might relate to varying care with personal hygiene dependent on the seasonal abundance or scarcity of water. With respect to rainfall and salinity, consideration might also be given to the implications of more favourable reservoirs for *V. cholerae* as wells become more saline but still potable during the dry season. This, combined with lower water levels and increased concentration of vibrios, might account for surges of cholera in coastal areas during a dry period.

Temperature has also been considered as a direct controlling factor. In the 19th century it was recognized that there were summer outbreaks of cholera between the isotherms at 60°F (16°C) and 80°F (27°C) and limits of 2–4 inches (5–10 cm) per month of rain (May, 1960). We are now able to consider the mechanisms behind this correlation, such as the possible role or synonymous behaviour of large algal and plankton blooms which increase with higher temperatures. The seasonal variation in nutrient levels in aquatic reservoirs caused by increased use of fertilizers and run-off from irrigation also play a role here. Seasonal differences in sunlight is a relevant factor if *V. cholerae* in drinking water is inactivated by direct ultraviolet (UV) irradiation (Acra *et al.*, 1989). However, a study by MacKenzie *et al.* (1992) carried out in Ecuador found that there was only a reduction in vibrios for bottled water at the high-altitude site of Quito and little change at the low-altitude site of Santo Domingo. The affects of UV radiation in the natural cholera reservoirs of the world is a complex factor of not only season, but also of altitude and availability of shade. The complexity of patterns of sunlight might therefore render it only minor in relevance as an identifiable controlling factor.

Cholera in the early 1990s became so widespread that there arises the possibility of more general climatic differences between zones north and south of the equator providing a continuum of favourable background conditions the whole year round, such as between the affected zones stretching the length of South and Central America or Africa. Variation of strains between regions and the relative lack of serotype immunity of people occupying one area against that of the other are likely to be subject to a seasonable dimension if vibrio types maintain different environmental sensitivities. A hint of this occurring is presented by the reciprocal pattern in prevalence of *V. cholerae* O139 with *V. cholerae* O1 in Bangladesh reported by Jesudason and Jacob John (1993).

The patterns of environmental factors outlined so far also need to be considered against seasonal variation in human well-being and changes in human behaviour that affect the susceptibility of the host.

8.3.1 SEASONALITY OF THE HOST

The correlation of incidence of cholera with poverty has been well documented and is referenced in the earlier part of this chapter. Seasonal poverty that impacts on cholera incidence relates to direct factors such as predisposing conditions of malnutrition and insufficient access to clean water at certain times of year, and indirect factors caused by a reduced coping mechanism in times of crisis. Further behavioural factors that are not necessarily linked to poverty might also include seasonal changes in diet and cultural practices that occur at specific times of year. Many of these factors are well covered in the edited work by Chambers *et al.* (1981) writing on seasonal dimensions to rural poverty.

In addition to the link between malnutrition and cholera infection through the pathophysiology of the lower intestine, there are further environmental and socioeconomic factors to be considered. In a multivariate study of association of malnutrition and diarrhoea in children aged under 5 years in a rural Sudanese community, El Samani *et al.* (1988) found that factors such as rainy season and a general socioeconomic effect were also consistent with this link. Arnold (1993) has illuminated how the seasonal onset of famine in 19th-century India resulted in the 'breakdown of normal social relations and produced a series of often dysfunctional behavioural responses', in terms of spreading epidemic diseases such as cholera. Survival strategies such as use of 'famine foods' and migration in search of food and work, facilitated epidemic diseases such as cholera in that the poor were further weakened. Weakening came about by the extra outlay of energy necessary to gather enough marginal foods and because these were coarse and unpalatable or poisonous unless thoroughly cooked. Interestingly, a rice-based diet, the indigenous staple diet of some of the world's cholera-prone regions is indicated as highly effective in the management of persistent diarrhoea in terms of improved nutrient absorption (Roy *et al.*, 1994). The optimum basis for this diet is rice powder, egg white, soya-bean oil, and glucose, with small additions of KCl, NaCl, $MgCl_2$ and $CaCl_2$. An extension of this discussion leads to a fundamental concept in holistic medicine in general, that the cause and remedies of sickness are found in the same place. Extraneous interference with this balance, such as occurred with the subsistence crises of 19th-century India, caused a breakdown of relative protection to certain communicable disease epidemics.

Non-availability of water is a major constraint in food hygiene and makes hand washing during food preparation and feeding a less frequent practice. During the dry season, contaminated surface streams and rivers often become the main sources of household water. This has been reported as a factor in Argentina near to the Bolivian Border (Chaudhary, 1992) and by Utsalo *et al.* (1992) in Calabar, Nigeria. Utsalo (1992) suggests an intensification in fishing and trade in fishery products carrying increased vibrio loads, occurs in the hot dry season when estuarine waters are shallow, and that this is a further factor in the seasonality of cholera. In Trujillo, Peru a multivariate study found that going to a fiesta was one of the associating factors as El Tor *V. cholerae* O1 grows rapidly in many cooked foods (Swerdlow *et al.*, 1992). Increased seasonal mobility at planting and at harvest time and religious ceremonies that attract large numbers of people into a small area have been further factors correlating with communicable disease transmission in general. Risks associated with seasonal ceremonies at spiritual sites on the River Ganges and migration to Mecca are well-known epidemic hazards of this type.

8.4 RECENT THOUGHTS ON DIFFUSION OF INCIDENCE OF CHOLERA

The spread of disease cannot be explained without understanding the spatial variation of any one moment. There are no simple diffusion models that can be easily applied to communicable diseases, such as cholera, which spread across space and through people. Non-uniformity of space influencing the diffusion of cholera includes differences in environmental reservoirs of *V. cholerae*, climate, settlements, social and economic well-being, and culture. Biological differences between the people who are exposed to the pathogen constitute a further non-uniformity guiding cholera transmission. The distribution of susceptibles, infectives and immunes is complex. Biological differences which may be considered as risk factors in contracting cholera are now thought to include lowered levels of gastric acid, not breast feeding, and presence of O blood group (Glass *et al.*, 1985, 1991, 1992; Van Loom, 1993). The nature of the pathogen is a third variable determining the spread of cholera in that different serotypes display a varying degree of virulence.

8.4.1 LOCATIONAL VULNERABILITY AND THE ECOLOGY OF RISK

As the location and persistence of cholera is determined by a combination of physical environmental phenomena, people, and pathogenesis, change to any one of these elements is likely to alter the distribution of cases. Diffusion of the disease therefore occurs against a backdrop of spatially defined environmental risk factors, a dynamic distribution of susceptible, infective and immune people, and a changing biology of the organism responsible. As mobility and local disruption in the late 20th century has increased globally, potential opportunity for communicable disease transmission has become complex. Consequently, there may be more relevance in identifying the nature of the terrain that the disease becomes established in rather than individual cases of transmission.

The role of terrain factors in the diffusion of cholera incidence was observed in the Quelimane epidemic (Collins, 1993). In its initial stages the main focus of incidence moved from one side to the other of the main city area. The shift was to areas identified as environmentally preferable to the longevity of *V. cholerae* in terms of a higher salinity and away from its first appearance in the zone with the highest population density. The new focus area became the one with the highest rate of incidence throughout the rest of the epidemic. Further analysis recorded areas with significantly higher than expected numbers of cholera cases for each month based on a method formerly employed by Ferguson (1975). This revealed that the majority of localized peaks in incidence were non-consecutive and non-contiguous and that 87% of all significant peaks, subsequent to the initial one, were in areas recording significantly higher salinity in the aquatic environment from which people drink. The controlling effect of the aquatic environment on cholera diffusion has also been hinted at by Marsden (1992) commenting on the spread of cholera down the Amazon basin. He suggests that the River Negro might be inimical to cholera as it has passed through large amounts of decomposed vegetable matter and is therefore acid in the range of pH 5–6, not suitable to *V. cholerae*. The importance of surface water in transmission in rural Bangladesh has been recorded by Hughes *et al.* (1982).

Susceptibility may be caused by factors such as hypochlorhydria in famine situations, as outlined above. However, it should be added here that it is not the only factor that involves changes to the human body. For example, alcoholism causes a loss of the resistant acids in the stomach. This has been suggested as an explanation of a particularly severe case of cholera in Belize (Hospedales, 1992). Baine *et al.* (1974) suggested that individuals with gastric hypoacidity from surgery, disease, or medication were particularly susceptible to cholera. Although investigations have shown that cholera can be transmitted by a variety of seafoods, a low pH in these same foods could inhibit transmission. St Louis *et al.* (1990) identified a lower risk of cholera in Conakry, Guinea when people ate the food with acidic tomato sauce (pH 3) rather than a neutral peanut-based sauce (pH 7). Breast-feeding in rural Bangladesh has been demonstrated as beneficial in protecting children up to the age of 36 months by Clemens *et al.* (1990). However, a study conducted by Bhattacharya *et al.* (1992) established that, in a cholera endemic area, about one-third of the admitted children up to the age of 2 years could have cholera and that even children below the age of 6 months were not spared, despite the fact that the average duration of predominant breast-feeding was about 4.7 months since birth.

Once established, there are numerous additional mechanisms through which **secondary transmission** may occur. They include factors such as drinking water contaminated during storage, going to festivals where food and beverages are prepared by multiple food-handlers, eating vegetables from farms which use untreated sewage to irrigate crops, and the consumption of raw seafood. Monitoring of diffusion of the disease through routes of secondary transmission is made difficult by the high proportion of symptom-free cases. This has been particularly prevalent with the El Tor strain. Symptom-free infections and mild to moderate cases of diarrhoea have been known to outnumber severe cases by as much as 100 to 1, particularly where the disease is endemic (Woodward and Mosley, 1971). Routine public health strategies aimed at controlling the diffusion of the disease via infected persons are not effective as most infected travellers are probably symptom-free.

8.4.2 HUMAN MOBILITY

The spread of cholera through and between nations by mobile populations is dependent on who, when and why people travel. For example, travel by people from developed countries into and out of endemic areas in the Third World has been proven to be relatively insignificant in terms of cholera transmission, as travellers such as these are generally socially and economically not so susceptible. One US report puts the risk of cholera to nationals travelling to affected areas at 1 per 500 000 travellers (Snyder and Black, 1982). Among the people living in the suburbs of Quelimane, many of whom were newcomers from the rural areas, the rate was more typically 200 per 10 000 people.

A part of the increased susceptibility of forceably displaced people undoubtedly relates to a decreased socioeconomic status during a period of instability. People forced to migrate due to war or famine usually need several growing seasons to re-establish adequate levels of food production to become self-sufficient again. The association of large-scale population displacement and subsequent resettlement with serious outbreaks of communicable diseases such as cholera is well documented (Prothero, 1977, 1994; Dick, 1984; Simmonds, 1984; Shears and Lusty, 1987; Guglielmetti *et al.*, 1992). In the 1970s and 1980s, millions of Africans became refugees often in densely populated camps and cholera may be diffused further when the refugees move on. Processes of interaction and the combined effects of changed environmental circumstances and population displacement on health have only been briefly indicated by some accounts (Dick, 1984, 1985; Meade, 1987; Prothero, 1994). The need for greater attention to be paid to this has become particularly urgent as more rapid change occurs to the natural environment and population stability in several parts of the world, not least sub-Saharan Africa. More specifically, there is a need for more informed planning strategies aimed at risk reduction in vulnerable regions and among vulnerable communities. A first stage might be in the identification of the geographical risk areas. A second, in identifying the role of population movements in influencing incidence of cholera operating either through: (i) changes in the local environment caused by the resettling of displaced people; and (ii) changes in environmental circumstance experienced by displaced people which make them more susceptible to health problems.

8.5 THE ECOLOGY OF CHOLERA IN CONTEXT

In addition to the specific determinants of disease incidence, wider contextual factors need to be considered. Particular emphasis is made here of war, famine, population displacement, and changing economic circumstances in developing countries. These are phenomena not entirely independent of each other; what happens at one scale is not only relevant to what is going on at another, but is also reciprocal. Major change, such as that caused by increased mass displacement of people from one zone to another and accelerated modifications to local environments through depletion of natural resources, often originates in structural developments occurring at national and international scale.

Structural phenomena determine the settlement patterns, congestion and environmental risk that impact on the distribution of cholera. The environmental factors of climate, soil, rainfall, temperature, altitude and seasonality that have been listed as possible proximate determinates of child mortality by Mosley and Chen (1984) and the added features of slope of terrain, population density and crowding within households listed by Blacker (1991) affect disease distributions in particular ecological settings, while simultaneously being affected by wider structural contexts.

Decisions of individuals to use clean water, observe basic **sanitational standards** and use appropriate health care – should these be available – are to a large part dependent on knowledge imparted through education. The rate of morbidity from cholera in any population may therefore be considered as the result of the interaction between hazards present in the environment and the ability of the population to defend itself against those hazards. Defences are generally those provided by the government or the community, such as provision of clean water and education, and those utilized by individuals, such as boiling drinking water and observing basic standards of hygiene. This also, however, depends on having the appropriate means and resources to take action, and therefore incidence of the disease is linked to wider-reaching economic considerations.

For example, if provision of basic infrastructures were good enough the world over, there would be little opportunity for the faecal–oral route of *V. cholerae* contamination, and incidence of cholera would be very minor or not occur at all. However, with a brief resume of the state of the world's water and sanitation supply, it can be quickly appreciated that contamination is not likely to be eradicated in the immediate future. In 1980, WHO estimated that only 20% of the world's population had access to totally safe drinking water, that 80% of all sick were suffering from diseases related to poor water and sanitation, and that 6 million children every year die from diarrhoeal diseases primarily associated with bad water, cholera being just one of the pathogens responsible.

During the United Nations Drinking Water and Sanitation Decade of the 1980s, access to uncontaminated water increased, with WHO figures suggesting that between 1980 and 1990 more than 1.6 million additional people were provided with access to water of reasonable quality. However, this barely kept pace with population growth and about 1 billion people still lack an adequate water supply with 1.7 billion people lacking adequate sanitation facilities (World Bank, 1992; Briscoe,

1993). There is also the harsh reality that many of those registered as officially having access to clean water still in fact drink polluted water. In short, the daunting scale of the effort required to eradicate the world's water and sanitation problems must mean that any additional preventative action that can be taken to reduce incidence should be explored with urgency. Greater appreciation of the ecological context of cholera, which is environmental, social, political and economic, is a step in the right direction for finding alternative and more sustainable solutions for controlling the disease.

Human conflict is a context of prime importance in understanding a proportion of today's cholera. Poignant demonstrations of its effects are the increased prevalence of refugee camps and associated cholera epidemics, as described by Mulholland (1985), Moren *et al.* (1991), Shears and Lusty (1987), and Sørenson and Dissler (1986), as with the cholera that has accompanied the aftermath of the Iraq war and among the survivors of the Rwandese genocide. The concentration of hundreds of thousands of Rwandan refugees into zones across the border in Tanzania and Zaire led to a rapid depletion of vegetation for firewood in those areas. Among the prevailing difficulties of environmental degradation, a continuous risk of cholera breaking out has been reported for the case of Tanzania. The long-term implications of this combination of disasters is not yet clear. Some 25 000 people a day arrived in Zaire from Rwanda. Among the over 1 million newcomers in the Goma area, Oxfam estimated that up to 3000 a day were dying of cholera. The total death toll by August 1994 was 13 000 (BBC, 1994). Conditions were so bad that it caused a return movement back across the border, despite continued security concerns. Many of those returning remained infected with cholera. Meanwhile, back in Rwanda other reports described a potentially good harvest, one of the best years for a while. Thus, disease and malnutrition relates specifically to human conflict in this region. In Angola, potentially one of the richest countries in Africa, harvests for 1994 would also have been good, but ongoing intensive fighting and displacement of peoples lives caused widespread malnutrition. Cholera and other communicable diseases abounded, in contrast to near disappearance of cholera in Mozambique during this period, and where a successful peace accord has been maintained since October 1992.

Thirdly, structural adjustment policies aimed at picking up the economies of some Third World countries have often been criticized as not giving priority to human well-being. UNICEF in Zimbabwe (1989) stated that 'adjustment programmes are rending the fabric of African society. Of the estimated half a million child deaths in 1988 which can be related to the reversal or slowing down of development, approximately two thirds were in Africa'. Loewenson (1993) argues that 'Structural Adjustment Programs have been associated with increasing food insecurity and undernutrition, rising ill-health, and decreasing access to health care in the two-thirds or more of the population of African countries that already lives below poverty levels'. The loss of a proactive health policy framework, a widening gap between the affected communities and policy makers, and the replacement of the underlying principle of equity in and social responsibility for health care are outlined in the same study.

Poorer countries' attempts to re-establish credit-worthiness with the International Monetary Fund, the World Bank and the international finance community after debt accumulation meant a tightening of public expenditure. Consequently, Peru's expenditure on health fell from $18.4 to $13.4 per head between 1980 and 1985 (Donovan, 1991). Average daily caloric intake dropped almost 30% since 1982, to well below accepted minimum standards, and in 1992 malnutrition rates approached 60% (Labonte, 1992). Meanwhile, the cost of cholera to Peru in 1991 was about 25% of its total budget. It has been suggested that cholera in the food and water of Peru found the population of that country to have little resistance as it was weakened by economic restructuring, and that policymakers had paid little attention to the public health impact of their policies.

At a minimum, socioeconomic factors can prove to be confounding variables when observing direct environmental links with cholera. This shortcoming in the general field of health and environment in urban areas of developing countries is highlighted by Stephens and Harpham (1992) who outline neglect of socioeconomic factors among household level studies. Outcomes determined by macroeconomics, settlement pattern, human connectivity, cost of transport, or concentrations of health care provision are generally visible and spatially definable phenomena and important in interpreting distributions of affected people and places. Intervention strategies therefore need to heed the relationship between the ecology of the areas and the social status of the inhabitants and this needs to be continuously monitored in the context of wider-reaching environmental changes and structural factors.

8.5.1 THE CASE OF STRUCTURAL CHANGE IN MOZAMBIQUE

The recent investigations carried out in Mozambique discovered that awareness of the causes, and ways of reducing the risk of getting cholera, among the inhabitants of Beira and Quelimane was generally good, but that people were unable to take action due to rising costs of living. Also, many interviewed said that on occasions they could not pay the 'nominal' fee that is charged for one bucket of water from the tap or protected well and were therefore forced to drink from the heavily contaminated ones. Very few people were found to be boiling the water as most could not afford to pay for the additional fuelwood or charcoal which is having to be transported increasingly large distances as more localized stocks dwindle.

Soaring prices in Mozambique, a result of structural adjustment policies implemented in the second half of the 1980s, hit the poorest and most vulnerable to infection more. For example, the rising cost of living caused by the Economic Reform Programme forced the poorer of those who had moved into the former colonial cement areas in the years following independence to move out to wattle-and-grass housing in the environmentally marginal and congested zones of the city. The new occupiers of the cement areas are often expatriates or part of Mozambique's small minority of emergent *nouveau riche.*

The water supply has been in need of renovation for many years. The relative 'boom town' scenario that has been unfolding in the last 3 years created a sense of urgency for it to be improved. The piped water quality is now much better and further work will ensure a more continuous supply. But water must now be paid for, properties containing taps have become privately owned, and the supply of water to the areas where the masses live in conditions of extreme poverty has not been extended. In many of these areas the existing public water sources have continued to deteriorate. Since independence was gained from Portugal, thousands have benefited from a sporadic supply to former colonial mansions occupied by their compatriots. However, the order of change has meant that taps are disappearing behind high walls where guard dogs patrol. This has not occurred in all cases; consequently the remaining taps in properties not yet bought out or in those occupied by people who did not catch the 'all mine' syndrome, have become inundated with long queues which begin before the supply is connected each dawn. Many people do not have time for this, and therefore continue to drink the notoriously contaminated water from the city's open concrete and oil drum wells. Thus in Beira, for some, improvements in the infrastructure are creating a good quality of life, but for the majority things at the best remain the same, or have deteriorated. Improvements in water supply structure that only benefit one sector of a community is likely to be fed back into only limited reduction in water-related health problems. To use the economic metaphor, 'it doesn't trickle down'.

Impacts of structural adjustment may become increasingly observable through the distribution of well-being measured in terms of health, and more specifically here, incidence of cholera. Meanwhile in Quelimane the main development in water provision since economic reform begun has been in the provision of more wells and bore holes with hand pumps, assisted by several NGOs. Sadly, many of these stand idle due to breaking down, there being a lack of adequate follow-up in pump maintenance. In some cases, and in particular with the bore-holes, bacterial contamination was reduced or removed, but the taste of the bore-hole water influenced by the pumping mechanism, is bad and people prefer to use the traditional wells, simple holes in the ground with 'sweet' water. Many of these are found to harbour high levels of contamination in terms of faecal coliforms. However, a few have been found to have low levels of contamination. More effort could be spent in determining how to make the best use of these. In Mozambique, as in many parts of the world, more effective strategies in reducing the amount of contaminated water that is actually consumed lies with the health and education sectors rather than solely the engineers. Unfortunately, in Mozambique, these are two sectors that have made only limited progress since their extensive disruption in the war of destabilization. Cliff and Noormahomed (1988) describe how after independence in 1975, Mozambique began successfully to implement a primary health care policy but how the preventative programmes became severely disrupted as units were attacked, transport destroyed and health workers kidnapped or killed. This report also records how the effects on health included an increase in mortality rates, famine and infectious disease epidemics including cholera.

8.6 A GEOGRAPHICAL RESEARCH STRATEGY

A recent increase in information on the epidemiology and ecology of cholera, improvements in the detection of *V. cholerae* in people and the environment, and the versatility of computerized information systems encourages a development in research approaches. This is coupled with a genuine need for surveillance of incidence of cholera, early warning systems, and a flexible preventative strategy appropriate to unique circumstances.

8.6.1 THE VALUE OF RECENT RESEARCH

Recent information on the ecology of *V. cholerae* assists in the identification of environmental zones with characteristics supporting higher disease incidence. This can aid implementation strategies for avoiding areas of primary transmission. Identification of processes of interaction leading to the creation of these unfavourable environmental conditions is necessary to avoid the development of further susceptible zones. With increased exposure of humans to varied environments caused by increased mobility and more rapidly changing local environments, the pattern of incidence of cholera will continue to alter. Understanding new patterns as they emerge may be particularly relevant in planning risk-reduction strategies by determining the availability of more favourable zones or areas of avoidance for communities that suffer from recurring epidemics and in the resettlement of refugees and displaced populations. Where an environmental reservoir is believed to assist endemicity of cholera, consideration should be given to the consequences of geophysical changes that create conditions that increase the survival period and toxigeneity of *V. cholerae*. In already settled areas, the identification of, and intervention in, processes of environmental change leading to the creation of physical conditions favourable to *V. cholerae*, such as salinization and alkalinization, should be considered as a factor in preventing the primary transmission and subsequent persistence of cholera.

8.6.2 MONITORING CHOLERA

Monitoring levels of *V. cholerae* in the physical environment using improved methods of detection, which allow for the identification of vibrios maintained in a temporary state of dormancy, has the capacity to assist in delineating the distribution of possible zones of primary transmission. Deriving patterns of secondary transmission may be assisted by more rapid techniques for detecting *V. cholerae* in cholera patients by use of coagglutination assay (Abbott and Janda, 1993) and rapid diagnostic kits (Andersson *et al.*, 1992; Carillo *et al.*, 1994). The importance of epidemiological surveillance for communicable disease monitoring has been emphasized in several recent disaster situations. Toole and Waldman (1988) indicated its relevance among refugees in Somalia, Sudan and Thailand, and Moren *et al.* (1991) among Mozambican refugees in Malawi. Siddique *et al.* (1992) outline the absence of reliable surveillance as one of the factors responsible for more deaths during the cholera epidemics in Bangladesh between 1985 and 1991. They call for 'effective simple and representative surveillance in the country which could provide early warning signals for impending outbreaks'. Labonte (1992) has extended the concept of surveillance to include community-based approaches in an account of the spread of cholera in South America in 1991. Quick *et al.* (1993), on epidemic cholera in the Amazon, suggests a long-term strategy which includes 'an effort to provide health training to residents of every village, and to develop a coordinated system of communication to provide a means of disseminating information rapidly'.

Babille *et al.* (1994), in monitoring health trends among 25 000 displaced Kurds, call for international agencies and NGOs in the field to make use of available technology to monitor health and disease among refugees, returnees and the permanently displaced. With their experience from work with Iraqi–Kurdish refugees in Iran, limitations are recognized in assessing changing risks and the proportion of risk attributable to the disaster and post-disaster experience. They suggest that more use should be made of existing information systems and complementary records so as to establish databases that could be used for epidemiological research, adding that 'Comparative data and data on trends from cross-sectional and prospective studies ... can help to identify and solve specific health problems when the international donors begin to lose interest'.

8.6.3 GEOGRAPHIC HEALTH INFORMATION SYSTEMS (GHIS)

Given the present status of health information systems what does a geographical approach have to offer? Firstly, cross-sectional and spatio-temporal techniques for environmental associative analyses are powerful tools of medical geographic research and their continued worth has been outlined in this chapter. The search for associations between variables using an essentially spatial dimension helps to reduce over simplified 'cause-and-effect' relationships.

Environmental zoning with indication of compatibility for human habitation with respect to health is a further development that could be applied in cholera-prone regions. Probability mapping indicating sublocations with significantly higher than expected numbers of cholera cases for individual periods of an epidemic has been included by Ferguson (1977) and Collins (1993). These can be used to indicate persistence of the disease in preferred environments and the movement of the epidemic focus to those areas. Analysis of combined factors or principal components includes use of what Jacquez and Kheifets (1993) refer to as 'synthetic variables' in their analysis of clustering of cancer; 'synthetic risk maps' can be generated from these. Residuals from multiple regression can also be spatially represented in sub-areas and then tested for contiguity. However, there are limitations in the use of sub-locations due to variation in disease clustering when the scale and size of sub-unit are altered. These are subjects explored in some detail by Openshaw (1977, 1984), Waller and Turnbull (1993) and Morris and Munasinghe (1993).

Spatial approaches such as those mentioned above currently benefit from the rapid development of Geographical Information Systems (GIS). These are data bases which have been combined with graphics software for mapping and spatial representation. There are many versions and space here does not allow a full discussion of their differences. However, in general, as a research tool one of their uses is in enabling integration of layers of spatially referenced and associated attribute data allowing analysis on phenomena where considerable amounts of disparate information is involved. For example, in the area

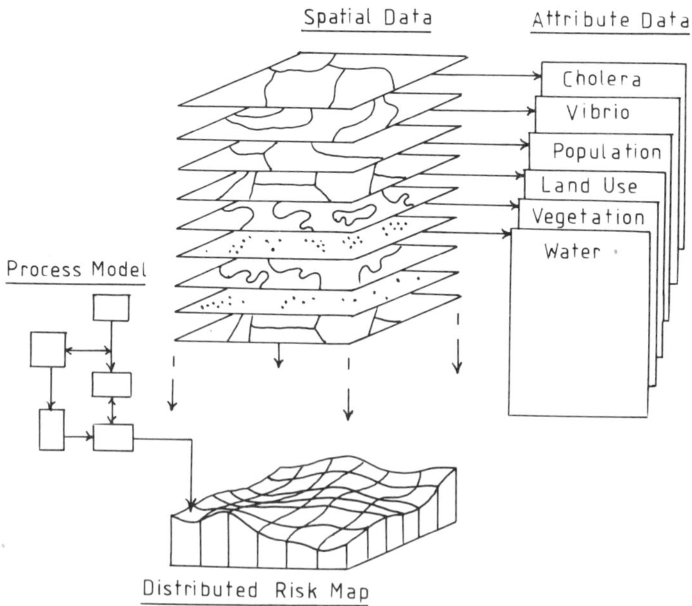

Fig. 8.4 Geographic Information System (GIS) overlay analysis producing risk map.

of environment and cholera, spatially distributed data can include attribute values such as rates of incidence in sublocations, levels of *V. cholerae* contamination and other indicator bacteria, population density, urbanization indexes, soils, water and more detailed geoecological measures of pH and salinity (Figure 8.4). Point positions and attributes, such as the location of individual households which are infected, can also be included. Wider investigations are able to use some of the same techniques to integrate empirical sociogeographic distributions, the GIS assisting as a tool for maximizing part of this procedure. The application of some developments in socioeconomic analysis using GIS to integrate conventional physical resource data with socioeconomic information in the context of benefiting land use planning is discussed by Martin and Bracken (1993).

Satellite imaging can also be integrated into GISs and, when combined with distribution maps of vector-borne diseases, has already been used. Examples are provided by Rogers and Randolph (1991), Rogers and Williams (1993) and Epstein *et al.* (1993b). A further aid in disease mapping is the use of Global Positioning Systems (GPS), such as that used by Snow *et al.* (1993) to map malaria cases on the coast of Kenya. This technology may be used rapidly to record the geographical locations of events and the stored information can be used interactively with a GIS. 'The GPS receiver becomes a digitizing cursor and the earth is the digitizer's table' (Van Demark, 1992). Abler (1993) suggests that a great attraction will be in the ability to 'create new maps interactively, in real time, amid the real phenomena the maps represent, while at the same time referring to and revising the background maps for the area contained in portable GIS's carried into the field'. In this manner GIS/GPS combinations could be applied to improving disease surveillance systems, particularly in emergencies such as those outlined in this chapter.

GIS also facilitates manipulation of data sets to produce new layers of information, for example, the intersection of overlays (Figure 8.4) to produce new representations of distributions not immediately apparent previously. The ready availability of integrated information in time series enables zones of environmental constraint to be identified and a three-dimensional distributed risk map can be produced (Figure 8.4). The distributed cells of risk are derived from accumulated values generated by the layers of information. Cells most typically represent geographical space, but could equally be made to represent sectors of a community. In other areas of physical geography cells in such a map become part of a larger distributed model within which processes of change can be analysed, as with catchment run-off modelling. The degree to which processes of change in each cell can be accurately represented is the degree to which the overall model can predict. The system is deterministic and it remains to be evaluated what such an extension to Geographic Health Information Systems (GHIS) might benefit given the range of possibilistic considerations associated with health, as have been outlined in this chapter.

8.6.4 A NOTE ON PREVENTION

An essentially geographical and holistic outline of the ecology of cholera has been presented through discussion of the organism, its physical and human environment, and links to the wider world. While incidence of cholera may relate to changing environmental factors affecting the ecology of the disease pathogen, changes in infrastructure and the wider development process are influencing the nature of local environments.

This contemporary geographical perspective on cholera is valuable in that it provides understanding of the issues borne out of a multifaceted research strategy with due attention to circumstances unique in place and time. On the one hand it provides a framework for empirical investigations of patterns and processes associated with cholera, while on the other it is able to combine contextual understanding of the wider issues involved.

It is exactly this perspective that is required in tackling the ongoing challenge that a rapidly changing world ecology of communicable disease presents. The most appropriate preventative strategies and interventive action will arise out of a programme mindful of this approach.

ACKNOWLEDGEMENT

Many thanks to Richard Black (King's College, London) for essential advice and comments on a draft of this chapter.

8.7 REFERENCES

Abbott, S.L. and Janda, J.M. (1993) Rapid detection of acute cholera in airline passengers by coagglutination assay. *Journal of Infectious Diseases*, **168**, 797–9.

Abler, R.F. (1993) Everything in its place: GPS, GIS, and Geography in the 1990s. *Professional Geographer*, **45**(2), 131–9.

Acra, A., Jurdi, M., Mu'Allem, H., Karahagopian, Y. and Raffoul, Z. (1989) Sunlight as disinfectant. *Lancet*, **i**, 280.

Adesina, H.O. (1987) The temporal and spatial analysis of cholera diffusion within Ibadan, Nigeria, 1971–1974, in *Health and Disease in Tropical Africa: Geographic and Medical Viewpoints*, (ed. R. Aktar), Harwood Academic Publishers, London, pp. 167–92.

Andersson, N., Morales, A. and Gillespie, S.H. (1992) Cholera: getting the message across. *British Medical Journal*, **304**, 1243–4.

Arnold, D. (1993) Social crisis and epidemic disease in the famines of nineteenth-century India. *Social History of Medicine*, **6**(3), 385–404.

Awad El Karim, M.A., El Hassan, B.M. and Hussien, K.K. (1985) Social and public health implications of water supply in arid zones in the Sudan. *Social Science and Medicine*, **20**, 393–8.

Babille, M., Colombani, P.De., Guerra, R., Zagaria, N. and Zanetti, C. (1994) Post-emergency epidemiological surveillance in Iraqi–Kurdish refugee camps in Iran. *Disasters*, **18**(1), 58–75.

Baine, W.B., Zampieri, A., Mazzotti, M. *et al.* (1974) Epidemiology of cholera in Italy in 1973. *Lancet*, **ii**, 1370–4.

Barrell, R.A.E. and Rowland, M.G.M. (1979) The relationship between rainfall and well water pollution in a West African (Gambian) village. *Journal of Hygiene (Cambridge)*, **83**, 143.

Barua, D. and Greenough, W.B. III (1992) *Cholera*, Plenum Medical Book Company, New York.

BBC (1994) *World Service for Africa, Focus on Africa News Reports*, London.

Bergstrom, S.K.D. and Ramalingaswami, V. (1992) Health, in *An Agenda of Science for Environment and Development into the 21st Century*, (eds J.C.I. Dooge, G.T. Goodman, J.W.M. la Rivière and J. Marton-Lefèvre *et al.*), Cambridge University Press, Cambridge, pp. 119–24.

Bhattacharya, S.K., Bhattacharya, M.K., Ramamurthy, T. *et al.* (1992) Cholera in young children in an endemic area. *Lancet*, **340**, 1549.

Bhattacharya, S.K., Bhattacharya, M.K., Balakrish Nair, G. *et al.* (1993) Clinical profile of acute diarrhoea cases infected with the new epidemic strain of

Vibrio cholerae O139: designation of the disease as cholera. *Journal of Infection*, **27**, 11–15.

Blacker, J.G.C. (1991) Infant and child mortality: development, environment, and custom, in *Disease and Mortality in Sub-Saharan Africa*, (eds R.G. Feachem and D.T. Jamison), Oxford University Press, Oxford, pp. 75–86.

Briscoe, J. (1993) When the cup is half full. Improving water and sanitation services in the developing world. *Environment*, **35**(4), 7–37.

Carillo, L., Gilman, R.H., Mantle, R.E. *et al.* (1994) Rapid detection of *Vibrio cholerae* O1 in stools of Peruvian cholera patients by using monoclonal immunodiagnostic kits. *Journal of Clinical Microbiology*, **32**(3), 856–7.

Cash, R.A., Music, S.I., Cibonati, J.P., Snyder, M.J., Wenzel, R.P. and Hornick, R.B. (1974) Response of man to infection with V. Cholerae I. Clinical, serologic and bacteriological responses to a known innoculum. *Journal of Infectious Diseases*, **129**, 45–52.

Chambers, R., Longhurst, R. and Arnold, P. (eds) (1981) *Seasonal Dimensions to Rural Poverty*, Francis Pinter, Exeter.

Chaudhary, V. (1992) Argentina waits for summer and cholera. *British Medical Journal*, **305**, 542–3.

Chongsa-nguan, M., Chaicumpa, W., Moolasart, P. *et al.* (1993) *Vibrio cholerae* O139 Bengal in Bangkok. *Lancet*, **342**, 430–1.

Clemens, J.D., Sack, D.A., Harris, J.R. *et al.* (1990) Breastfeeding and the risk of severe cholera in rural Bangladesh children. *American Journal of Epidemiology*, **3**, 400–11.

Cliff, J. and Noormahomed, A.R. (1988) Health as a target: South Africa's destabilization of Mozambique. *Social Science and Medicine*, **27**(7), 717–22.

Collins, A.E. (1993) Environmental influences on the distribution of the incidence of cholera: a case study in Quelimane, Mozambique. *Disasters*, **17**(4), 321–40.

Colombo, M.M., Francisco, M., Ferreira, B.D., Rubino, S. and Cappuccinelli, P. (1993) The early stage of the recurrent cholera epidemic in Luanda, Angola. *European Journal of Epidemiology*, **9**(5), 563–5

Colwell, R.R. and Spira, W.M. (1992) The ecology of *Vibrio cholerae*, in *Cholera*, (eds D. Barua and W.B. Greenough III), Plenum Medical Book Company, New York, pp. 107–27.

Colwell, R.R., Brayton, P., Grimes, D.J., Roszak, D.R., Huq, S.A. and Palmer, L.M. (1985) Viable, but non-culturable *Vibrio cholerae* and related pathogens in the environment: implications for release of genetically engineered micro-organisms. *Bio/Technology*, **3**, 817–20.

Council of Environmental Quality (C.E.Q.) (1981) *Global Future – Time to Act*, (ed. G. Speth), Government Printing Office, Washington DC.

Craig, M. (1988) Time-space clustering of *Vibrio cholerae* O1 in Matlab, Bangladesh, 1970–1982. *Social Science and Medicine*, **26**(1), 5–13.

Dick, B. (1984) Diseases of refugees – causes, effects and control. *Transactions of the Royal Society of Tropical Medicine and Hygiene*, **78**, 734–41.

Dick, B. (1985) The impact of refugees on the health status and health services of host communities: compounding bad with worse? *Disasters*, **9**(4), 259–69.

Donovan, P. (1991) 'Collateral Damage', *South Magazine*, August, p. 28.

Doran, M., Shillan, P., Hoffman, R.E., MacFarland, L.M. (1989) Toxigenic Vibrio Cholerae O1 infection acquired in Colorado. *Morbidity and Mortality Weekly Report*, **38**(2), 19–20.

Drasar, B.S. (1992) Pathogenesis and ecology: the case of cholera. *Journal of Tropical Medicine and Hygiene*, **95**, 365–72.

El Samani, E.F., Willett, W.C. and Ware, J.H. (1988) Association of malnutrition and diarrhea in children aged under five years. *American Journal of Epidemiology*, **128**(1), 93–105.

Epstein, P.R. (1992) Cholera and the environment. *Lancet*, **339**, 1167–8.

Epstein, P.R., Ford, T.E. and Colwell, R.R. (1993a) Marine ecosystems. *Lancet*, **342**, 1216–19.

Epstein, P.R., Rogers, D.J. and Sloof, R. (1993b) Satellite imaging and vector-borne disease. *Lancet*, **341**, 1404–6.

Feachem, R.G., Burns, E., Cairncross, A. *et al. (1978) Water, Health and Development: an Interdisciplinary Evaluation*, Tri-Med, London.

Ferguson, A.G. (1977) Probability mapping of the 1975 Cholera epidemic in Kisumu District, Kenya. *Journal of Tropical Geography*, **44**, 23–32.

Glass, R.I., Holmgren, J., Haley, C. *et al.* (1985) Predisposition for cholera of individuals with O blood group: possible evolutionary significance. *American Journal of Epidemiology*, **121**, 791–6.

Glass, R.I., Claeson, M., Blake, P.A., Waldman, R.J. and Pierce, N.F. (1991) Cholera in Africa: lessons on transmission and control for Latin America. *Lancet*, **338**, 791–5.

Glass, R.I., Libel, M. and Brandling-Bennett, A.D. (1992) Epidemic cholera in the Americas. *Science*, **256**, 1524–5.

Guglielmetti, 'P., Bartoloni, A., Roselli, M. *et al.* (1992) Population movements and cholera spread in Cordillera Province, Santa Cruz Department, Bolivia. *Lancet*, **340**, 113.

Haines, A. and Parry, M. (1993) Climate change and human health. *Journal of the Royal Society of Medicine*, **86**, 707–11.

Higa, N., Honma, Y., Albert, M.J. and Iwanaga, M. (1993) Characterization of *Vibrio cholerae* O139 synonym Bengal isolated from patients with cholera-like disease in Bangladesh. *Microbiology and Immunology*, **37**(12), 971–4.

Hospedales, C.J. (1992) Cholera in Belize. *West Indian Medical Journal*, **41**, 88–9.

Howe, G.M. (1982) Disease and environment, in *Disease and the Environment: Proceedings of the Inaugural Conference of the Society for Environmental Therapy*, (eds A.R. Rees and H.J. Purcell), Oxford, 21–23 March 1981, Wiley, Chichester, pp. 1–9.

Hughes, C.C. and Hunter, J.M. (1970) Disease and 'development' in Africa. *Social Science and Medicine*, **3**, 443–93.

Hughes, J.M., Boyce, J.M., Levine, R.J. *et al.* (1982) Epidemiology of El Tor cholera in rural Bangladesh: importance of surface water in transmission. *Bulletin of the World Health Organization*, **60**, 395–404.

ICDDR,B (1993) International Center for Diarrhoeal Diseases Research, Bangladesh. Large epidemic of cholera-like disease in Bangladesh caused by *Vibrio cholerae* O139 synonym Bengal. *Lancet*, **342**, 387–90.

Islam, M.S., Hasan, M.K., Miah, M.A. *et al.* (1993) Isolation of *Vibrio cholerae* O139 Bengal from water in Bangladesh. *Lancet*, **342**, 430.

Jacquez, G.M. and Kheifets, L.I. (1993) Synthetic cancer variables and the construction and testing of synthetic risk maps. *Statistics in Medicine*, **12**, 1931–42.

Jesudason, M.V. and Jacob John, T. (1993) Major shift in prevalence of non-O1 and El Tor *Vibrio cholerae*. *Lancet*, **341**, 1090–1.

Kaysner, C.A., Abeyta, C. Jr, Wekell, M.M., De Paola, A. Jr, Stott, R.F. and Leitch, J.M. (1987) Incidence of *Vibrio cholerae* from estuaries of the United States West Coast. *Applied and Environmental Microbiology*, **53**(6), 1344–8.

Kovda, V.A. (1972) Soil preservation, in *The Environmental Future*, (ed. N. Polunin), Macmillan, London.

Labonte, R. (1992) South America's cholera pandemic provides lesson in public health, politics. *Canadian Medical Association Journal*, **147**(7), 1052–6.

Learmonth, A. (1988) *Disease Ecology*, Blackwell, Oxford.

Lida, T., Shrestha, J., Yamamoto, K. and Honda, T. (1993) *Lancet*, **342**, 926.

Lindskog, R.U.M. and Lindskog, P.A. (1988) Bacteriological contamination of water in rural areas: an intervention study from Malawi. *Journal of Tropical Medicine and Hygiene*, **91**, 1–7.

Loewenson, R. (1993) Structural adjustment and health policy in Africa. *International Journal of Health Services*, **23**(4), 717–30.

MacKenzie, D.T., Ellison, R.T. III and Mostow, S.R. (1992) Sunlight and cholera. *Lancet*, **340**, 367.

Mandal, B.K. (1993) Epidemic cholera due to a novel strain of *Vibrio cholerae* non-O1: the beginning of a new pandemic. *Journal of Infection*, **27**, 115–17.

Marsden, P.D. (1992) Cholera. *British Medical Journal*, **304**, 1170–1.

Martin, D. and Bracken, I. (1993) The integration of socioeconomic and physical resource data for applied land management information systems. *Applied Geography*, **13**, 45–53.

Martins, M.T., Sanchez, P.S., Sato, M.I.Z., Brayton, P.R. and Colwell, R.R. (1993) Detection of *Vibrio cholerae* O1 in the aquatic environment in Brazil employing direct immunofluorescence microscopy. *World Journal of Microbiology and Biotechnology*, **9**, 390–2.

May, J.M. (1950) Medical geographical: its methods and objectives. *Geographical Review*, **40**, 9–41; reprinted *Social Science and Medicine*, **11**, 715–30.

May, J.M. (1960) *Disease Ecology*, Hafner, New York.

Mbwette, T.S.A. (1987) Cholera outbreaks in Tanzania. *Journal of the Royal Society of Health*, **4**, 134–6.

McCormack, W.M., Mosley, W.H., Fahimuddin, M. and Benenson, A.S. (1969) Endemic cholera in rural East Pakistan. *American Journal of Clinical Nutrition*, **25**, 1236–42.

Meade, M.S., Florin, J. and Gesher, W. (1987) *Medical Geography*, The Guildford Press, London, New York.

Merson, M.H., Black, R.E., Moslemuddin, K. and Huq, I. (1978) Epidemiology of cholera and enterotoxigenic *Escherichia coli* diarrhoea, in *Cholera and Related Diarrhoea, 43rd Nobel Symposium, Stockholm 1978*, Karger, Basel, pp. 34–45.

Miller, C.J., Drasar, B.S. and Feachem, R.G. (1982) Cholera and estuarine salinity in Calcutta and London. *Lancet*, **i**, 1216–18.

Miller, C.J., Drasar, B.S. and Feachem, R.G. (1984) Response of toxigenic *Vibrio cholerae* O1 to stresses in aquatic environments. *Journal of Hygiene (Cambridge)*, **93**, 475–95.

Miller, C.J., Feachem, R.G. and Drasar, B.S. (1985) Cholera epidemiology in developed and developing countries: new thoughts of transmission, seasonality, and control. *Lancet*, **ii**, 261–3.

Miller, C.J., Feachem, R.G. and Drasar, B.S. (1986) The impact of physiochemical stress on the toxigenicity of *Vibrio cholerae*. *Journal of Hygiene (Cambridge)*, **96**, 49–57.

Miyaki, K., Iwahara, S., Sato, K., Fujmoto, S. and Aibara, K. (1967) Basic studies on the viability of El Tor Vibrios. *Bulletin of the World Health Organization*, **37**, 773–8.

Moe, C.L., Sobsey, M.D., Samsa, G.P. and Mesolo, V. (1991) Bacterial indicators of risk of diarrhoeal disease from drinking-water in the Philippines. *Bulletin of the World Health Organization*, **69**(3), 305–17.

Moore, H.A., de la Cruz, E. and Vargas-Mendez, O. (1965) Diarrhoeal disease studies in Costa Rica. *American Journal of Epidemiology*, **82**, 162–4.

Moren, A., Stefanagge, S., Antona, D., Bitar, D., Gastellu Etchegorry, M., Tchatchioka, M. and Lungu, G. (1991) Practical field epidemiology to investigate a cholera outbreak in a Mozambican refugee camp in Malawi, 1988. *Journal of Tropical Medicine and Hygiene*, **94**, 1–7.

Morris, R.D. and Munasinghe, R.L. (1993) Aggregation of existing geographic regions to diminish spurious variability of disease rates. *Statistics in Medicine*, **12**, 1915–29.

Mosley, W.H. and Chen, L.C. (1984) An analytical framework for the study of child survival in developing countries. *Population and Development Review*, **10** (suppl.), 25–45.

Mulholland, K. (1985) Cholera in Sudan: an account of an epidemic in a refugee camp in Eastern Sudan, May–June 1985. *Disasters*, **9**(4), 247–8.

Nalin, D.R., Levine, R.S., Levine, M.M. and Hoover, D. (1978) Cholera non-vibrio cholera and stomach acid. *Lancet*, **ii**, 856–9.

Openshaw, S. (1977) A geographical study of scale and aggregation problems in region-building, partitioning, and spatial modelling. *Transactions of the Institute of British Geographers*, **2**, 459–72.

Openshaw, S. (1984) *The modifiable areal unit problem*, CATMOG 38. Geo Books, Norwich.

Owen, M., Headworth, H.G. and Morgan-Jones, M. (1991) Groundwater in basin management, in *Applied Groundwater Hydrology: a British Perspective*, (eds R.A. Downing and W.B. Wilkinson), Clarendon Press, Oxford, pp. 16–34.

Pan American Health Organization (1991) Cholera situation in the Americas. *Epidemiological Bulletin*, 12:2.

Pandit, C.G., Pal, S.C., Murti, G.V.S., Misra, B.S., Murty, D.K. and Shrivastav, J.B. (1967) Survival of *Vibrio cholerae* biotype El Tor in well water. *Bulletin of The World Health Organization*, **37**, 681–5.

Pavlovskiy, E.N., Petrishcheva, P.A., Zarukhin, D.N. and Olsofiev, N.G. (eds) (1955) Natural nidi of human diseases and regional epidemiology. Medgiz, Leningrad.

Prescott, L.M. and Bhattacharjee, N.K. (1969) *Bulletin of The World Health Organization*, **40**, 980–2.

Preston, N.W. (1993) Cholera isolates in relation to the 'eighth pandemic'. *Lancet*, **342**, 925–6.

Prothero, R.M. (1977) Disease and human mobility: a neglected factor in epidemiology. *International Journal of Epidemiology*, **6**, 259–67.

Prothero, R.M. (1994) Forced movements of population and health hazards in tropical Africa. *International Journal of Epidemiology*, **23**(4), 657–64.

Quick, R.E., Vargas, R., Moreno, D. *et al.* (1993) Epidemic cholera in the Amazon: the challenge of preventing death. *American Journal of Tropical Medicine and Hygiene*, **48**(5), 597–602.

Reed, D. (ed.) (1993) *Structural Adjustment and the Environment*, Earthscan, London.

Rich, B. (1994) *Mortgaging the Earth. The World Bank, Environmental Impoverishment and the Crisis of Development*, Earthscan, London.

Robinson, E. (1991) Peru battles 12,600 cases of cholera. *The Washington Post*, 14 February.

Rogers, D.J. and Randolph, S.E. (1991) Mortality rates and population density of tsetse flies correlated with satellite imagery. *Nature*, **351**, 739–41.

Rogers, D.J. and Williams, B.G. (1993) Monitoring trypanosomiasis in space and time. *Parasitology*, **106**, S77–92.

Roy, S.K., Akramuzzaman, S.M., Haider, R. *et al.* (1994) Persistent diarrhoea: efficacy of a rice-based diet and role of nutritional status in recovery and nutrient absorption. *British Journal of Nutrition*, **71**, 123–34.

Rozenzweig, C., Parry, M., Fischer, G. and Frohberg, K. (1992) *Climate Change and World Food Supply – a Preliminary Report*, Environmental change unit, University of Oxford, Oxford.

Rutherford, G.W. and Mahanjane, A.E. (1985) Morbidity and mortality in the Mozambican famine of 1983: prevalence of malnutrition and causes and rates of death and illness among dislocated persons in Gaza and Inhambane provinces. *Journal of Tropical Pediatrics*, **31**, 143–9.

Salazar-Lindo, E. (1993) Recent developments in gastrointestinal infections with a focus on cholera. *Current Opinion in Infectious Diseases*, **6**, 41–7.

Sarkar, B.L., De, S.P., Sircar, B.K. *et al.* (1993) Polymyxin B sensitive strains of *Vibrio cholerae* non-O1 from recent epidemic in India. *Lancet*, **341**, 1090.

Scrimshaw, N.S., Taylor, C.E. and Gordon, J.E. (1968) Interactions of nutrition and infection, *WHO Monograph Series, no. 57*, WHO, Geneva.

Shears, P. and Lusty, T. (1987) Communicable disease epidemiology following migration: studies from the African famine. *International Migration Review*, **21**(3), 783–95.

Siddique, A.K., Baqui, A.H., Eusof, A. *et al.* (1991) Survival of classic cholera in Bangladesh. *Lancet*, **337**, 1125–7.

Siddique, A.K., Zaman, K., Baqui, A.H. *et al.* (1992) Cholera epidemics in Bangladesh: 1985–1991. *Journal of Diarrhoeal Disease Research*, **10**(2), 79–86.

Simmonds, S. (1984) Refugees, health and development. *Transactions of the Royal Society of Tropical Medicine and Hygiene*, **78**, 726–33.

Singleton, F.L., Attwell, R.W., Jangi, M.S. and Colwell, R.R. (1982) Influence of salinity and nutrient concentration on survival and growth of *Vibrio cholerae* O1 in aquatic microcosms. *Applied and Environmental Microbiology*, **43**, 1080–5.

Snow, J. (1855) *On the Mode of Communication of Cholera*, London.

Snow, R.W., Schellenberg, J.R.M., Peshu, N. *et al.* (1993) Periodicity and space–time clustering of severe childhood malaria on the coast of Kenya. *Transactions of the Royal Society of Tropical Medicine and Hygiene*, **87**, 386–90.

Snyder, J.D. and Black, P.A. (1982) Is cholera a problem for US travellers? *Journal of the American Medical Association*, **247**, 1495–9.

Sørensen, E. and Dissler, K. (1986) Practical experience with the management of a cholera outbreak in a refugee camp in Eastern Sudan, 1985. *Disasters*, **12**(3), 274–81.

Stephens, C. and Harpham, T. (1992) Health and environment in urban areas of developing countries. *Third World Planning Review*, **14**(3), 267–82.

St Louis, M.E., Porter, J.D., Helai, A. *et al.* (1990) Epidemic cholera in West Africa: the role of food handling and high-risk foods. *American Journal of Epidemiology*, **131**, 719–28.

Stock, R. (1986) 'Disease and development' of 'The underdevelopment of health': a critical review of geographical perspectives on African health problems. *Social Science and Medicine*, **23**(7), 689–700.

Swerdlow, D.L. and Ries, A.A. (1993) *Vibrio cholerae* non-O1: the eighth pandemic? *Lancet*, **342**, 392–3.

Swerdlow, D.L., Mintz, E.D., Rodriguez, M. *et al.* (1992) Waterborne transmission of epidemic cholera in Trujillo, Peru: lessons for a continent at risk. *Lancet*, **340**, 28–32.

Tamplin, M.L. and Parodi, C.C. (1991) Environmental spread of *Vibrio cholerae* in Peru. *Lancet*, **338**, 1216–17.

Thomason, H., Burke, V. and Gracey, M. (1981) Impaired gastric function in experimental malnutrition. *American Journal of Clinical Nutrition*, **34**, 1278–80.

Timberlake, L. (1985) *Africa in Crisis*, Earthscan, London.

Tivy, J. and O'Hare, G. (1989) *Human Impact on the Ecosystem*, Oliver and Boyd, Edinburgh.

Toole, M.J. and Waldman, R.J. (1988) An analysis of mortality trends among refugee populations in Somalia, Sudan, and Thailand. *Bulletin of the World Health Organization*, **66**(2), 237–47.

UNICEF (1989) in *Situation of Women and Children in Zimbabwe 1985–1990*, UNICEF/Government of Zimbabwe, Jongwe Printers, Harare.

Utsalo, S.J., Eko, F.O. and Antia-Obong, E.O. (1992) Features of cholera and *Vibrio parahaemolyticus* diarrhoea endemicity in Calabar, Nigeria. *European Journal of Epidemiology*, **8**(6), 856–60.

Van Demark, P. (1992) Exhibits redux: what I saw at GIS/LIS '91. *URISA News*, **119**, 5–6.

Van Loom, F.P.L. (1993) Cholera: developments in prevention and cure. *Tropical and Geographic Medicine*, **45**(6), 269–73.

Ventura, G., Roberts, L. and Gilman, R. (1992) *Vibrio cholerae* non-O1 in sewage lagoons and seasonality in Peru cholera epidemic. *Lancet*, **339**, 937–8.

Wachsmuth, I.K., Evins, G.M., Fields, P.I. *et al.* (1993) The molecular epidemiology of cholera in Latin America. *Journal of Infectious Diseases*, **167**, 621–6.

Waller, L.A. and Turnbull, B.W. (1993) The effects of scale on tests for disease clustering. *Statistics in Medicine*, **12**, 1869–84.

WHO (1948) *The Constitution*, WHO, Geneva.

WHO (1970) *Principles and Practice of Cholera Control*, WHO, Geneva.

WHO (1986) *Guidelines for cholera control*, WHO/CDD/SER/80.4RE.1.

WHO (1991) *Environmental Health in Urban Development*, WHO Technical Report Series No. 807, Geneva.

WHO (1993) *Weekly Epidemiological Record*, 21.

WHO (1995) *The World Health Report*, Geneva.

Woodward, W.E. and Mosley, W.H. (1971) The spectrum of cholera in rural Bangladesh. Comparison of El Tor, Ogawa and classical Inaba infection. *American Journal of Epidemiology*, **96**, 342.

World Bank (1992) *The World Bank Development Report 1992: Development and the Environment*, World Bank, Washington DC.

Wright, R.C. (1986) The seasonality of bacterial quality of water in a tropical developing country (Sierra Leone). *Journal of Hygiene (Cambridge)*, **96**, 75.

Xu, H.S., Roberts, N., Singleton, F.L., Attwell, R.W., Grimes, D.J. and Colwell, R.R. (1982) Survival and viability of non-culturable *Escherichia coli* and *Vibrio cholerae* in the estuarine and marine environment. *Microbial Ecology*, **8**, 313–23.

The fate of *Vibrio cholerae* in wastewater treatment systems

9

T. Curtis

9.1 INTRODUCTION

Contaminated wastewater can, and has, contributed to the spread of cholera. Yet, despite the enormous amount of literature devoted to *Vibrio cholerae*, what happens to the organism in wastewater treatment plants (sewage works) has been largely overlooked. It is a shame, verging on a scandal, that although the molecular biology and physiology of the organism is known in considerable detail, the ecology of the organism in wastewater treatment plants can barely be sketched. The appearance of a new pandemic strain (Cholera Working Group, 1993) – a strain that appears to subvert the 'high-tech' preventative strategy of vaccination – places renewed emphasis on the need to control the organism in the environment and underscores our ignorance in this area.

Why is relatively little attention paid to the fate of *V. cholerae* in wastewater treatment works? This oversight may be attributed to a variety of misconceptions. For example, some believe that wastewater treatment works are a barrier to pathogenic organisms and that therefore sewage treatment will prevent the organism contaminating the environment. Others have assumed that *V. cholerae*'s behaviour in wastewater may be inferred from studies of *Escherichia coli* and other pathogen indicators; also the assumption – common until recently – that 90% of cholera is water-borne has led to public health workers forgetting the importance of other routes of transmission. In reality, most wastewater treatment plants are not barriers to pathogens. *V. cholerae* can be spread by many routes and its behaviour may be quite different from that of *E. coli* and other bacterial indicators.

Cholera and the Ecology of Vibrio cholerae. Edited by B.S. Drasar and B.D. Forrest.
Published in 1996 by Chapman & Hall, London. ISBN 0 412 61220 8.

I do not wish to be alarmist about the lack of information. Therefore, this chapter seeks to set the role of wastewater treatment in a wider context before discussing our knowledge (or rather ignorance) in more detail. The term sewage and wastewater are used interchangeably.

9.2 WHY WE PRODUCE WASTEWATER

To get a clear picture of the role of wastewater treatment it is instructive to take a step back and think why we produce and treat wastewater.

The use of water to carry excreta away, in modern times, has its roots in the cholera outbreaks in the rapidly expanding cities of western Europe in the 1800s. The English experience is archetypal, but not unique. At the dawn of the 19th century English sewers were intended solely for the drainage of surface water. Indeed, in Westminster, London it was a penal offence to put excreta in a sewer until 1815 (Sheppard, 1971). Cholera and other diseases were widespread and concern at the cost of supporting those affected by the disease combined with compassion for those most affected led to the commissioning of a *Report on the Sanitary Conditions of the Labouring Population of Great Britain* (Chadwick, 1842). Written in ignorance of the microbial causes of disease, the report identified the miasma produced by the large amounts of stagnant human excrement as a principal cause of disease. At the time, the bulk of the population was obliged to use pits dug in the ground (middens and cesspits) as toilets. This inadequate system was further strained by the rapid urbanization and the introduction of the water closet (flush toilet). This latter innovation was credited with encouraging frequent overflows of cesspits. The inquiry concluded that self-cleansing sewers were the best and cheapest way of removing this waste and that 'for these purposes, as well as for domestic use, better supplies water were absolutely necessary' (Chadwick, 1842).

The authors of the report were aware that sewage would pollute rivers. However, they calculated that level of pollution in the River Thames would only be doubled and reasoned that river pollution was a lesser evil than an excreta-ridden environment. Since wastewater treatment is still the exception rather than the rule, even in many developed countries (World Bank, 1992), this latter reasoning appears to be used to this day.

9.3 WHY SEWAGE TREATMENT WAS INTRODUCED

The widespread use of sewage to transport wastes led to a very large increase in the pollution of watercourses, leading to the terrible smells and, with hindsight (Sheppard, 1971), even worse water-borne outbreaks of cholera because drinking water was pumped, untreated, from the vicinity of the outfalls. Ignorant of the microbial nature of disease, but powerfully motivated by the stench of the River Thames, the problem in London was initially overcome by laying more sewers to carry the waste away from the town and controlling the discharge of the sewage. After 10 years the river became overloaded and sewage treatment was introduced to prevent the organic pollution of the receiving waters. In the succeeding years a wide variety of methods have evolved, by accident, design and experimentation (Sidwick, 1977). However, the earliest standards were framed in the first years of this century (Royal Commission on Sewage Disposal, 1912). Microbiology was in its infancy and the emphasis was on the removal of organic pollutants rather than pathogenic organisms. Consequently, most standard designs of wastewater treatment will not remove significant numbers of pathogens unless specific and costly disinfection procedures are included in the design (the exception being a system known as a waste stabilization pond or WSP).

Nevertheless, it is self-evident that the relative inability of these systems to remove cholera and other pathogenic organisms has not prevented the sanitary revolution from reducing the excreta related diseases to acceptable levels in western European countries.

9.4 THE ROLE OF WASTEWATER TREATMENT IN CHOLERA PREVENTION

Belief in a disease causing miasma has been replaced by a sophisticated understanding of the causes of infectious disease in general and cholera in particular. Nevertheless, the spirit of the recommendations of the early sanitary reformers have been endorsed by recent studies confirming the benefit of adequate excreta disposal facilities (Esrey *et al.*, 1991). The lessons of the 19th century have been well learnt, perhaps too well learnt, since the existence of alternatives to sewerage (Cairncross and Feachem, 1993) is sometimes overlooked and the early epidemiological work on cholera has been reduced almost to a myth by simplification and generalization (Feachem, 1982). Outbreaks of cholera are often accompanied by calls for improved sanitation and wastewater treatment (Mcabe, 1970; Nichols, 1991; Tauxe *et al.*, 1994). Unfortunately, the role of wastewater treatment in the prevention of cholera is often misunderstood. For example, in response to the disastrous outbreak of cholera in Peru, the Commission of the European Community required all Peruvian factories producing fruit and vegetables for export to the community to **chlorinate** their wastewater (EEC, 1991). For some reason it was felt that chlorinating the water **after** it had been used for food processing would protect the quality of the food. Pathogens are also found in wastewater in the European Community, but no such requirements are demanded of European food processors; indeed, in many European countries a significant proportion of all wastewater is not even properly treated, much less chlorinated.

Although raw sewage should not be returned to a water course or reused without adequate treatment, the complete removal of *V. cholerae* may not be required. In those countries where cholera persists, only a limited amount of money is available for the provision of sanitation. Therefore, it is important to consider in what circumstances the removal of *V. cholerae* is important. To spend money to remove the cholera vibrio when there is no need to do so may divert funds and administrative expertise from more pressing problems.

It is only necessary to remove cholera from wastewater if the failure to do so will result in the transmission of the disease. Typically, this occurs if the treated wastewater contaminates food that is eaten raw. For example, the irrigation of salad vegetables with wastewater was strongly implicated in the cholera outbreak in Jerusalem in the 1970s (Shuval *et al.*, 1986) and case control studies in the Peruvian outbreak (Swerdlow *et al.*, 1992) have implicated cabbages which may have been irrigated with contaminated water. Raw or inadequately cooked seafood has also been strongly implicated in the spread of cholera (Merson *et al.*, 1977).

To prevent the transmission of disease by these routes a number of organizations have suggested guidelines or set standards for unrestricted irrigation (the irrigation of food crops eaten raw) and for waters used for harvesting shellfish. For example, current WHO guidelines for unrestricted irrigation advise that wastewater may be used for this purpose if the numbers of faecal coliforms are reduced to 1000/100 ml and there is less than 1 nematode egg/litre (Mara and Cairncross, 1989; WHO, 1989). The validity of the use of faecal coliforms as indicators is discussed below. Of course guidelines and standards are only as valid as the process by which they were derived. The WHO guidelines for the reuse of wastewater in irrigation (Mara and Cairncross, 1989; WHO, 1989) are based on a rigorous review of the available epidemiological data (Shuval *et al.*, 1986) which sought to distinguish between the actual risks of disease and the potential or hypothetical risks. The review concluded that the irrigation of non-food crops with wastewater had no impact on the transmission of cholera or other bacterial diseases. However, the review did find there to be a moderate risk of the transmission of cholera transmission when food crops were irrigated with wastewater. Accordingly, the guideline of 1000 faecal coliforms/100 ml was proposed. The assumption underlying this guideline is that, since there are typically 10^7 faecal coliforms/100 ml in raw sewage, the 4-log reduction in faecal coliforms will be associated with sufficient reduction in the bacterial pathogens in the wastewater to eliminate the organism. This rationale has been corroborated for organisms other than *V. cholerae* O1 (Oragui *et al.*, 1987).

Treatment guidelines of this type have the disadvantage of being based on the use of faecal coliforms; this approach will be discussed later.

The thought of wastewater, even if treated, being used for irrigation may seem alarming to some. However, in many areas of the world water resources are limited and there is competition between agriculture and urban populations for water. The use of wastewater for irrigation can release water supplies for use by an urban population whose health standards were previously compromised by lack of water (World Bank, 1992).

It has become apparent that in some areas of the world certain waters may act as a reservoir for *V. cholerae* (Islam *et al.*, 1993). If conditions in a particular water course favour the survival of *V. cholerae* then wastewater treatment may reduce the numbers of the organism present. However, it is unlikely that wastewater treatment, no matter how rigorous, will be able to prevent a water body being colonized by *V. cholerae*. This is in part because no treatment method can guarantee to remove 100% of *V. cholerae* 100% of the time, and sooner or later one organism will 'get through' and partly because contamination may occur from uncontrolled sources such as passing boats, casual defecation and sewer overflows that operate during high rainfall. Moreover, a casual comparison of the impact of colonization of waters off the Gulf Coast of the USA (where a small number of cases occur sporadically) (Shandera *et al.*, 1983), and the waters of Bangladesh (where large outbreaks are common) (Glass *et al.*, 1982) would suggest that the impact of this colonization is a reflection of the prevailing level of hygiene and sanitation rather than the presence of the organism.

9.5 A WORD ABOUT WATER TREATMENT

Wastewater treatment before disposal is no substitute for adequate treatment of water abstracted for drinking-water supplies. However, the collection and treatment of wastewater may also help protect sources of drinking water, even if the cholera vibrio is not completely removed from the sewage. This is because treatment will reduce the pathogen load on the water treatment works. All too often, such water treatment systems are poorly designed and badly run, dogged by the use of inappropriate technologies and poor supplies of vital chemicals. There is even less information on the fate of cholera in water treatment works than there is in sewage treatment plants (Dixo, 1994).

9.6 HOW THE REMOVAL OF PATHOGENS FROM WASTEWATER TREATMENT WORKS IS ASSESSED

Pathogen removal from wastewater treatment plants has traditionally been expressed in terms of the removal of bacterial faecal indicators. This group of organisms has been borrowed from the water supply industry where they were used to indicate the presence of faecal contamination. Sewage is, of course, implicitly faecally contaminated so nothing can be inferred from the presence of bacterial faecal indicators in wastewater. However, the organisms are easy to enumerate, usually non-pathogenic, and occur in large numbers. The organisms are therefore monitored on the assumption that they will act as analogues of the less numerous and more difficult to count pathogenic bacteria. Feachem and Mara (1979) have suggested the term 'pathogen indicator' to describe the use of these organisms to monitor the numbers of pathogens in water.

Ideally, pathogen indicators should be used in situations where there is evidence that the indicator is a valid analogue of the pathogens of concern. However, because the pathogens of concern are often difficult or impossible to enumerate, this condition is rarely fulfilled. In the absence of such evidence some indication of the validity of the indicator may be sought by determining how closely the pathogen and the indicator are related. For example, the observation that indicator bacteria are poor analogues of enteric viruses is common place (Bitton, 1994). Since bacteria and viruses are such completely different biological entities, this is not surprising. However, bacteriophage (bacterial viruses) which do resemble enteric-viruses have been found to be reasonable indicators of viruses (Bitton, 1994).

The most widely used pathogen indicator is the faecal coliform, which is also referred to as a thermotolerant coliform. In tropical (north-east Brazil) wastewater, most (>90%) faecal coliforms are *E. coli* (Curtis, 1987, 1991). The degree of similarity between two bacteria can be assessed by comparing the degree of similarity between the cell's phenotype (growth characteristics, shape and so on), DNA and ribosomal RNA (rRNA). A detailed comparison of the relationships between different groups of organisms requires a polyphasic approach (the use of phenotypic and genetic information). However, with respect to *E. coli* and *V. cholerae* the overall picture is approximately represented by comparisons of the rRNA sequences in the two organisms.

Ribosomal RNA performs the same function in all living things, guiding the translation of mRNA (RNA transcripts of DNA) into protein. Consequently its sequence is highly conserved and a comparison of the rRNA sequences helps to define the evolutionary relationship between different organisms. On this basis it can be seen that *E.*

coli and *V. cholerae* have a common evolutionary ancestry as they both fall within the same grouping of the gamma subdivision of the purple-photosynthetic phyla of the eubacteria (Woese *et al.*, 1985). However, a more detailed analysis of the vibrio-enteric group shows that considerable divergence has occurred from the indicators and pathogens common ancestors and that *E. coli* and *V. cholerae* are members of distinct and varied bacterial families: the Enterobacteriacae and the Vibrionacae (MacDonell *et al.*, 1986). Thus it appears that *E. coli* is no more related to *V. cholerae* than it is to any of the Vibrionaceae, a diverse family that includes marine vibrios and photobacteria. Interestingly, *V. cholerae* O1 and another organism (*V. mimicus*) have recently been shown to form a distinct group (at the genus level) when compared with members of the family Vibrionacae, including other marine vibrios (Kita-Tsukamoto *et al.*, 1993). This picture is corroborated by phenotypic studies (West *et al.*, 1986).

What is the significance of these evolutionary relationships to sanitary engineering? These relationships show that there is no reason to suppose that the most common pathogen indicator (*E. coli*) will behave like *V. cholerae*. Comparisons of the fate of *E. coli* and *V. cholerae* in estuarine waters (Peterson *et al.*, 1984) and microcosms of estuarine waters (Hood and Ness, 1982; Guthrie and Scovill, 1984) have not found the former to be an adequate indicator of the latter. However, some of these studies were examining *V. cholerae* non-O1, which begs the question: is non-O1 an adequate indicator of O1, or indeed O139?

9.7 AQUATIC ECOLOGY OF *V. CHOLERAE*

A knowledge of the behaviour of *V. cholerae* in aquatic systems may help in the understanding of the behaviour of *V. cholerae* in wastewater treatment plants.

Our understanding of the aquatic ecology of *V. cholerae* has changed over the past 15 years with the realization that the organism was an aquatic bacterium which infected humans rather than a human pathogen that occasionally entered bodies of water (Feachem, 1982; West, 1989). Moreover, it has become apparent that natural waters, particularly estuarine waters, may act as a reservoir for the spread of *V. cholerae* (Islam *et al.*, 1993). It has been implied that wastewater treatment systems (Anonymous, 1992) may also act as a reservoir for *V. cholerae* and that changes in the numbers of the *V. cholerae* non-O1 were associated with rise and fall of the levels of cholera in the surrounding population. The aquatic environment in wastewater treatment plants will vary within and between designs (Table 9.1). Nevertheless, they are, almost invariably, relatively high nutrient fresh water environments in which there may well be intense competition for nutrients.

Table 9.1 A summary of the generalized differences between the four main types of wastewater treatment mentioned in the text. Details of salinity, temperature and concentrations of sodium ions and organic carbon typical of wastewater are given in the text

System	Dissolved oxygen (mg/l)	pH	Higher organisms	Retention times (days)	Comments
Anaerobic	0	5.5–7.0	Not reported	0.5–15	H_2S present
Suspended growth	1–3	7.5	Protozoa	0.2–1.5	
Attached growth	0–6	7.5	Protozoa, algae	0.05–0.3	Large areas of biofilm
Waste stabilization pond (WSP)	0–25	7.5–9.5	Protozoa, algae, zooplankton	6–30	Intense photooxidation, H_2S present

Depending on the specific design, conditions may include presence of algae, protozoa, higher plants and large areas of biofilm.

V. cholerae is usually associated with brackish water (Kaper *et al.*, 1979; Lee *et al.*, 1982) and has an absolute requirement for sodium ions (Singleton *et al.*, 1982a; Miller *et al.*, 1984) while treated wastewater is typically of a relatively low salinity (0.05–1.8 dS/m, equivalent to 0.0023–0.1% NaCl) and sodium levels of between 5 and 300 mg/l (Sawyer and McCarty, 1978; Asano *et al.*, 1985; APHA, 1989; Ahmed and Chughtai, 1991). Wastewater has total organic carbon levels which range from 300 mg/l or more for raw sewage to about 15 mg/l in a final effluent. There is evidence that nutrient levels typical of wastewater may permit growth in fresh water (Singleton *et al.*, 1982b; Nair *et al.*, 1988). High temperatures (<10°C), the norm in tropical wastewater, have been reported to offset suboptimal salinities (Singleton *et al.*, 1982b), though other workers have found that this interaction may not occur at the very low (<0.1% NaCl) salinities usually found in wastewater (Miller *et al.*, 1984). Of course, higher temperatures alone favour the survival and growth of *V. cholerae* (Kaper *et al.*, 1979; Lee *et al.*, 1982).

The effect of pH on survival has also been shown to interact with salinity. Unfortunately, this phenomenon has not been studied at the salinities typical of wastewater. The optimum pH for survival was found to increase with decreasing salinity. At salinities equivalent to about 0.1% NaCl the survival is best at pH values of between 7.5 and 9.0 (Miller *et al.*, 1984). This range of pH values is typical of WSPs.

Though pH values as low as 6.5 have been found to permit survival, this was only observed at salinities >0.3% (Huq *et al.*, 1984; Miller *et al.*, 1984). It is not clear whether the nutrient levels typical of wastewater might ameliorate the deleterious effect of low pH on survival. Low pH (<6.5) values can be created in specially designed anaerobic digesters (Alexiou and Anderson, 1995). The effect of high oxygen concentrations on *V. cholerae* does not appear to have been studied; however, the anaerobic conditions found in some parts of wastewater treatment plants does not appear to be a barrier to growth (Youngren-Grimes *et al.*, 1988).

Sunlight, which kills through the intermediacy of toxic forms of oxygen (Curtis *et al.*, 1992) does appear to be able to remove *V. cholerae* in certain circumstances (Lema *et al.*, 1979; MacKenzie *et al.*, 1992). Protozoal predation has also been shown to be capable of reducing the number of *V. cholerae* in seeded river water (Chao *et al.*, 1988).

The realization that *V. cholerae* is an aquatic organism, and an interest in possible reservoirs, has led to suggestions that the bacteria may be associated with higher organisms, some of which may be found in wastewater treatment plants including: phytoplankton (Tamplin *et al.*, 1990; Islam *et al.*, 1994a), water hyacinth (Spira *et al.*, 1981) and freshwater amoebae (Thom *et al.*, 1992).

9.8 BARRIERS TO THE STUDY OF *V. CHOLERAE* IN WASTEWATER TREATMENT PLANTS

When considering what is, and is not, known about the fate of *V. cholerae* in wastewater treatment works (or indeed any pathogen in any ecosystem) it is best to be aware of the methodological and conceptual limitations of current knowledge.

The observation that the cholera vibrio may be detectable by a number of microscopical techniques when it is not detectable on standard culture media led to the hypothesis of the viable but not culturable (VBNC) bacterial cell. In essence this proposes that, under certain conditions, some or all bacteria may enter into a state where they remain alive (and if pathogenic, infective) but are unable to grow on standard culture media. The details of this debate (Barer *et al.*, 1993) are beyond the scope of this chapter. However, the VBNC hypothesis has highlighted the absence of a definition of a dead bacterial cell and that survival information based on culturability may not necessarily be taken at face value.

A less frequently cited limitation of our understanding of the fate of *V. cholerae* in water is our reliance on artificial, laboratory-based experimental systems. In the case of the ecology of *V. cholerae* it is best to conduct experiments that reflect the normal physiological state of the organism in question in the environment of interest (Brock, 1987). However, much of what we know about the behaviour of *V. cholerae* is based on studies that, for practical reasons, depart from this ideal because they employ bacteria cultured in the laboratory (for example, Huq *et al.*, 1984), often in solutions prepared in the laboratory (for example, Miller *et al.*, 1984). A bacterium's physiological condition is known to be affected by both the conditions in which it is grown (Klamen and Tuveson, 1982; Chamberlain and Moss, 1987) and the nutrient levels of its surroundings (Matin *et al.*, 1989; Watson, 1990). A cholera vibrio grown in a laboratory culture may be quite different from one that has grown in the gut and then passed through a sewerage system into a wastewater treatment system. Moreover, chemicals inherently present in wastewater may have an unsuspected role in the survival of an organism; for example, humic substances play a key role in the effect of light on microorganisms in WSPs (Curtis *et al.*, 1992). Departing from the ideal of studying naturally occurring microorganisms in the environment of interest does not invalidate the experiment. However, it does mean that the results of the study must be interpreted with caution. For example, Miller *et al.* (1984) suggested – on the basis of laboratory-based experiments – that salinity would exert a strong influence on growth and survival at low salinities, irrespective of the nutrient concentration and temperature. However, when Nair *et al.* (1988) actually looked at freshwater environments in and around Calcutta they found that *V. cholerae* was present and evidently growing at very high concentrations in waters of very low salinity. If lives and economies depend on a proper understanding of the ecology of an organism in a certain ecosystem, there is no trustworthy substitute for studying that organism in that ecosystem.

Since the detection of *V. cholerae* by culture (that is, by growth on media) is suspect and the study of organisms in the laboratory is inadequate, truly authoritative studies on the fate of *V. cholerae* in wastewater treatment requires that real systems are studied and that both culture and non-culture-based enumeration methods are employed. No such studies have been published so far.

9.9 *V. CHOLERAE* CONCENTRATIONS IN WASTEWATER

The amount of pathogenic *V. cholerae* in raw sewage is an important consideration when designing for the removal of pathogens. The numbers of pathogenic *V. cholerae* will clearly be related to the numbers of carriers and sick people using the sewerage system. There is little published information on the numbers of organisms in raw sewage in endemic and epidemic situations. Fortunately, the numbers of organisms excreted by carriers and the sick has been studied (Dizon *et al.*, 1967). This may be related to the point prevalence rate (the numbers of sick and carriers at one moment in time), water use and the weight of faeces excreted daily to give an indication of the numbers of *V. cholerae* O1 present in sewage by using the following formula:

$$V.\ cholerae/100\,\text{ml of sewage} = ((Vc \times f)/w \times 10) \times p$$

where Vc is a weighted mean of *V. cholerae*/g faeces, f is the mean weight of faeces excreted/person/day, w is the mean volume (litres) of water used/person/day, and p is the proportion of the population excreting the organism; 10 is a correction factor.

The values of the variables will depend on whether the excreters being considered are carriers or patients. For the carriers: $Vc = 7 \times 10^4$ organisms/g (Dizon *et al.*, 1967), f = 100 g/day (Feachem *et al.*, 1983) and p is unlikely to exceed 0.01 (a point prevalence of 1%; Feachem, 1982), w = 50 l/day. Experimental infections have suggested that most cholera patients have mild symptoms (Levine *et al.*, 1979), for such patients: $Vc = 10^8$ organisms/g (Dizon *et al.*, 1967), f = 4000 g/day (that is, 4 l of diarrhoea, assumed to have a density of 1; Levine *et al.*, 1979), w = 50 l/day, and p is unlikely to exceed 0.002 (a point prevalence of 0.2%).

Cholera patients with severe symptoms are unlikely to use the toilet and may be discounted in the calculations.

The equation and information outlined above have been used to estimate (Figure 9.1) the number of *V. cholerae* in raw sewage in populations with differing rates of sick and carriers.

The large estimated range of *V. cholerae* concentrations is consistent with the few published accounts of the numbers of organisms in raw sewage. Daniel and Lloyd (1980a) studying sewage in a refugee camp in Bangladesh and Martins *et al.* (1993) in urban São Paulo, Brazil failed consistently to isolate any *V. cholerae* O1 by culture. In north-east Brazil, Oragui *et al.* (1993) and Arridge (1995) found geometric means of 4.85 and 3.85 organisms/100 ml respectively. These reports are consistent with a low prevalence of asymptomatic infections, though it is possible that these workers have underestimated the numbers of *V. cholerae* O1 due to the inadequacy of the methods for the quantitative isolation of this organism from sewage. By contrast, Tamplin and Parodi (1991) working in Peru at the height of the recent epidemic, reported 10^5 and 10^2 organisms/100 ml in municipal sewage in Lima and Puno respectively. The levels observed in Lima are consistent with the lower end of the estimates above, probably reflecting genuine low point prevalence in the sewered population of Lima. The lower levels observed in Puno are attributed to the lower levels of disease found in this cooler city. This appears to corroborate the common sense assumption that the numbers of organisms in sewage reflects the numbers of excreters in

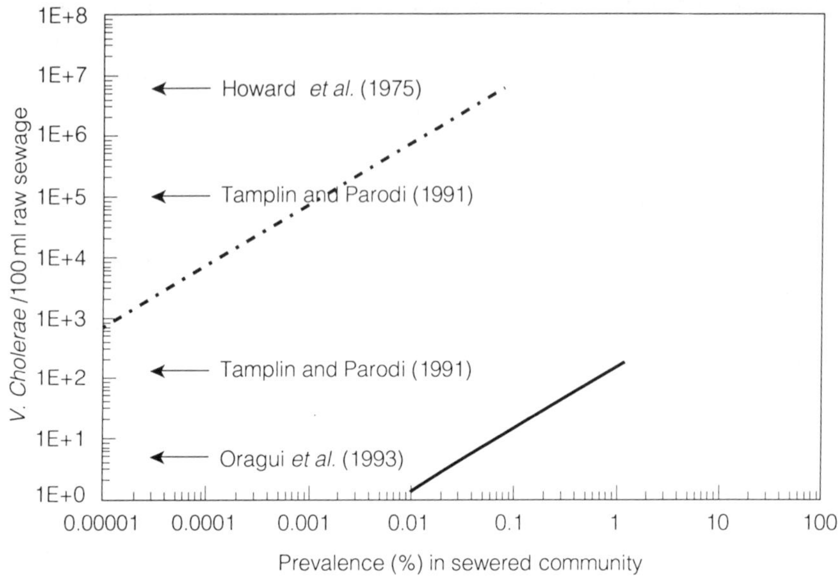

Fig. 9.1 Estimated (lines: ——, carriers; -·-·-·-, patients) and recorded (arrows) levels of *V. cholerae* O1 in raw sewage.

the population. When Howard *et al.* (1975) examined the sewage produced by a hospital admitting 2–40 confirmed cholera cases per day they found between 5×10^5 and 5×10^7 *V. cholerae* O1/100 ml – there were more cholera vibrios than coliforms! If such sewage is mixed with municipal wastewater the concentrations reaching the sewage works will probably be a great deal lower. Nevertheless, undiluted hospital wastewater is clearly a potent source of *V. cholerae* O1 and as such would require very high levels of treatment.

The estimated cholera concentration shows that a very low point prevalence, perhaps 1 or 2 people suffering from mild symptoms in a population of 100 000, could cause a vast increase in the numbers of *V. cholerae* in raw sewage and thus the load on the wastewater treatment system (if one is in use). If the sewage contaminated seafood, a food crop eaten raw, or a water supply, a large number of people could become infected quite quickly. Moreover, depending on the hydraulic flow regime of the sewerage system, the organisms may not be present in the sewage for 24 hours a day but appear sporadically at very high concentration, reflecting the defaecation of the small numbers of individuals suffering from the disease. Thus, grab samples may fail to detect or underestimate the numbers of organisms present in the sewage. There are no reports of anyone sampling and processing raw sewage samples over a 24-hour period (the logical alternative to grab sampling). Moore swabs and vaginal tampons function as crude non-quantitative 24-hour sampling devices, which may explain why they have been used successfully to monitor sewers for the presence of *V. cholerae* (Isaacson *et al.*, 1974; Barrett *et al.*, 1980; Pazzaglia *et al.*, 1993).

Feachem *et al.* (1983) estimated that tropical sewage would contain 1000 *V. cholerae*/100 ml. The key assumptions in this calculation appear to have been: (i) that those ill with cholera did not use the sewerage system; (ii) a prevalence of carriers of 1%; (iii) an average faecal mass of 100 g; (iv) water use of 100 l per person per day; and (v) that 90% of the organisms died shortly after defaecation. The discrepancy between this estimate and the reports of Tamplin and Parodi (1991) probably lies in the neglection of the mildly ill, whose faeces contain up to 10^8 vibrios (Dizon *et al.*, 1967), and may be more numerous than in the severely ill. Kott and Betzer (1972) suggest that there may be 10–10^4 organisms/100 ml, though the basis of this estimation is not stated. For design purposes, engineers would do well to assume a mean cholera concentration of about 10^5 organisms/100 ml.

As all the studies and estimates above are based on studies of culturability, one should consider whether the presence of VBNC *V. cholerae* in raw sewage has been overlooked. One might think that fresh raw sewage is one of the few environments where the use of culturability in detection may be valid. Since cholera sufferers may be expected to excrete viable *V. cholerae* and sewage has been reported to support culturability (Mukerjee *et al.*, 1961; Gerichter *et al.*, 1975) and even growth in warm (19–25°C) sewage (Feachem *et al.*, 1981, citing Altukhov *et al.*, 1975 and Zaidenov *et al.*, 1976). However, Martins *et al.* (1993) have used immunofluorescence (Brayton and Colwell, 1987; Huq *et al.*, 1990) to screen 1-day enrichment broths of sewage samples in São Paulo, Brazil. They reported that in all samples the presence of *V. cholerae* O1 was detected, even though sub-cultures of the broths were negative for this organism. Unfortunately, this paper is not quantitative, has not rigorously excluded the possibility of false positives, or sought to corroborate the apparent presence of the organism by using phage (Sechter *et al.*, 1975) or molecular-based detection techniques. Nevertheless, it raises the possibility that more sophisticated detection methods may yield new information about pathogenic *V. cholerae* in raw sewage.

In addition to the reports and estimates above, there are a number of reports of the presence and concentration of *V. cholerae* non-O1 in raw sewage in Europe (Muller, 1977; Martin and Bonnefont, 1989), South America (Corrales *et al.*, 1989; Hofer and Ernandez, 1990; Martins *et al.*, 1993), Bangladesh (Daniel and Lloyd, 1980a) and North Africa (Lesne *et al.*, 1991). These organisms appear to be almost ubiquitous in sewage and give no clue to the likely concentration of their more feared pathogenic relatives. In view of the limited public health significance of these organisms, their presence is of only academic interest, though they may act as indicators of the behaviour of the toxigenic *V. cholerae*.

9.10 THE FATE OF *V. CHOLERAE* IN WASTEWATER TREATMENT WORKS

As was noted above, most discussions of the prevention of cholera assume that sewage treatment systems are an integral part of any preventative strategy 'Can Safe ... Sewage Treatment Systems Be Built Efficiently throughout the Developing World' (Tauxe *et al.*, 1994). Yet in the words of Feachem *et al.* (1983):

> 'Those whose job is to select and design appropriate systems for the collection and treatment of sewage in developing countries must bear in mind that European and North American practices do not represent the zenith of scientific treatment, nor are they the product of a logical and rational and design process. Rather, treatment practices in developed countries are the products of history, a history that started about 100 years ago when little was known about the fundamental physics and chemistry of the subject and when practically no applicable microbiology had been discovered.'
>
> 'These practices are not especially clever, nor logical, nor completely effective – and it is not necessarily what would be done today if these countries had the chance to start again.'

The conventional sewerage systems (Figure 9.2) can be divided into: anaerobic systems (septic tanks, anaerobic ponds and anaerobic digesters), aerobic systems with attached growth (typically trickling filters) and aerobic suspended growth systems (typically activated sludge). Aerobic systems may be followed by some form of wastewater disinfection. A well-established alternative technology is the WSP (Figure 9.2), a system which is relatively simple to construct yet biologically sophisticated, containing both aerobic and anaerobic zones. A less well-established alternative is the reed bed. Of these different designs, only one, WSP, may be constructed to ensure the efficacious removal of *V. cholerae*; though it is possible that reed beds may also prove to be able to remove pathogens satisfactorily.

9.10.1 ANAEROBIC SYSTEMS

A number of systems are characterized by anaerobic conditions. Septic tanks consist of one or more sedimentation chambers through which sewage passes before being discharged into a soak away, a local water course or a sewerage system. Such systems typically have retention times of less than 3 days and are often used to provide on-site sanitation where flush toilets are used but there is no sewerage system. Anaerobic digesters are typically single-chambered reactors that are, ideally, designed and operated to optimize the production and collection of gas. Although they may be used to provide continuous pretreatment for waste stabilization ponds with short (<1 day) retention times (Dixo *et al.*, 1995), they are more usually fed with sewage sludge or other high strength wastes and have much higher (10–20 days) retention times. Anaerobic WSP are considered below with other WSP. Anaerobic conditions are not inimical to *V. cholerae* which has been shown to grow in the absence of oxygen (Youngren-Grimes, 1988). Nevertheless, the volatile fatty acids and hydrogen sulphide (Arridge, 1995) produced in anaerobic conditions may discourage the survival of *V. cholerae.*

Some of the earliest work on the removal of cholera vibrios was conducted by Flu (1921) in what is now Indonesia. In this study, five different septic tanks were challenged with cultures containing an unstated number of *V. cholerae.* 'The comma bacillus' was only found in the effluent of one tank on one occasion 24 hours after the challenge. The tanks were simultaneously challenged with a 'typhoid culture' and this organism was almost invariably present in the effluents. One must doubt the efficacy of the detection method employed and the author does not state the number of bacteria used or detected or give any details of the hydraulic flows in the tank. The central conclusion of the paper, was that septic tanks are not an efficient barrier to *V. cholerae.*

In a relatively recent study (a mere 14 to 20 years ago) the removal of *V. cholerae* O1 (Howard *et al.*, 1975) and *V. cholerae* non-O1 (Daniel and Lloyd, 1980a) in Oxfam Sanitation Units was examined. The unit was in essence a double-chambered septic tank consisting of two flexible butyl rubber containers (Daniel and Lloyd, 1979) with a total mean hydraulic retention time of 12–15 days. The study of the removal of *V. cholerae* O1 employed the effluent from the wards of the Cholera Research Laboratory in Dacca, Bangladesh. In view of the long retention time, removal was relatively poor, with only 98% being removed in the first tank and 91% being removed in the second tank (overall removal 99.8%). The unit was monitored for 2 months (November and December); in this time the temperature of the wastewater fell by 10°C without affecting the removal of the cholera vibrio. The conditions in the tanks were relatively inhospitable (anaerobic, with pH values of 6.3–6.7 and 40–90 mg/l of ammonia) so the poor removal may have been attributable to short-circuiting. The removal of *V. cholerae* non-O1 was studied in units in refugee camps. The removal of the *V. cholerae* non-O1 was comparable with the results of the earlier study, with geometric mean removals of 99.8% and 96.7% in the two systems

Fig. 9.2 Top left: a stream of raw sewage entering an estuary in north-east Brazil. The sewage was diverted from a defunct sewage works. Top right: a trickling filter in north-east Brazil; the trickling filter is inoperative due to blocked pipe work. Bottom left: a suspended growth system (oxidation ditch), not functioning due to the failure of the grit removal stage at the inlet leading to blockage in the pipework. Bottom right: Dandora Phase I waste stabilization ponds, Nairobi, Kenya (two series, each comprising a facultative pond and three maturation ponds). Further series have been added in Phase II. In Phase III, anaerobic ponds will be constructed at the head of each series. (Photograph of Dandora courtesy of Prof. Duncan Mara.)

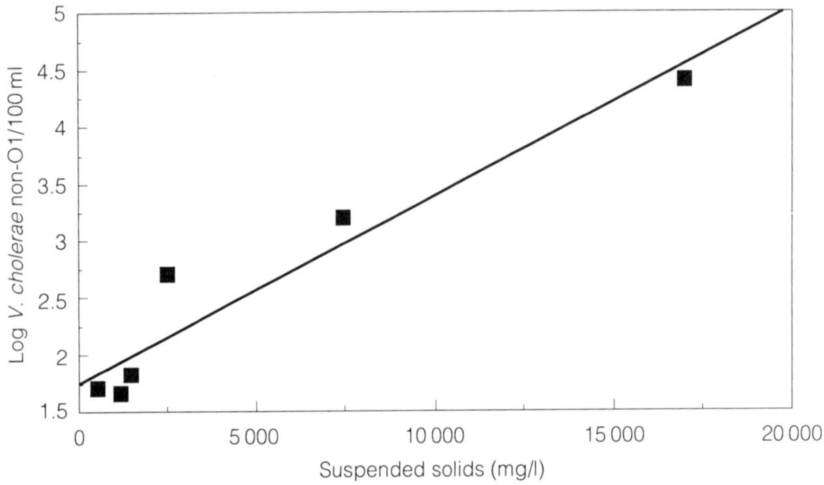

Fig. 9.3 Suspended solids (ss) and the numbers of *V. cholerae* non-O1 in the influent and effluent of septic tanks in refugee camps in Bangladesh. $r = 0.96$, log *V. cholerae* non-O1/100 ml = 0.00016(ss) + 1.73. (Data from Daniel and Lloyd, 1980.)

studied. There is evidence that sedimentation played a role in the removal observed, since there was a correlation between the amount of suspended solids and the numbers of the *V. cholerae* non-O1 (Figure 9.3). However, other mechanisms may also be involved, the secondary tank in one location removing 87% of the *V. cholerae* non-O1 in the influent and an identical secondary tank at another location removing nothing, though removal of the suspended solids was similar in both tanks (Figure 9.4). This variation could be due to the presence of vibriocidal conditions in one system and short-circuiting in the other. Interestingly, coliform and *Salmonellae* spp. removal in the Oxfam Sanitation Units was found to be very poor and inferior to the removal of *V. cholerae* O1 and *V. cholerae* non-O1. This appears to corroborate the observation of Flu (1921) that *S. typhi* was found more frequently in the effluents than pathogenic *V. cholerae*. A septic tank receiving warm, saline liquid (artificial kidney dialysate) was reported to have been colonized by toxigenic *V. cholerae* (Fukimi, 1979). Such unusual circumstances have little bearing on the wastewater treatment except that these findings reinforce the concept of *V. cholerae* as a pathogen that will colonize suitable aquatic ecosystems.

There appear to be no studies of the fate of *V. cholerae* in anaerobic digesters. However, Rojas and Hazen (1989) did study the survival of *V. cholerae* non-O1 in the seeded effluents of a digester treating rum

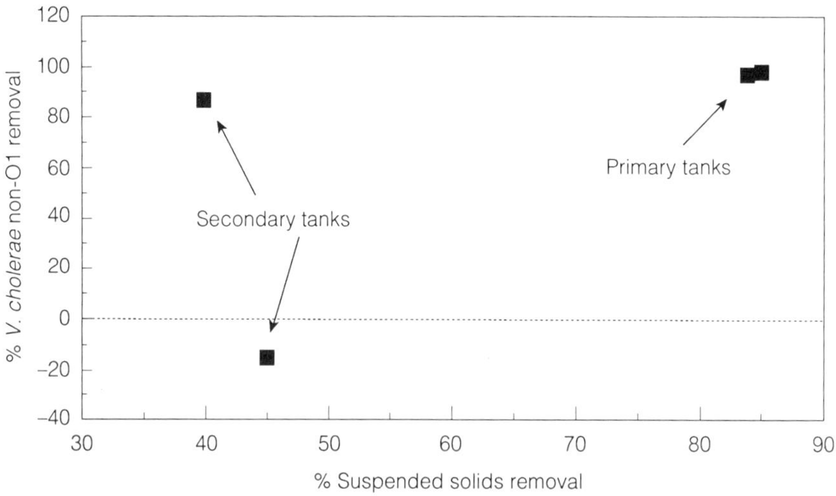

Fig. 9.4 Percentage removal of suspended solids and *V. cholerae* non-O1 from the primary and secondary tanks in Oxfam sanitation units (septic tanks) in refugee camps in Bangladesh (r = 0.58).

distillery waste. This form of industrial waste would not normally contain pathogenic bacteria and, even after digestion, contains very high levels of organic waste [biochemical and chemical oxygen demands (BOD, COD) of 27500 and 82680 mg/l respectively]. Moreover, a number of the microcosms used were constantly shaken in air. Consequently it is not clear to what extent their findings reflect the conditions inside a digester treating domestic sewage or sewage sludge. Both microcosms (shaken bottles) and pure cultures in dialysis bags in larger containers of effluent were studied. In the former, some loss of culturability was observed, a process that was enhanced by high temperatures. Whilst a range of pH values of between 4.6 and 8.6 had no detectable effect on the numbers of culturable organisms that could be recovered, it is possible that the high nutrient levels protected the bacteria from adverse pH values. However, high (45°C) temperatures appeared to interact with high pH to lower the numbers of culturable organisms. When the survival of the organisms was evaluated using total cell counts and measures of respiration and cell activity, such clear-cut patterns were not observed. The studies of pure cultures in dialysis bags evaluated survival using coulter counts, whole-cell counts and respiration rates, extended survival and even growth was observed. The authors concluded that anaerobic digestion was not a barrier to *V. cholerae*.

Although the anaerobic systems reviewed have not been found to completely remove the cholera vibrio, there is some evidence that the anaerobic processes produce the conditions that affect the survival of *V. cholerae*. Anaerobic conditions will permit the growth (Youngren-Grimes *et al.*, 1988) and presumably survival of the organisms. However, some members of the microbial consortia characteristic of anaerobic systems, produce acids (through acetogenosis) and hydrogen sulphide (from sulphate-reducing bacteria) which may discourage the survival of *V. cholerae*. Arridge (1995) found that the numbers of culturable *V. cholerae* O1 declined in a microcosm (bucket) of anaerobic WSP effluent faster than comparable microcosms of aerobic WSP. She postulated that elevated levels of hydrogen sulphide (8 mg/l) observed in the anaerobic pond may have accelerated the decline of the seeded organism. As part of the study mentioned above Rojas and Hazen (1989) compared the survival of seeded *V. cholerae* non-O1 in untreated and anaerobically digested rum distillery effluents using culturable counts, direct counts and measures of respiration. Surprisingly the BOD, COD and pH of the treated and untreated wastes were reported to be similar. Nevertheless, the untreated effluents allowed the growth of the seeded organism while, as noted above, treated effluents appeared to encourage a decline of the organism. This difference could be because some, as yet undefined, product of anaerobic digestion was promoting the decline of the organism. Engineers already have considerable expertise in the manipulation of the conditions in anaerobic processes. If these putative vibriostatic conditions could be identified, engineers could adapt designs to ensure that such conditions occurred in the anaerobic processes where *V. cholerae* removal was a consideration.

9.10.2 AEROBIC SYSTEMS

Conventional aerobic systems typically consist of a primary treatment stage, in which a proportion of the wastes are removed by sedimentation and a secondary biological treatment stage comprising either a suspended growth or attached growth system. There are numerous variations on this theme in operation, all of which are designed to remove organic waste or nutrients rather than pathogens. There is no reliable information on the fates of *V. cholerae* in such systems, which can be assumed to be, at best, poor (90–99%). Mean retention times are typically a matter of hours and conditions in the reactors are not known strongly to affect *V. cholerae* or indeed any of the other pathogenic bacteria or viruses found in sewage.

In the primary stage, (primary) sedimentation tanks are used to settle out the larger particles and 50–60% of the settleable solids can be removed (quite cheaply) in this manner. The degree of *V. cholerae* removal associated with primary sedimentation is unknown. However, since the removal is a physical, as opposed to biological process, it may be valid to infer the behaviour of the organisms from the behaviour of faecal coliforms, in which case no more than 30–50% removal can be expected (Feachem *et al.*, 1983). Indeed, some workers have reported an increase in the numbers of indicator bacteria during primary sedimentation (Kawamura and Kaneko, 1986).

Fixed film secondary treatment processes work by passing the wastewater over a biofilm which is capable of removing the oxygen demand. Units in which the water is distributed over stationary media are usually called trickling filters (though no filtration occurs), while units in which the biofilm is passed through the wastewater are called rotating biological contactors. Factors likely to affect the fate of *V. cholerae* in such systems include binding to the biofilm, predation by protozoa and competition for nutrients. But the short retention times characteristic of such systems militate against any significant reduction in numbers. Daniel and Lloyd (1980b) working in Bangladesh reported a 3-log increase in the numbers of *V. cholerae* non-O1 when they passed the effluent from a dual-chambered septic tank over an improvised trickling filter. The authors showed that *V. cholerae* non-O1 would reproduce in batch samples of the trickling filter effluent at 20–25°C and suggested that ponding in the filter was permitting the growth of the organism. However, when the survival of *V. cholerae* El Tor was studied in identical batches of effluent, the number of culturable organisms was observed to decline by a factor 10 over a period of 7 hours and then to maintain itself at the reduced level for a further 41 hours. The apparent discrepancy between the two forms of *V. cholerae* could not be readily explained. It is conceivable that the organisms thought to be *V. cholerae* non-O1 was in fact some other bacteria which behaves quite differently from the subject of this paper or that, under the conditions described, that *V. cholerae* O1 and non-O1 behave quite differently.

Suspended growth secondary treatment systems, such as activated sludge, are probably the most widely used form of sewage treatment. Typically they comprise an aeration basin in which the growth of floc-forming bacteria is promoted and a subsequent sedimentation tank or clarifier which is employed to separate the flocculated bacteria (and much of the oxygen demand) before disposal of the liquid. A portion of the flocculating bacteria (the return activated sludge) is returned to the aeration basin. Factors that may affect the survival of *V. cholerae* in suspended growth systems include the competition for nutrients, binding to the flocs and protozoal grazing. Retention times are short, typically 6–8 hours in a conventional plant, rising to 30 hours in certain designs. This review of the literature has found no studies of the fate of *V. cholerae* in working plants. Streeter (1930) reported that Courmont and Rochaix (1922) had seeded a batch of activated sludge and sewage with *V. cholerae* are found that they could not recover the organisms after 5–6 hours' aeration. The same review stated that Bruns and Sierp (1927) had conducted analogous experiments and found that 98% of the seeded *V. cholerae* were removed. It is doubtful if even this level of removal could be achieved in practice since most suspended growth systems operate on a continuous flow basis and consequently a proportion of the influent will leave the aeration basin before the mean hydraulic retention time. More recently, a short report by Valkov *et al.* (1981) stated that an inoculum of an environmental isolate of *V. cholerae* O1 El Tor of unstated size survived for 4–6 days in aerated sewage. Overall, it would be wise to assume that a well-operated and well-designed primary sedimentation tank and suspended growth system could remove 90% of *V. cholerae* in the influent. It would be wiser still to obtain more up-to-date information on the fate of *V. cholerae* in such systems.

9.10.3 WASTEWATER DISINFECTION

One response to the poor pathogen removal observed in conventional effluent treatment is to use some form of wastewater disinfection. However, this should be approached with caution. It is one thing to design and fit some form of disinfection unit to a wastewater treatment unit; it is quite another to ensure that that unit operates 24 hours a day, 365 days a year and is replaced at the end of its design life. The effluent from the treatment plant must be of a consistent quality, the disinfection equipment must be regularly maintained and the chemicals and spares required for continuous operation must be available at all times. These conditions are sometimes difficult to fulfil in many water companies and municipalities in the countries most afflicted by cholera. Feachem *et al.* (1983) have recommended that effluent chlorination should be rejected unless the highest level of process control and management can be guaranteed. Public health officials and engineers would do well to heed their advice. Having said the above, the ability of chlorine (Kott, 1973) and ultraviolet lamps (Mazoit *et al.*, 1975) to kill *V. cholerae* in wastewater has been established and it seems likely that other forms of disinfection will also be able to kill the organism.

9.11 WASTE STABILIZATION PONDS (WSP)

WSP are a very good form of sewage treatment. They combine physical simplicity, which means they are cheap to build and maintain, with biological sophistication. This means that they are ecologically stable and able to remove pathogens efficiently. Consequently WSP are particularly suited to developing countries though they are widely used in Western Europe, the USA and Australia and New Zealand. WSP systems are usually a series of shallow (1.5m) lagoons and should ideally consist of an anaerobic pond, a facultative pond and a number of maturation ponds. An anaerobic pond, if used, is the first pond in a series; anaerobic conditions are maintained by virtue of a very heavy organic load and retention times are between 1 and 5 days. The biology of anaerobic ponds is probably analogous to that of anaerobic digesters or septic tanks. Facultative ponds may be used as the first pond in the system (primary facultative ponds) or after an anaerobic pond (secondary facultative ponds). Such ponds sustain very large algal (1000–3000 μg/l of chlorophyll a) populations, the photosynthetic action of which oxygenates the upper layers while the lower zones remain anaerobic. The coexistence of the aerobic and anaerobic zones in the pond led to the use of the term 'facultative pond'. Because of the intense photosynthetic activity the oxygen levels in the pond may become very high (>20 mg/l) and the concomitant take up of carbon dioxide may lead to high pH values (8.5–9.5) due to shifts in the carbonate–bicarbonate buffering system. The retention time of facultative ponds should depend on the organic load, and can be anything between 5 and 25 days. Maturation ponds also support large algal populations, have high pH and dissolved oxygen values, but are aerobic throughout. The function of maturation ponds is to remove pathogenic organisms not eliminated in earlier stages. Retention times are about 3–5 days and sufficient ponds are added onto the system to achieve the level of pathogen removal desired (Mara and Pearson, 1986).

Unlike other sewage treatment systems WSP, when correctly designed, can achieve very high levels of pathogen removal; 99.9999% faecal coliform removal being readily achievable if sufficient land is available (Mara and Pearson, 1986). Moreover, the reduction in the indicator organisms has been shown to be associated with the reduction of a variety of pathogenic bacteria and viruses (Oragui *et al.*, 1987) including *V. cholerae* O1 (Oragui *et al.*, 1993). However, there is an apparent discrepancy between the known mechanisms of pathogen removal and the ecology of the *V. cholerae*. Studies of pathogen removal based on faecal coliforms have suggested that fast light and algae-based processes, photooxidation (Curtis *et al.*, 1992; Curtis and Mara, 1994) and high pH (Parhad and Rao, 1974), combine with slower and less well understood processes such as protozoal grazing and adsorption to eliminate pathogens. Variations in the faster, light and algae-based effects probably account for much of the observed geographical and seasonal variation in pond performance. The changes in removal efficiency reflect the rise or falls in the amount of sunlight and the algal population (Troussellier *et al.*, 1986). Yet the studies, reviewed above, of the aquatic ecology of *V. cholerae* have shown that the growth and survival of the organism is favoured by both the presence of algae and high pH. Consequently, there have been calls for research to clarify this situation (Anonymous, 1992).

Noting the absence of information in this area Kott and Betzer (1972) seeded a model scale waste stabilization pond with *V. cholerae* O1 El Tor. The model plant consisted of a single, baffled 70-l aquarium that was fed for 10 hours daily with sufficient diluted sewage to give a 5-day retention time. The authors did not state whether the WSP was indoors or outdoors, or the concentration of the algal population. However, the pH of the effluent was said to be between 8 and 8.5, which implies some algal activity in the pond. Removal was assessed in two ways: (i) the reduction in numbers between the influent and effluent; and (ii) the survival of the seeded organisms in dialysis bags suspended in the pond. The former assessment found a 3.1-log (99.94%) reduction in *V. cholerae* O1; by comparison, only 1.62-log (97%) of the coliforms were removed in the same system. In the dialysis bag studies an initial inoculum of 10^3 organisms/100 ml could not be recovered after 24 hours. Kott and Betzer (1972) concluded that, notwithstanding the elevated pH of the pond, *V. cholerae* O1 was rapidly removed from WSPs. Although these results are encouraging they cannot be assumed to apply to all WSPs. This is because of the limited nature of the experimental set-up (a small-scale, single pond of an atypical design) and the paucity of the information about the ecology and location of the pond.

A relatively conventional series of WSP in northeast Brazil removed naturally occurring (49/100 ml in the raw sewage) *V. cholerae* El Tor (Oragui *et al.*, 1993) after a mean hydraulic retention time of 11 days (Figure 9.5). The pilot-scale experimental series consisted of a 1-day anaerobic pond, a 2-day facultative pond and a further eight, 2-day maturation ponds. The order of the ponds and the total retention time of the system in the series is conventional good practice. However, the retention times of the individual facultative and maturation ponds is shorter than the usual 4–6 days, this could have resulted in some short-circuiting. These findings were confirmed by subsequent studies at the same site over a 12-month period, though the numbers of *V. cholerae* O1 in the raw sewage was lower at 4 organisms/100 ml (Arridge, 1995).

The WSP were not colonized by the cholera vibrio. This should allay fears (Anonymous, 1992) that associations between the algae present in WSP and *V. cholerae* O1 could cause ponds to act as a source of, rather than a barrier to, the organism. There does not appear to be a relationship between the amount of algae and the survival of the organism

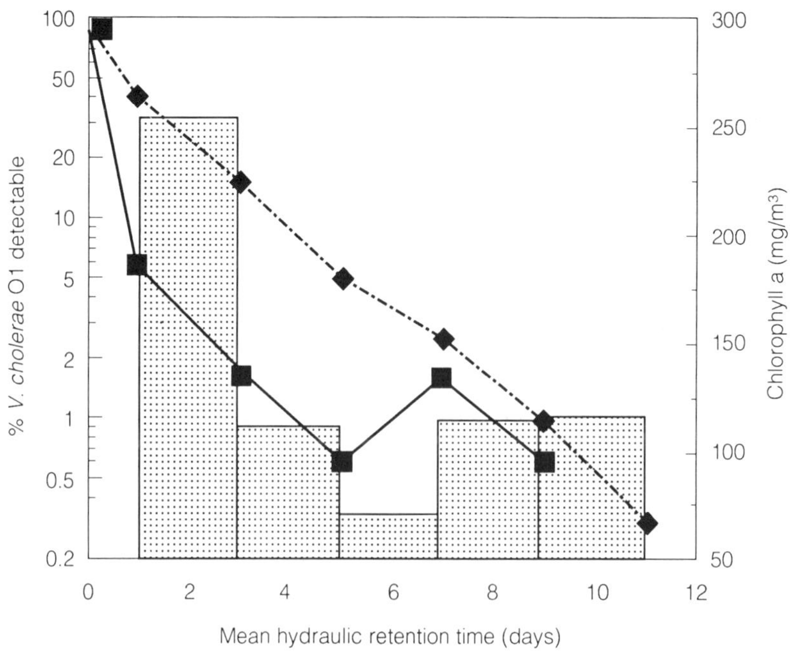

Fig. 9.5 A comparison of faecal coliform (◆) and *V. cholerae* O1 (■) removal in a series of waste stabilization ponds (WSP) in northeast Brazil. The first pond was anaerobic with a retention time of 1 day; subsequent ponds in the series had retention times of ca. 2 days. No *V. cholerae* O1 could be detected after a mean hydraulic retention time of 9 days, by which time the count was reduced to 0.3 organisms/100 ml. The striped bars indicate chlorophyll a concentrations. (Data from Oragui *et al*., 1993.)

(Figure 9.5). The same team subsequently reported the removal of *V. cholerae* O1 El Tor from facultative ponds with up to 700 $\mu g\,l^{-1}$ of chlorophyll a (Arridge, 1995). As part of the same study, column samples taken from facultative and maturation ponds reported to contain 435 and 369 $\mu g\,l^{-1}$ of chlorophyll a respectively were placed in buckets suspended in WSP and seeded with 10^8 organisms. Seven-log removal was observed after 6 days in the facultative pond and 8 days in the maturation pond. Although there was a large decrease in the numbers of seeded organisms, it is possible that the presence of algae afforded some protection to *V. cholerae* O1 since analogous experiments with raw sewage, anaerobic pond effluent and Ringer's solution removed the organism more quickly. The failure of WSP to support the cholera vibrio is probably because the relationship between *V. cholerae* O1 and algae is species- or genus-specific (Islam *et al*., 1994a) and that the two algal genera which typically predominate in WSP, *Euglena* and *Chlorella* (Mara and Pearson, 1986), do not support the organism. Ideally, this hypothesis should be tested by examining microscopically the flora of WSP using immunofluorescence microscopy in a manner analogous to the studies of Brayton and Colwell (1987), Tamplin *et al.* (1990) and Huq *et al.* (1990). However, work by Mezrioui *et al.* (1994) employing culture-based methods appears to support this hypothesis. Three different kinds of algae, *Chlorella sorokiniana* (a green algae) and *Synechosistis* and *Synechococcus* (cyanobacteria) were isolated from a WSP and grown in batches of treated wastewater in the presence of *V. cholerae* O1. In the dark the presence of the algae had little effect on the rate of reduction of the colony count of the *V. cholerae* O1, while in the light the presence of the algae enhanced the rate of removal of the cholera vibrio. The lack of effect of the dark control suggests that passive adsorption may be excluded as a removal mechanism. This work does not prove conclusively that there is no active association between the algae found in WSP and the cholera vibrio, since a number of other algal species are found in WSP and it is conceivable that the enhanced reduction observed may be due to active adsorption to the algae. Nevertheless, it does suggest that such algae do not support the growth of *V. cholerae* O1. This area merits further systematic study.

Not surprisingly Oragui *et al.* (1993) also reported that the rate of removal of faecal coliforms was different from the rate of removal of *V. cholerae* (Figure 9.5). In the anaerobic pond the *V. cholerae* was removed more effectively (94%) than the indicator organism (60%). Though this result was not confirmed by later studies at the same site (with much lower cholera concentrations) (Arridge, 1995). Arridge (1995) has postulated that the enhanced removal observed in the anaerobic pond is due to the elevated sulphide concentrations (11.3 mg/l) found in the pond. This phenomenon deserves further investigation in order to discover whether it could be incorporated into design on a more systematic basis. However, the available information is sufficient to commend the use of anaerobic ponds in WSP series to remove the cholera vibrio, in addition to this pond's more usual role in removal of organic matter. In the algae-containing facultative and maturation ponds (Oragui *et al.*, 1993) the rate of removal of *V. cholerae* O1 appears to differ from that of faecal coliforms in so much as the rate of decline in the former appears to change while the latter declines at a steady rate (Figure 9.5). However, the values for the numbers of *V. cholerae* O1 for the last two or three ponds in the series will be based on a low number of positives detected by the most probable number (MPN) method. Such results are subject to a great deal of variation and it is doubtful if the apparent rise and fall in removal rate is statistically significant. Unfortunately, no confidence intervals were presented with the data. Mezrioui *et al.* (1994)

compared the fate of *E. coli* and *V. cholerae* O1 in cultures of green algae and cyanobacteria isolated from WSP. They found that the cholera vibrio declined more quickly than the indicator organism in the presence of *Chlorella sorokiniana* and less quickly than the indicator in the presence of *Synechosistis* and *Synechococcus*. *Chlorella* spp. are more usual in WSP than the cyanobacteria, the latter being associated with very underloaded systems (Konig, 1984). The genera of algae present in a WSP system depends on the organic load, climate depth and design; all these factors may, theoretically, influence the relative removal rates of faecal coliforms and *V. cholerae* O1.

Although all the studies of the fate of *V. cholerae* O1 in WSP have suggested that the pathogen is rapidly removed, studies of *V. cholerae* non-O1 in WSP present a very different picture. Lesne *et al.* (1991) studied an experimental WSP consisting of a facultative and maturation pond in series (no anaerobic pond was used) each with a retention time of 25 days. This system is probably underloaded and short-circuiting is inevitable. In the cooler months the influent contained 1700 organisms/100 ml and there was a 2-log reduction in the numbers of *V. cholerae* non-O1; however, in the warmer months 51 000 organisms/100 ml were present in the raw sewage and no removal took place at all. There was a positive correlation between the air temperature and the numbers of *V. cholerae* non-O1 in the raw sewage and effluent of both ponds. Significantly, the number of *V. cholerae* non-O1 in the first pond also correlated positively with the algal concentration. The number of *V. cholerae* was negatively correlated with the number of protozoa in the first pond. The number of faecal coliforms was also negatively correlated with the numbers of *V. cholerae* non-O1. Interestingly, these findings are corroborated by Ventura *et al.* (1992) who reported work carried out in 1986. Studying full-scale systems in Lima, Peru they also found that the numbers of *V. cholerae* non-O1 rose and fell seasonally, with the peak coinciding with the warmer months of the year. Details of the pond system are sketchy; it was said to consist of three ponds in series. These studies echo other work in the freshwater environment (Nair *et al.*, 1988) where *V. cholerae* non-O1 was readily isolated but *V. cholerae* O1 was not.

It appears that the *V. cholerae* O1 is removed in WSP while the non-O1 forms are not. However, no one has reported successfully simultaneously studying *V. cholerae* non-O1 and O1; Lesne *et al.* (1991) sought, but did not find, the cholera vibrio. It is not easy to reconcile the different reports, since it is not clear why different forms of the same species of bacteria should behave so differently. There are a number of possible explanations: the *V. cholerae* non-O1 could have been misidentified; the difference could reflect differences in the ecology of the different serotypes of bacteria; the *V. cholerae* O1 could be entering a viable but not culturable state while the non-O1 bacteria are not; or, there could be differences in the ecology of different WSP.

The first and simplest explanation, that other organisms were misidentified as *V. cholerae* non-O1, does not appear to bear examination. Both Lesne *et al.* (1991) and Ventura *et al.* (1992) report checking yellow colonies on TCBS (thiosulphate-citrate-bile salt agar) using the API system and polyvalent antisera.

An alternative explanation could be that the ecology of the O1 and non-O1 organisms differ in some crucial way that permits the survival of the latter but not the former. There is circumstantial evidence to support this hypothesis. First, both Kott and Betzer (1972) and Arridge (1995) commented that the detection of *V. cholerae* O1 was particularly difficult because of the large numbers of yellow colonies that were found on TCBS agar after enrichment in alkaline-buffered saline. Arridge (1995) reported that many of these colonies were *V. cholerae* non-O1, which suggests that these organisms were ecologically more successful than *V. cholerae* O1. Secondly, when Mezrioui *et al.* (1994) observed the prolonged survival of *V. cholerae* non-O1 in the summer months recorded by Lesne *et al.* (1991) they attempted to further examine the problem by investigating the effect of green bacteria and cyanobacteria isolated from the WSP on *V. cholerae* O1 isolated from a local patient. They presumably expected the cyanobacteria, blooms of which are associated with the summer months when *V. cholerae* non-O1 survival is at its highest, to be associated with the extended survival of the seeded organism. Instead, the *V. cholerae* O1 declined at a significant rate in the presence of both the green bacteria and cyanobacteria. It is interesting to speculate that had the workers used *V. cholerae* non-O1 isolated from the pond the behaviour of the seeded organisms may have resembled the organism in the pond. Also, as noted above, when Daniel and Lloyd (1980b) compared a clinical isolate of *V. cholerae* O1 and a 'non-cholera vibrio' in a tropical wastewater effluent, culturable counts of the former declined while the latter grew. These putative ecological differences presumably relate, in some way, to differences in the surface characteristics of the different serotypes; for example, different serotypes of *V. cholerae* may bind preferentially to different species of phytoplankton and the growth and survival of a given serotype may depend on the species of algae present.

An alternative possibility is that the *V. cholerae* O1 could be entering a viable but not culturable state while the non-O1 organisms are not. This hypothesis requires that the conditions in the WSP are sufficiently unfavourable to induce the VBNC state in the O1 serotype and yet sufficiently favourable to allow other serotypes to remain culturable. This hypothesis could only be tested by using molecular and direct microscopic methods to enumerate *V. cholerae* O1 in the pond system.

Colwell and Huq (1994) have compared the VBNC state to spore formation. It is therefore interesting to note that *Clostridium perfringens* spores are removed by WSP (Oragui *et al.*, 1987; Pearson *et al.*, 1995) and that the observed removal of pathogens from WSP can be attributed to the interaction of a number of mechanisms (Curtis and Mara, 1994) some of which, such as oxygen radical formation, will affect all living things. A further consideration is the well-documented (Huq *et al.*, 1990) association between algae and putative non-culturable *V. cholerae*, since the speciation of the algae changes in succession in a treatment series. Thus, an association formed between a cell of the genus *Euglena* and a cell of *V. cholerae* in a facultative pond is unlikely to sustain the organisms until it reaches the effluent, since the majority of the *Euglena* algae will have been replaced by other species in subsequent ponds (Konig, 1984).

It is conceivable that the ecology of *V. cholerae* O1 and non-O1 in WSP is the same but that the conditions in the different ponds were not, coincidentally favouring reduction when the O1 serotype was studied and survival when the non-O1 serotypes were studied. This would be quite a coincidence. However, the climate (two Mediterranean, two just south of the South American equator), salinity (potable water) and nutrient levels in the different studies are likely to have been comparable. The putative differences would therefore have to be related to some other aspect of the pond ecology such as the species and activity of algae present. This hypothesis is supported by earlier microcosm studies comparing the O1 and non-O1 types in other environments (Muller, 1977; Singleton *et al.*, 1982b; Huq *et al.*, 1983) that found no strong differences in the ecology of *V. cholerae* O1 and non-O1. It is known that design, particularly organic load, affects algal speciation (Konig, 1984) and activity. Therefore, if the observed differences were attributable to some aspect of the pond ecology it should be possible to adapt the design appropriately.

The answers to the questions posed by the few available studies on the fate of *V. cholerae* are well within the ingenious scope of modern microbial ecologists. These questions should be addressed. Wastewater treatment undoubtedly has a role in the prevention of *V. cholerae* (Tauxe *et al.*, 1994) and a WSP system is the wastewater treatment system of choice in developing countries (Mara *et al.*, 1992). It is about time that the sophistication of our knowledge of *V. cholerae* in wastewater treatment matched the sophistication of our knowledge about the physiology and molecular ecology of this organism. The appearance of *V. cholerae* O139 should lend further impetus to this call for work in this area. Preliminary investigations in this area suggest that the ecology of the new organism may be different from previous pandemic strains (Islam *et al.*, 1994b).

9.11.1 MACROPHYTE PONDS AND REED BEDS

Floating macrophyte WSP contain plants that float on the water, but their use is only recommended in certain specialized situations (Mara *et al.*, 1992). The genera of macrophyte employed include *Salvinia, Lemna, Eichhornia* and *Pistia*. *Eichhornia* (Spira *et al.*, 1981) and *Lemna* (Islam *et al.*, 1990) have been shown to support *V. cholerae* O1. It would therefore seem prudent to avoid macrophyte ponds when *V. cholerae* removal is a consideration. The routine maintenance of conventional ponds should include the energetic removal of any macrophytes that invade the pond system. I have not found any information on the fate of *V. cholerae* in reed bed systems.

9.12 SUMMARY

1. The need for adequate wastewater treatment is frequently cited as a prerequisite for the prevention of cholera. However, such sentiments are probably not based on a rational analysis of the most cost-effective interventions in this context. Nevertheless, untreated wastewater can and has caused significant outbreaks of the disease, notably through the irrigation of food crops and the contamination of seafood. It is therefore necessary, in certain circumstances, to remove the cholera vibrio from treated wastewater before the water can be reused or returned to the environment. Even if the wastewater treatment employed does not remove all of the cholera vibrios present or prevent the colonization of a particular water course by the organism, it may reduce the load on any subsequent drinking-water treatment systems abstracting water from the water course.
2. The vast majority of studies on the fate of pathogenic organisms in wastewater have relied on faecal coliforms (which are 90% *E. coli* in tropical sewage) in the expectation that such organisms will act as analogues of bacterial and viral pathogens. Neither the evolutionary relationship between *E. coli* and *V. cholerae* nor the limited published empirical data support this assumption. There is very little information on the removal of *V. cholerae* from wastewater treatment plants. The data which are available are not in keeping with current good practice in microbial ecology and some of the most 'recent' data are over 70 years old!
3. A consideration of known information on the excretion of pathogenic *V. cholerae* by the sick and carriers and the few published reports in the literature suggest that the numbers of organisms in the raw sewage will range from over 10^5/100 ml in an epidemic situation to less than 10 organisms/100 ml if a low number of

carriers are present in the sewered population. However, even one mildly ill person in a population of 100 000 can excrete enough organisms to give a mean concentration of 10^3 cholera vibrios/ 100 ml in a city's raw sewage.

4. For the purposes of this review wastewater treatment systems were divided into: anaerobic systems such as septic tanks, suspended growth systems such as activated sludge, attached growth systems, also called trickling filters, and waste stabilization ponds. Only waste stabilization ponds appeared to be able to remove high levels of *V. cholerae* from raw sewage. Other systems removed 90–99% of the organisms with which they were challenged. Growth of *V. cholerae* non-O1 was reported in one makeshift trickling filter. There is some evidence that the vibriocidal conditions exist within anaerobic systems and that this aspect of their operation could be optimized.
5. Not withstanding the reports of the satisfactory removal of *V. cholerae* O1 from waste stabilization ponds others have reported that *V. cholerae* non-O1 will survive and even grow in such systems. This discrepancy could reflect: poor experimental techniques, genuine differences in the ecology of the pathogenic and non-pathogenic forms of the organism or differences in the waste stabilization ponds used in the different studies.
6. Our ignorance concerning the fate of *V. cholerae* in sewage treatment works is unfortunate. Engineers and microbiologists must work together to rectify this gap in our knowledge to ensure that water-borne sanitation systems – which can so effectively prevent the disease at a domestic level – can be used without spreading the disease by contaminating the water resources upon which we all depend for our health and well-being.

9.13 REFERENCES

Ahmed, K. and Chughtai, M.I.D. (1990) *Microbiology of Waste Stabilization Ponds*, Pakistan Society of Biochemists, Lahore, Pakistan.

Alexiou, I.E. and Anderson, G.K. (1995) Pre-acidification: concepts, guidelines and parameters for application. *Water Science and Technology*, **31**, 1–00.

Altukhov, A.A., Ivanov, S.I., Semiotrochev, V.L., Bulegenov, E.I., Lebedev, K.K., Filimonova, L.A., Zakharov, N.I. and Rybak, E.I. (1975) Prolonged detection of El Tor cholera vibrio in sewage in a bathhouse. *Zhurnal Mikrobiologii, Epidemiologii i Immunobiologii*, **2**, 41–4.

Anonymous (1992) Of cabbages and chlorine: cholera in Peru. *Lancet*, **340**, 20–1.

APHA (1989) *Standard Methods for the Examination of Water and Wastewater*, 17th edn, American Public Health Association, Washington, DC.

Arridge, HJ. (1995) The influence of geometry on the performance of waste stabilisation ponds with special reference to pathogen removal, PhD Thesis, University of Liverpool.

Asano, T., Smith, R.G. and Tchobanoglous, G. (1985) Municipal wastewater: treatment and reclaimed water characteristics, in *Irrigation With Reclaimed Municipal Wastewater – Guidance Manual*, (eds G.S. Pettygrove and T. Asano), Lewis Publishers Inc., Michigan, USA, pp. 2-1–2-26.

Barer, M.R., Gribbon L.T., Harwood, C.R. and Nwoguh, C.E. (1993) The viable but non-culturable hypothesis and medical bacteriology. *Reviews in Medical Microbiology*, **4**, 183–91.

Barrett, T.J., Blake, P.A., Morris, G.K., Puhr, N.D., Bradford, H.B. and Wells, J.G. (1980) Use of Moore swabs for isolating *Vibrio cholerae* from sewage. *Journal of Clinical Microbiology*, **11**, 395–8.

Bitton, G. (1994) *Wastewater Microbiology*, Wiley-Liss Inc., New York.

Brayton, P.R. and Colwell, R.R. (1987) Fluorescent antibody staining method for enumeration of viable environmental *Vibrio cholerae* O1. *Journal of Microbiological Methods*, **6**, 309–14.

Brock, T.D. (1987) The study of microorganisms *in situ*: progress and problems, in *Ecology of Microbial Communities*, (eds M. Fletcher, T.R.G. Gray, and J.G. Jones), Cambridge University Press, Cambridge, pp. 1–17.

Bruns, H. and Sierp, F. (1927) Einflub der schlammbelebung des abwassers auf pathogene keime. *Zeitschrift fur Hygiene*, Leipzig (later Berlin), **107**, 571–84.

Cairncross, S. and Feachem, R. (1993) *Environmental Health Engineering In The Tropics*, 2nd edn, J. Wiley, Chichester, England.

Chadwick, E. (1842) *Report on the Sanitary Conditions of the Labouring Population of Great Britain*, Her Majesty's Stationery Office, London.

Chamberlain, J. and Moss, S.H. (1987) Lipid peroxidation and other membrane damage produced in *Escherichia coli* K1060 by near U.V. radiation and deuterium oxide. *Photochemistry and Photobiology*, **45**, 625–30.

Chao, W.-L., Chen, R.-S. and Tai, C.-L. (1988) Factors affecting the survival of pathogenic bacteria in subtropical river water. *Chinese Journal of Microbiology and Immunology*, **21**, 85–92.

Cholera Working Group (1993) Large epidemic of cholera-like disease in Bangladesh caused by *Vibrio cholerae* O139 synonym Bengal. *Lancet*, **342**, 387–90.

Colwell, R.R. and Huq, A. (1994) Vibrios in the environment: viable but non-culturable *Vibrio cholerae*, in *Vibrio cholerae and cholera* (eds I.K. Wachsmuth, P.A. Blake and O. Olsvik), American Society for Microbiology, Washington, USA, pp. 117–33.

Corrales, M.T., Fronchkowsky, G.B. and Eiguer, T. (1989) Aislamento en Argentina de *Vibrio cholerae* no O1 en liquidos cloacales. *Revista Argentina de Microbiologia*, **21**, 71–7.

Courmont, P. and Rochaix, A. (1922) Sur l'epuration bacteriènne des eaux d'egouts par le procede des 'boues activiees'. *Revue d'Hygiene*, **44**, 907–19.

Curtis, T.P. (1987) Pathogen removal from pilot scale deep waste stabilisation ponds. M.Eng Thesis, University of Leeds, England.

Curtis, T.P. (1991) Mechanisms of removal of faecal coliforms from waste stabilisation ponds. PhD Thesis, University of Leeds, England.

Curtis, T.P. and Mara, D.D. (1994) *The Effect of Sunlight on Mechanisms for the Die-off of Faecal Bacteria in Waste Stabilisation Ponds*, University of Leeds/ ODA Research Monograph No. 1. Leeds, UK.

Curtis, T.P., Mara, D.D. and Silva, S.A. (1992) Influence of pH, oxygen and humic substances on ability of sunlight to damage fecal coliforms in waste stabilization pond water. *Applied and Environmental Microbiology*, **58**, 1335–43.

Daniel, R.R. and Lloyd, B. (1979) Sanitation and disease in Bangladesh urban slums and refugee camps. *Progress in Water Technology*, **11**, 191–200.

Daniel, R.R. and Lloyd, B.J. (1980a) Microbiological studies on two Oxfam sanitation units operating in Bengali refugee camps. *Water Research*, **14**, 1567–71.

Daniel, R.R. and Lloyd, B.J. (1980b) A note on the fate of El Tor cholera and other vibrios in percolating filters. *Journal of Applied Bacteriology*, **48**, 207–9.

Dixo, N.G.H. (1994) The removal of *V. cholerae* by water treatment systems in developing countries. MSc Thesis, University of London.

Dixo, N.G.H., Gambrill, M.P., Catunda, P.F.C. and van Haandel, A. (1995) Removal of pathogenic organisms from the effluent of an upflow anaerobic digester using waste stabilisation ponds. *Water Science and Technology*, **31**(12), 275–84.

Dizon, J.J., Fukimi, H., Barua, D. *et al.* (1967) Studies on cholera carriers. *Bulletin of the World Health Organization*, **37**, 737–43.

EEC (1991) Commission Decision of 19 March 1991 adopting protective measures against cholera in Peru in respect of imports (91/146/EEC). *Official Journal of the European Communities*, L 73, 35.

Esrey, S.A., Potash, J.B., Roberts, L. and Shiff, C. (1991) Effects of improved water supply and sanitation on ascariasis, diarrhoea, dracunculiasis, hookworm infection, schistosomiasis and trachoma. *Bulletin of the World Health Organization*, **69**, 609–21.

Feachem, R.G. (1982) Environmental aspects of cholera epidemiology III. Transmission and control. *Tropical Diseases Bulletin*, **79**, 1–47.

Feachem, R.G. and Mara, D.D. (1979) A reappraisal of the role of faecal indicator organisms in tropical waste treatment processes. *The Public Health Engineer*, **7**(1), 31–3.

Feachem, R.G., Miller, C. and Drasar, B. (1981) Environmental aspects of cholera epidemiology II. Occurrence and survival of *Vibrio cholerae* in the environment. *Tropical Diseases Bulletin*, **78**, 865–80.

Feachem, R.G., Bradley, D.J., Garelick, H.D. and Mara, D.D. (1983) *Sanitation and Disease, Health Aspects of Excreta and Waste Management and Disease*, World Bank Studies in Water Supply and Sanitation 3, John Wiley & Sons Ltd, Chichester, UK.

Flu, P.C. (1921) Investigations of the duration of life of cholera vibriones and typhoid in septic tanks at Batavia. *Mededeelingen van den Burgerlijken Geneeskundigen Dienst in Nederlandsch-Indie*, **3**, 178–87.

Fukimi, H. (1979) Brief summary report of cholera happening in Tsurimi River, Kanagawa, Japan, in *Proceedings of the 14th Joint Conference, U.S.–Japan Cooperative Medical Science Program, Cholera Panel Symposium on Cholera*, (eds K. Takeya and Y. Zinnaka), Toho University, Tokyo, pp. 18–22.

Gerichter, C.B., Sechter, I., Gavish, A. and Cahan, D. (1975) Viability of *Vibrio cholerae* biotype El Tor and of cholera phage on vegetables. *Israel Journal of Medical Sciences*, **11**, 889–95.

Glass, R.I., Becker, S., Huq, M.I., Stoll, B.J., Khan, M.U., Merson, M.H., Lee, J.V. and Black, R.E. (1982) Endemic cholera in rural Bangladesh, 1966–1980. *American Journal of Epidemiology*, **116**, 959–70.

Guthrie, R.K. and Scovill, M.A. (1984) Recovery of *Escherichia coli* and *Vibrio cholerae* from aquatic microcosms. *Water Research*, **18**, 1055–7.

Hofer, E. and Ernandez, D. (1990) Incidencia de *Vibrio cholerae* nao O1 em afluentes de estacoes de tratamento de esgotos da cidade do Rio de Janeiro, RJ. *Revista Microbiologica de Sao Paulo*, **21**, 31–4.

Hood, M.A. and Ness, G.E. (1982) Survival of *Vibrio cholerae* and *Escherichia coli* in estuarine waters and sediments. *Applied and Environmental Microbiology*, **43**, 578–84.

Howard, J., Lloyd, B. and Webber, D. (1975) *Oxfams Sanitation Unit, The design and testing of a sanitation and sewage treatment unit for disasters and long term use*, 2nd edn, Oxfam Technical Paper, Oxfam, Oxfam House, 274 Banbury Road, OX2 7DZ, UK.

Huq, A., Small, E.B., West, P.A., Huq, M.I., Rahman, R. and Colwell, R.R. (1983) Ecological relationships between *Vibrio cholerae* and planktonic crustacean copepods. *Applied and Environmental Microbiology*, **45**, 275–83.

Huq, A., West, P.A., Small, E.B., Huq, M.I. and Colwell, R.R. (1984) Influence of water temperature, salinity and pH on survival and growth of toxigenic *Vibrio cholerae* serovar O1 associated with live copepods in laboratory microcosms. *Applied and Environmental Microbiology*, **48**, 420–4.

Huq, A., Colwell, R.R., Rahman, R., Ali, A., Chowdhury, M.A.R., Parveen, S., Sack, D.A. and Russek-Cohen, E. (1990) Detection of *Vibrio cholerae* O1 in the aquatic environment by fluorescent-monoclonal antibody and culture methods. *Applied and Environmental Microbiology*, **56**, 2370–3.

Isaacson, M., Clarke, K.R., Ellacombe, G.H., Smit, W.A., Smit, P., Koornhof, H.J., Smith, L.S. and Kriel, L.J. (1974) The recent cholera outbreak in the South African gold mining industry, a preliminary report. *South African Medical Journal*, **48**, 2557–60.

Islam, M.S., Drasar, B.S. and Bradley, D.J. (1990) Survival of toxigenic *Vibrio cholerae* O1 with a common duckweed, *Lemna minor*, in artificial aquatic ecosystems. *Transactions of the Royal Society of Tropical Medicine and Hygiene*, **84**, 422–4.

Islam, M.S., Drasar, B.S. and Sack, R.B. (1993) The aquatic environment as a reservoir of *Vibrio cholerae*: a review. *Journal of Diarrhoeal Disease*, **11**, 197–206.

Islam, M.S., Drasar, B.S. and Sack, R.B. (1994a) The aquatic flora and fauna as reservoirs of *Vibrio cholerae*: a review. *Journal of Diarrhoeal Disease*, **12**, 87–96.

Islam, M.S., Hasan, M.K., Miah, M.A., Yunus, M., Zaman, K. and Albert, M.J. (1994b) Isolation of *Vibrio cholerae* O139 synonym Bengal from the aquatic environment in Bangladesh: implications for disease transmission. *Applied and Environmental Microbiology*, **60**, 1684–6.

Kaper, J., Lockman, H., Colwell, R.R. and Joseph, S.W. (1979) Ecology, serology and enterotoxin production of *Vibrio cholerae* in Chesapeake Bay. *Applied and Environmental Microbiology*, **37**, 91–103.

Kawamura, K. and Kaneko, M. (1986) Microbial quality of human wastes and treatment of plant effluent. *Water Science and Technology*, **18**, 257–65.

Kita-Tsukamoto, K., Oyaizu, H., Nanba, K. and Simidu, U. (1993) Phylogenetic relationships of marine bacteria, mainly members of the family *Vibrionaceae*, determined on the basis of 16S rRNA sequences. *International Journal of Systematic Bacteriology*, **43**, 8–19.

Klamen, D.L. and Tuveson, R.W. (1982) The effect of membrane fatty acid composition on the near U.V. (300–400 nm) sensitivity of *Escherichia coli*. *Photochemistry and Photobiology*, **35**, 167–73.

Konig, A. (1984) Ecophysiological Studies on Some Algae and Bacteria of Waste Stabilization Ponds, PhD Thesis, University of Liverpool.

Kott, Y. (1973) Hazards associated with the use of chlorinated oxidation pond effluents for irrigation. *Water Research*, **7**, 853–62.

Kott, Y. and Betzer, N. (1972) The fate of *Vibrio cholerae* (El Tor) in oxidation pond effluents. *Israel Journal of Medical Sciences*, **8**, 1912–16.

Lee, J.V., Bashford, D.J., Donovan, T.J., Furniss, A.L. and West, P.A. (1982) The incidence of *Vibrio cholerae* in water, animals and birds in Kent, England. *Journal of Applied Bacteriology*, **52**, 281–91.

Lema, O., Ogwa, M. and Mhalu, F.S. (1979) Survival of *El Tor* cholera vibrio in local water sources and beverages in Tanzania. *East African Medical Journal*, **56**, 504–8.

Lesne, J., Baleux, B., Boussaid, A. and Hassaini, L. (1991) Dynamics of non-O1 *Vibrio cholerae* in experimental sewage stabilization ponds under arid Mediterranean climate. *Water Science and Technology*, **24**, 387–90.

Levine, M.M., Nalin, D.R., Rennels, M.B., Hornick, R.B., Sotman, S., Van Blerk, G., Hughes, T.P., O'Donnell, S. and Barua, D. (1979) Genetic susceptibility to cholera. *Annals of Human Biology*, **6**, 369–74.

MacDonell, M.T., Swartz, D.G., Ortiz-Conde, Last, G.A. and Colwell, R.R. (1986) Ribosomal RNA phylogenies for the vibrio-enteric group of eubacteria. *Microbiological Sciences*, **3**, 172–8.

MacKenzie, T.D., Ellison, R.T. and Mostow, S.R. (1992) Sunlight and cholera. *Lancet*, **340**, 367.

Mara, D.D. and Cairncross, S. (1989) *Guidelines for the Safe Use of Wastewater and Excreta in Agriculture and Aquaculture*, World Health Organization, Geneva.

Mara, D.D. and Pearson, P.P. (1986) Artificial freshwater environment: waste stabilization ponds, in *Biotechnology*, **8**, (eds H.J. Rehm and G. Reed), VCH Verlagsgesellschaft, Weinheim, pp. 177–206.

Mara, D.D., Alabaster, G.P., Pearson, G.P. and Mills, S.W. (1992) *Waste Stabilisation Ponds, A Design Manual for Eastern Africa*, Lagoon Technology International, Leeds, UK.

Martin, Y.P. and Bonnefort, J.-L. (1989) Variations annuelles et identification des vibrions cultivant a 37°C dans un effluent urbain, dans des moules et dans l'eau de mer en rade de Toulon (Méditerranée, France). *Canadian Journal of Microbiology*, **36**, 47–52.

Martins, M.T., Sanchez, P.S., Sato, M.I.Z., Brayton, P.R. and Colwell, R.R. (1993) Detection of *Vibrio cholerae* O1 in the aquatic environment in Brazil employing direct immunofluorescence microscopy. *World Journal of Microbiology*, **9**, 390–2.

Matin, A., Anger, E.A., Blum, P.H. and Shultz, J.E. (1989) Genetic basis of starvation survival in non-differentiating bacteria. *Annual Reviews in Microbiology*, **43**, 293–316.

Mazoit, L.P., Maurin, J., Dodin, A. and Escallier, G. (1975) Use of ultraviolet irradiation for the disinfection of water supplies. *Tribune du Cebedeau*, **28**, 21–30.

Mcabe, D.B. (1970) Water and waste-water systems to combat cholera in East Pakistan. *Journal of Water Pollution Control Federation*, **42**, 1968–81.

Merson, M.H., Martin, W.T., Craig, J.P., Morris, G.K., Blake, P.A., Craun, G.F., Feeley, J.C., Camacho, J.C. and Gangarosa, E.J. (1977) Cholera on Guam, 1974, epidemiologic findings and isolation and non-toxinogenic strains. *American Journal of Epidemiology*, **105**, 349–61.

Mezrioui, N., Oudra, B., Oufdou, K., Hassani, L., Loudiki, M. and Darley, J.

(1994) Effect of microalgae growing on wastewater batch culture on *Eschericia coli* and *Vibrio cholerae* survival. *Water Science and Technology*, **30**(8), 295–302.

Miller, C.J., Drasar, B.S. and Feachem, R.G. (1984) Response of toxigenic *Vibrio cholerae* O1 to physico-chemical stresses in aquatic environments. *Journal of Hygiene (Cambridge)*, **93**, 475–95.

Mukerjee, S., Rudra, B.C. and Guha Roy, U.K. (1961) Observations on cholera endemicity in Calcutta and survival of *Vibrio cholerae* in the water sources. *Annals of Biochemistry and Experimental Medicine*, **11**, 31–40.

Muller, G. (1977) Non agglutinable cholera vibrios (NAG) in sewage, riverwater, and seawater. *Zentralblatt fur Bakteriologie, Parasitenkunde, Infektionskrankheiten und Hygiene I*, **167B**, 272–84.

Nair, G.B., Sarkar, B.L., De, S.P., Chakrabarti, Bhandra, R.K. and Pal, S.C. (1988) Ecology of *Vibrio cholerae* in the freshwater environs of Calcutta, India. *Microbial Ecology*, **15**, 203–15.

Nichols, A.B. (1991) Sanitation does contribute to Peruvian epidemic. *Water Environment and Technology*, **7**, 13–14.

Oragui, J.I., Curtis, T.P., Silva, S.A. and Mara, D.D. (1987) The removal of excreted bacteria and viruses in deep waste stabilization ponds. *Water Science and Technology*, **19**, 569–73.

Oragui, J.I., Arridge, H., Mara, D.D., Pearson, H.W. and Silva, S.A. (1993) *Vibrio cholerae* O1 (El Tor) removal in Waste Stabilization Ponds in Northeast Brazil. *Water Research*, **27**, 727–8.

Parhad, N.M. and Rao, N.U. (1974) Effect of pH on survival of *Escherichia coli*. *Journal of Water Pollution Control*, **46**, 980–6.

Pazzaglia, G., Lesmana, M., Tjaniadi, P., Subekti, D. and Kay, B. (1993) Use of vaginal tampons in sewer surveys for non-O1 *Vibrio cholera*. *Applied and Environmental Microbiology*, **59**, 2740–2.

Peterson, M.E., Yokel, B.J. and Lim, D.V. (1984) Recovery of selected pathogens from Naples bay, Florida, and associated waterways. *Estuaries*, **7**, 133–8.

Rojas, Y.A. and Hazen, T.C. (1989) Survival of *Vibrio cholerae* in treated and untreated rum distillery effluents. *Water Research*, **23**, 103–13.

Royal Commission on Sewage Disposal (1912) *Eighth Report*, His Majesty's Stationery Office, London.

Sawyer, C.N. and McCarty, P.L. (1978) *Chemistry for Environmental Engineering*, 3rd edn, McGraw-Hill Inc., New York.

Sechter, L., Gerichter, C.B. and Cahan, D. (1975) Method for detecting small numbers of *Vibrio cholerae* in very polluted substrates. *Applied Microbiology*, **29**, 814–18.

Shandera, W.X., Hafkin, B., Martin, D. *et al.* (1983) Persistence of cholera in the United States. *American Journal of Tropical Medicine and Hygiene*, **32**, 812–17.

Sheppard, F. (1971) *London 1808–1870: The Infernal Wen*, Secker and Warburg, London.

Shuval, H.I., Adin, A., Fattal, B., Raawitz, E. and Yekutiel, P. (1986) *Wastewater Irrigation in Developing Countries. Health Effects and Sanitation Solutions*, World Bank Technical Paper 51, World Bank, Washington DC, USA.

Sidwick, J.M. (1977) *A Brief History Of Sewage Treatment*, Thunderbird Enterprises, Harrow, Middlesex.

Singleton, F.L., Attwell, R., Jangi, S. and Colwell, R.R. (1982a) Effects of temperature and salinity on *Vibrio cholerae* growth. *Applied and Environmental Microbiology*, **44**, 1047–58.

Singleton, F.L., Attwell, R.W., Jangi, M.S. and Colwell, R.R. (1982b) Influence of salinity and organic nutrient concentration on survival and growth of *Vibrio cholerae* in aquatic microcosms. *Applied and Environmental Microbiology*, **43**, 1080–5.

Spira, W.M., Huq, A., Ahmed, Q.S. and Saeed, Y.A. (1981) Uptake of *Vibrio cholerae* biotype *eltor* from contaminated water by water hyacinth (*Eichhornia crassipes*). *Applied and Environmental Microbiology*, **42**, 550–3.

Streeter, H.W. (1930) The effects of activated sludge on pathogenic organisms. *Sewage Works Journal*, **2**, 292–5.

Swerdlow, D.L., Mintz, E.D., Rodriguez, M. *et al.* (1992) Waterborne transmission of epidemic cholera in Trujillo, Peru: lessons for a continent at risk. *Lancet*, **340**, 28–32.

Tamplin, M.L. and Parodi, C.C. (1991) Environmental spread of *Vibrio cholerae* in Peru. *Lancet*, **338**, 1216–17.

Tamplin, M.L., Gauzens, A.L., Huq, A., Sack, D.A. and Colwell, R.R. (1990) Attachment of *Vibrio cholerae* serogroup O1 to zooplankton and phytoplankton of Bangladesh waters. *Applied and Environmental Microbiology*, **56**, 1977–80.

Tauxe, R., Blake, P., Olsvik, O. and Wachsmuth, I.K. (1994) The future of cholera: persistence, change, and an expanding research agenda, in *Vibrio cholerae and cholera*, (eds I.K. Wachsmuth, P.A. Blake and O. Olsvik), American Society for Microbiology, Washington, USA, pp. 443–53.

Thom, S., Warhurst, D. and Drasar, B.S. (1992) Association of *Vibrio cholerae* from fresh water amoebae. *Journal of Medical Microbiology*, **36**, 303–6.

Troussellier, M., Legendre, P. and Baleux, B. (1986) Modelling the evolution of bacterial densities in an eutrophic ecosystem (sewage lagoons). *Microbial Ecology*, **12**, 355–79.

Valkov, B.G., Malygin, A.V., Petrova, L.S., Malysheva, G.G. and Saleeva, V.N. (1981) *Zhurnal Mikrobiologii, Epidemiologii i Immunobiologii*, **10**, 114–15.

Ventura, G., Roberts, L. and Gilman, R. (1992) *Vibrio cholerae* non-O1 in sewage lagoons and seasonality in Peru cholera epidemic. *Lancet*, **339**, 937–8.

Watson, K. (1990) Microbial stress proteins. *Advances in Microbial Physiology*, **31**, 183–223.

West, P.A. (1989) The human pathogenic vibrios – a public health update with environmental perspectives. *Epidemiology and Infection*, **103**, 1–34.

West, P.A., Brayton, P.R., Bryant, T.N. and Colwell, R.R. (1986) Numerical taxonomy of vibrios isolated from the aquatic environments. *International Journal of Systematic Bacteriology*, **36**, 531–43.

WHO (1989) *Guidelines for the use of wastewater in agriculture and aquaculture*, WHO Technical Report Series 778, WHO, Geneva.

Woese, C.R., Weisburg, W.G., Hahn, C.M., Paster, B.J., Zablen, L.B., Lewis, B.J., Macke, T.J., Ludwig, W. and Stackebrandt, E. (1985) The phylogeny of purple bacteria: the gamma subdivision. *Systematic and Applied Microbiology*, **6**, 25–33.

World Bank (1992) *World Development Report 1992, Development and the Environment*, Oxford University Press, New York.

Youngren-Grimes, B.L., Grimes, D.J. and Colwell, R.R. (1988) Growth of *Vibrio cholerae, Vibrio parahaemolyticus,* and *Vibrio vulnificus* under strict anaerobic conditions. *Systematic and Applied Microbiology,* **11**, 13–15.

Zaidenov, A.M., Sayamov, R.M., Maloletkov, I.S. *et al.* (1976) Prolonged survival of El Tor cholera vibrio in naturally infected sewage. *Zhurnal Mikrobiologii, Epidemiologii i Immunobiologii,* **12**, 61–9.

Problems of the epidemiology of cholera and the ecology of *Vibrio cholerae* 10

B.S. Drasar

10.1 INTRODUCTION

The past 25 years have seen the development of major understandings as to how *Vibrio cholerae* causes cholera. From the point of view of the patient, widespread recognition of the value and efficacy of oral rehydration has revolutionized treatment. The current series of epidemics in South America has shown how the advent of adequate systems for rehydration can reduce the death rates from cholera. Until recently this was among the most dreaded of diseases. Now, a patient is unlucky if they die.

The application of molecular biology to the study of the pathogenic mechanisms and virulence factors produced by *V. cholerae* has led to major advances in our understanding. Indeed, our understanding of cholera can be used as a yardstick against which to evaluate our understanding of other diseases. The understanding of the pathogenesis of the disease has had tangible benefits through the development of genetically engineered live oral vaccines. Though these do not yet provide the complete means of protecting people, such vaccines represent a major advance on the parenteral vaccines that were in use until recently.

The recognition that *V. cholerae*, including those *V. cholerae* associated with the disease, are normal members of the autochthonous flora of

Cholera and the Ecology of Vibrio cholerae. Edited by B.S. Drasar and B.D. Forrest.
Published in 1996 by Chapman & Hall, London. ISBN 0 412 61220 8.

natural waters has improved our understanding of the epidemiology of cholera and gives hope for more rationally based environmental control measures. It is important to realize when speaking of cholera that we must think in terms of control, and not of eradication. A consequence of the ability of *V. cholerae* to colonize aquatic environments is that it cannot be eliminated. The ecology of *V. cholerae* is such that the protection of the human population by means of vaccine and environmental measures related to water supplies would not lead to the elimination of the organism. *V. cholerae* has undoubtedly persisted for many years in a variety of aqueous habitats and will continue to do so.

The physicochemical conditions for survival have been examined in some detail (Singleton *et al.*, 1982; Miller *et al.*, 1984, 1986; Tamplin and Colwell, 1986). Though there is considerable strain-to-strain variation, many strains can persist for many months in 0.1% saline. Survival is better when balanced mixtures of salts like those that occur in seawater are used. When data from outbreaks are considered in the light of these studies, it can be seen how survival in brackish water can enhance disease transmission and explain epidemiological relationships (Miller *et al.*, 1982, 1984).

The physicochemical conditions for the survival of *V. cholerae* O1 are now well defined and the possibility of survival in an estuarine environment and other brackish water is widely accepted (Glass *et al.*, 1991). Doubts exist, however, as to the nature of this survival and the niche occupied in aquatic microecology. *V. cholerae* is probably best regarded as having two inter-linked ecologies, the first related to its persistence in the environment, and the second to the colonization of human beings.

In spite of recent advances major questions remain to be answered about the epidemiology and ecology of cholera. Three questions seem to be of particular importance.

1. How does *V. cholerae* survive in the inter-epidemic periods in endemic areas? That is, why are there cholera seasons?
2. What is the nature of pandemic spread and by what means are *V. cholerae* spread about the world?
3. How are new pandemic strains developed?

10.2 INTER-EPIDEMIC SURVIVAL

It seems likely as mentioned above that *V. cholerae* is a normal member of the autochthonous flora of natural aqueous habitats. A variety of members of the aquatic flora and fauna have been suggested as likely to produce the right habitat. These include the plants, crustaceans, amoebae and various forms of zooplankton/phytoplankton. These studies assume that the vibrios persist as active members of a biofilm microbiota. It is likely that the occurrence of starved, and viable but not culturable forms, is also important. These studies are particularly pertinent when considering the maintenance of existing, and the development of new, endemic foci. If it is postulated that in endemic areas vibrios persist as part of the normal flora of a number of bodies of water it can be readily seen how local epidemics will arise. Seasonal variation can be explained in terms of seasonal variations in the aquatic microbiota, e.g. algal blooms linked to an increase in the numbers of *V. cholerae* in water. Further, colonization of lakes and watercourses in non-endemic areas may result in the development of new endemic areas and the range of areas in which cholera is endemic. Endemic foci associated with colonized water sources probably occur in Bangladesh, in Louisiana in the US, in parts of Africa, and perhaps in Australia. It is likely that during a current South American outbreak, new endemic foci have been developed.

Under this scenario the seasonality of cholera in an endemic area is explained in terms of seasonal variations in the local ecology. It is assumed that seasonal variations in human activity bring people and vibrios into contact. This theory fits in well with what we know about the potential for survival of *V. cholerae* biotype El Tor. However, our knowledge of the survival of the Classical biotype is more fragmentary and we cannot at this stage be sure that a similar explanation applies. It may be that the comparatively short duration of the Classical pandemics was a reflection of the limited number of ecological niches available for Classical vibrios when compared with El Tor. Any speculation as to mechanisms involved in the early historic pandemics must be restricted by our lack of knowledge as to the causal agent.

Thus, we have an attractive hypothesis that explains many features of the seventh pandemic. This should not, however, be extrapolated too far. Though the ecology of *V. cholerae*, whether of the El Tor or Classical biotype, or indeed of other types, must explain the seasonal variation in disease, it is likely that different strains of *V. cholerae* have different ecologies and indeed different requirements of the habitats within the aqueous environment. There are two important questions that need to be addressed in some detail:

(a) the relationship of the different ecologies of *V. cholerae*; and
(b) the form in which *V. cholerae* survives in different ecological niches.

10.3 ECOLOGICAL MODELS FOR *V. CHOLERAE* SURVIVAL

V. cholerae is able to survive in a number of different ecological systems. Best understood is the human intestinal ecosystem, the colonization of which by the vibrio can produce cholera and increases the numbers of toxigenic vibrio. Survival also occurs in the aquatic ecosystem and in a variety of biological microhabitats associated with this. Transition from the intestine to the aquatic environment and from the aquatic environment to a biological microhabitat involves environmental changes.

10.3.1 ENVIRONMENTAL TRANSITIONS

It is known that *V. cholerae* responds to environmental signals. This phenomenon has been most extensively studied in terms of the virulence cascade (DiRita *et al.*, 1991; DiRita, 1992; Mekalanos, 1992). However, it is likely that environmental signals are also important when the vibrio enters the aqueous environment and when it forms a biological association with other members of the aquatic microbiota. The ecological transitions are set out in Figure 10.1. Questions related to the interaction with the human ecosystem and the control of virulence are dealt with in Chapter 5. The production of the specific survival phase in the aquatic environment is dealt with below.

Possible biological microhabitats are detailed in Chapters 6 and 7. It seems likely that different microhabitats will involve different biological transitions and different forms of association. Some possible models are presented here.

(a) Model 1

Model 1 (Figure 10.2) postulates that only minimal changes occur through vibrios when they enter the aquatic environment. In this model

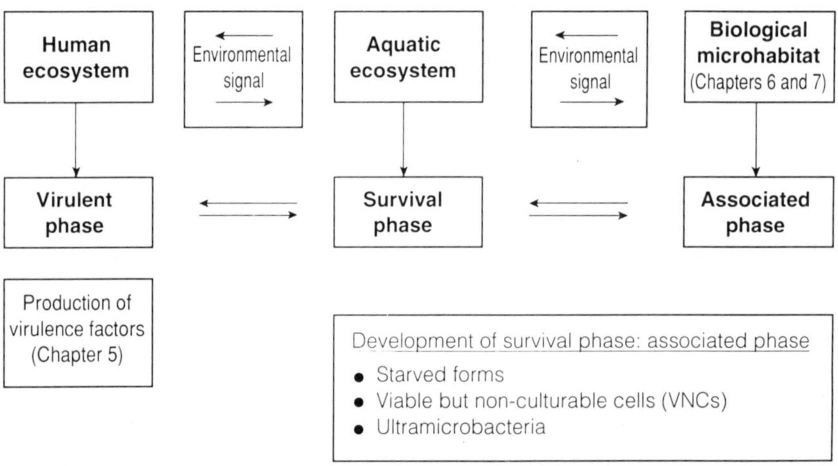

Fig. 10.1 A simplified ecological model for *V. cholerae* survival.

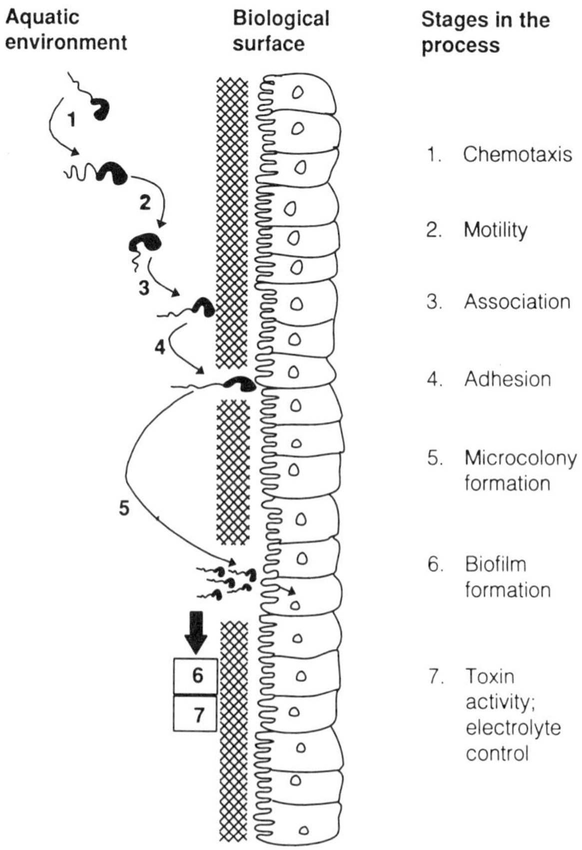

Fig. 10.2 Model 1: survival of *V. cholerae* in biological microhabitats; normal cell type.

when the vibrio enters the aquatic environment it seeks out and forms an association with a suitable biological host. This might be an alga, a plankton or any of the other possible hosts detailed elsewhere. The purpose of this association is to provide the vibrio with an osmotically protected ecological niche. It is suggested that participation in a biofilm may in part solve this problem. It may be that the production of cholera toxin (CT) and CT-like molecules enables the vibrio to alter the electrolyte balance within this microenvironment. Under this scenario the main purpose of these molecules is to enable the vibrio to control the osmolarity of its environment rather than to cause cholera. Some support for this idea is provided by the observation that toxin productions may be up regulated at 25°C (Parsot and Mekalanos, 1990). Obviously, consideration of the specificity of toxin action would indicate that the number of niches in which such an interaction is possible must be limited. However, this need for specific interaction with particular hosts may explain the variety of molecules produced by vibrios, particularly the non O1 *V. cholerae*.

(b) Model 2

This possibility (Figure 10.3) is in effect a variant of Model 1. On entering the aquatic environment, the vibrio seeks out a biological microhabitat but on forming an attachment to a suitable host, is then transformed/changed into an associated or survival phase in response to environmental signals. It is this type of association that has been described in studies of blue-green algae.

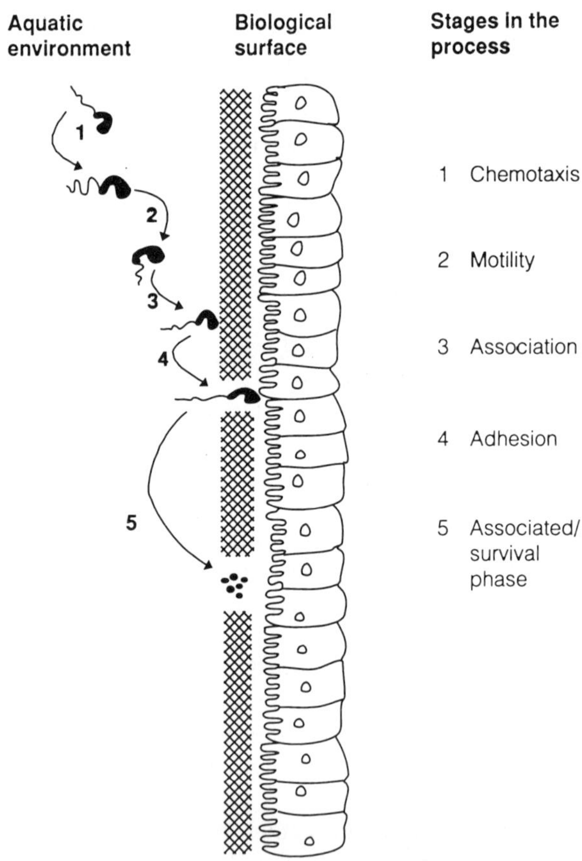

Fig. 10.3 Model 2: survival of *V. cholerae* in biological microhabitats, cell type changed on association.

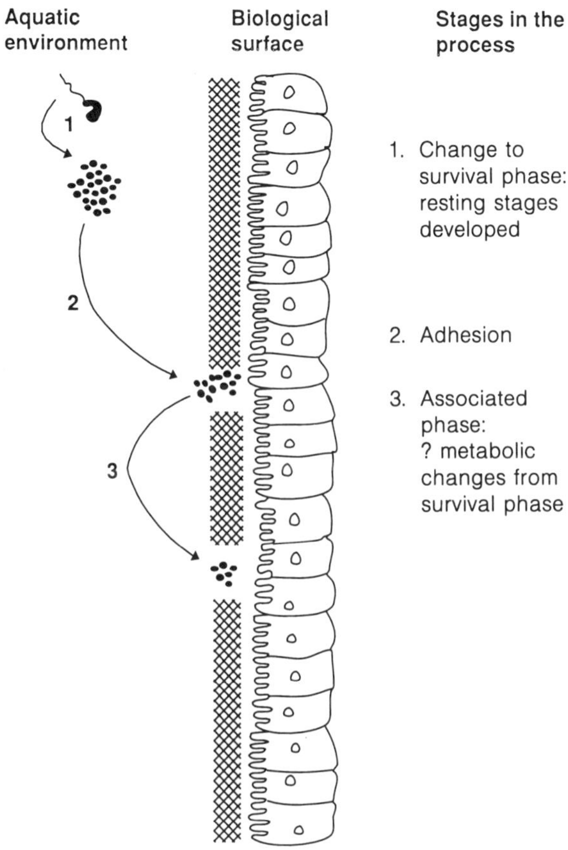

Fig. 10.4 Model 3: survival of *V. cholerae* in biological microhabitats, cell type changed by exposure to aquatic environment and an association.

(c) Model 3

Model 3 (Figure 10.4) represents a further possible variant. It is likely that when vibrios enter the aquatic environment, this transition stimulates the change to a starvation/stress response. This may result in the production of starved forms/ultramicrobacteria, or the viable but nonculturable forms discussed below. However, these different phenotypes are not precluded from entering into an interaction with various biological hosts. Indeed, it may be that the increased surface hydrophobicity associated with these changes facilitates these interactions. It must also be considered that the starvation form produced in the aquatic phase might undergo further changes when associated with a suitable host.

The variety of interactions outlined here by no means exhausts the possibilities. Intimate associations such as those that result when vibrios are ingested by amoebae may be a further variant. The interactions outlined in these models are not exclusive and it may be that all these types of association – and indeed combinations of them – occur under different ecological circumstances. The variety of biological interactions and possible ecological niches that have been described for *V. cholerae* suggest that it is an extremely ecologically versatile organism. If this is true, we should not be surprised that it is able successfully to colonize new endemic areas and to survive in the environment during inter-epidemic periods.

10.3.2 SURVIVAL PHENOTYPES OF *V. CHOLERAE*

Central to this discussion of the survival of *V. cholerae* in the environment is the concept of the survival phenotype. The earlier literature on survival was reviewed by Roszak and Colwell (1987). Particular attention has been paid to the concept of viable and non-culturable forms (VNCs). Thus, it has been postulated that under stress conditions the vibrios had converted to a form which cannot be recovered by standard culture techniques but remains viable. Such VNCs can be detected by fluorescent microscopy, by biochemical techniques and remain able to produce infection. Other resting stages of viable forms that have been considered include ultramicrobacteria and starved cells (Table 10.1).

The relationship of these various forms one to another is uncertain. It is clear however that in most natural environments and in laboratory microcosms, the number of bacterial cells that can be demonstrated by culture techniques is much lower than the total count demonstrable by microscopy. From this it must be concluded that either most bacteria in the environment are non-viable, or that the methods normally used to detect bacterial viability are inadequate. As a working hypothesis it seems reasonable to assume that starved bacteria, ultramicrobacteria and viable but non-culturable forms are all aspects of the same biological phenomenon, different definitions being a consequence of the different methods used for their study.

Table 10.1 Postulated resting stages

- Ultramicrobacteria
- Viable but non-culturable (VNCs) (Xu *et al.*, 1982)
- Starved bacteria

Conceptually and perhaps practically, starved forms can be distinguished from VNCs. Starvation responses can be regarded as individual examples of the well-described stress or heat-shock response. Though starvation may be a more complex phenomenon.

Culture strategies devised for the recovery of stressed cells are often successful for starved cells. The starved form of *V. cholerae* rapidly reverts to the vibrio form on the addition of nutrients (Baker *et al.*, 1983).

The VNC hypothesis postulates that VNCs are a specific quiescent or dormant phase. *V. cholerae* enters a viable but not non-culturable phase which is thought to facilitate its association with a wide range of marine life. Under favourable conditions they revert to a culturable and infectious form. Such conditions are thought to arise in association with algal blooms. Though it may be that the same conditions stimulate both phenomena (Epstein *et al.*, 1993).

As might be expected, considerable attention has focused on the growth or resuscitation of VNCs. However, growth of VNCs in the absence of culturable cells has not so far been convincingly demonstrated (Barer *et al.*, 1993; Ravel *et al.*, 1995). This is not to say that VNCs are not viable, but rather underlies the distinction between viability and culturability.

In part at least, resting stages may be explained in terms of the properties of starved cells. When bacteria are placed in a non-nutrient environment, i.e. under starvation conditions, they undergo a variety of changes as a result of the re-metabolism and conversion of the cell structure in an attempt to prolong cell viability. When compared with bacteria grown on laboratory culture media they are different in a number of ways (Siegele and Kolter, 1992) (Table 10.2). Cells are much smaller, usually spherical and may be referred to as ultra-microcells. Further, cells are more resistant to autolysis and more adhesive to surfaces as a result of the greater hydrophobicity of the cell wall. In the starved state, bacteria appear to undergo more spontaneous mutations than when they are fully fed. This is probably the effect of the down-regulation of DNA repair systems under starvation conditions as a part of energy and nuclear tied conservation rather than an increase in mutation rate.

V. cholerae is able to exist in very different ecological situations. As a

Table 10.2 Starved forms of bacteria

- Ultramicrocells
- More resistant to autolysis
- More adhesive (hydrophobic)
- More spontaneous mutations

result of this versatility, it must be able to express a variety of phenotypes and to exist in a number of forms. This is an area in which much further work is needed in that it will enable us to understand not only the ecology of *V. cholerae* and important aspects of the epidemiology of cholera, but also the basic biology of the environmental survival of many other bacteria.

10.4 PANDEMIC SPREAD

The occurrence of pandemics first brought cholera to the attention of European medicine. Though the movement of infected individuals is undoubtedly important this is by no means the whole of the story. In an epidemic most people will be colonized by *V. cholerae* but not all those so infected will develop clinical cholera. It has been estimated that about half of those infected with the Classical biotype and about 90% of those infected with the El Tor biotype remain asymptomatic. Thus, it is clear how in the short term such individuals can move the vibrios from place-to-place and thus spread the epidemic. However, in the past the efficacy of this means of spread must have been limited. People did not travel far and they travelled slowly. Furthermore, intestinal carriage of *V. cholerae* seems to be usually short term.

The reasons for, and the mechanisms of, pandemic spread remain obscure. In the most recent pandemic rapid travel of persons by air helped spread. The movement of ballast water by ships can be seen as a means of moving local microbiota around the world and possibly enhancing the range of colonization of *V. cholerae*. However, in past pandemics this did not occur.

Carriers in cholera – in the long-term sense that they occur in typhoid – are very rare; so, spread by humans until the advent of rapid air transport does not seem to be an adequate explanation. Thus, although the epidemiology of cholera has been studied since before the advent of microbiology and the pioneering studies of John Snow established many of the principles of epidemiology, our increased knowledge of the survival of the vibrio, its pathogenicity and the advent of molecular typing methods have not made our understanding that much clearer.

It has long been assumed that pandemics result from the spread of strains of *V. cholerae* from a source. The Ganges delta has been considered to be the ancestral home of vibrios and it has long been believed that pandemics spread from this area. Under this theory, a new, highly virulent strain of *V. cholerae* would develop in an endemic focus and then spread widely. However, application of ribotyping systems to cholera vibrios isolated during the seventh pandemic cast doubts on this. A variety of ribotypes have been demonstrated, each in part restricted to a particular area (Koblavi *et al.*, 1990). Thus, different types of *V. cholerae* El Tor occur on the Indian sub-continent, Africa and in South America. It has been suggested that this diversity of types represents the acquisition of virulence factors by *V. cholerae* normally resident in aquatic environments in the different areas. Studies on the possible eighth pandemic caused by *V. cholerae* O139 suggests a single type is responsible for all the cases so far studied.

The explanatory value of the results of the ribotyping is limited in that they relate only to the seventh pandemic. The occurrence of different types in different epidemic areas may represent the ability of strains to colonize local aquatic reservoirs. The problem remains; either we can explain the spread of virulence factors and their acquisition by local vibrios world-wide, or alternatively, the spread of the virulent vibrios. A further complication stems from the different degrees of susceptibility to cholera of different groups of people. This matter is dealt with more fully in Chapter 6. For example, cholera had not been reported in South America for nearly 100 years and it was assumed for that reason it would not occur there and that perhaps the population was not susceptible. The recent extension of the seventh pandemic to South America has made us realize that there is no area of the World in which cholera cannot become established.

The pandemic spread of cholera can be best explained by changes in the ecology of *V. cholerae*. These changes can result from colonization of new areas with strains of *V. cholerae* carried in and adapted to the human intestinal ecosystem. This mechanism allows *V. cholerae* to become established in new ecological niches and to extend the range of biological microhabitats that it uses. Pandemic spread can also occur when whole ecologies are moved around the world, as for example in the bilge water of ships. This results not only in the spread of the vibrios, but also in the spread of suitable biological microhabitats. It is likely that both of these mechanisms play a role.

10.5 THE DEVELOPMENT OF PANDEMIC STRAINS

Until recently it was regarded as a truism to say that there had been seven pandemics of cholera, the first six caused by the Classical biotype and the seventh by the El Tor biotype. A little reflection will show this to be false. It was not until the fifth pandemic that *V. cholerae* was isolated and identified. Thus, we have no way of knowing what biotype caused the first four pandemics. However, we may safely say that the fifth and sixth pandemics were caused by the Classical biotype and the seventh by the El Tor biotype.

V. cholerae O139 is the most recent type of *V. cholerae* capable of causing epidemic cholera; indeed, it may be about to cause the eighth pandemic. If this is the case, of the four pandemics of which a cause would be known, three would have been caused by different organisms. It is important therefore to consider how and by what mechanisms new pandemic strains occur. Is the history of cholera the history of the biological evolution of a particular bacterial species to produce virulent organisms capable of pandemic spread?

Molecular biology techniques have been used to study the degree of relatedness of *V. cholerae* from patients and from the environment. These studies suggest that the sixth and seventh pandemic strains may be evolved from separate lineages of environmental non-O1 *V. cholerae* (Karaolis *et al.*, 1995).

Ribotype studies of vibrio from Bangladesh have shown that O1 and non-O1 clinical and environmental isolates may be very closely related. Indeed, an environmental isolate of non-O1 *V. cholerae* has been described that seems to be closely related to a clone of classical biotype of *V. cholerae* (Faruque *et al.*, 1995).

These studies further illustrate the survival in the environment, in Bangladesh, of strains having genetic characteristics associated with previous pandemic strains. This suggests that such characteristics are important for survival in the environment and environmental strains could form a reservoir of genes needed for the development of pandemic strains.

The diversity of strains of *V. cholerae* must reflect the complex ecology with its several means of generating mutation and selecting for different characteristics (Figure 10.5). The intestinal environment of a person with cholera is the most dramatic evidence of the amplification of bacterial numbers. The immune response of the host and the stresses consequent on entry into the aquatic environment provide potent selective pressures. Within the aquatic environment the starvation response may provide a means by which spontaneous mutations are conserved. If DNA repair is closed down and damage accumulates, many different possible variants will be available for selection when re-growth occurs. Similarly, macromolecular scavenging may increase the range of genes available to the bacterial cell. This would be particularly the case if restriction systems were disabled as a result of cell starvation.

Mutations are probably not simple context-free events involving simple changes to nucleic acid sequences. In reality, mutations occur within cells and intercellular environments may influence their occurrence. Cairns *et al.* (1988) suggested that mutations arise more frequently when the organism is under selective pressure that would favour that mutation. In the present context, this would entail *V. cholerae* responding to the external environment by changes in the internal environment that would favour mutation. In this context it may be important that the repeated sequence (RS1) flanking the cholera

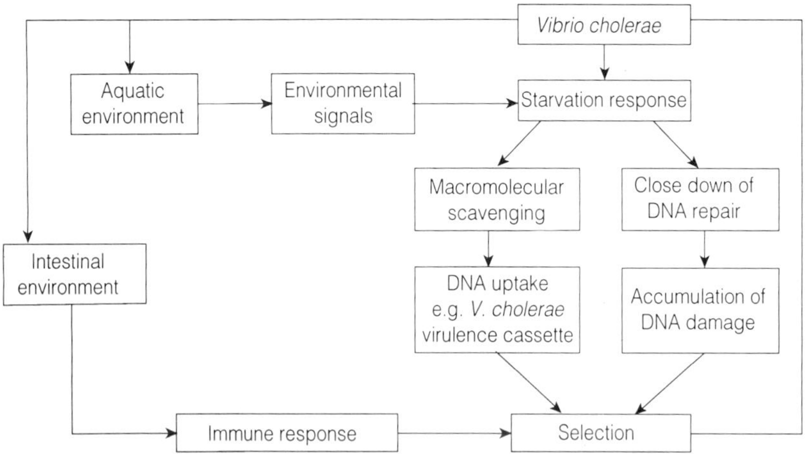

Fig. 10.5 Model for the development of pandemic strains of *V. cholerae* by selection in intestinal and aquatic environments.

toxin gene includes a site-specific recombination system (Pearson *et al.*, 1993) that might be affected by environmental changes.

The question of adaptive mutation is the subject of active research and the results should illuminate the complex problem of adaptation to a range of dissimilar ecological niches. The idea that 'organisms use their perception of environment to modify their genetic metabolism' (Thaler, 1994) would make the possibility of complex changes understandable. However, the increased frequency of mutation in the starved state may also give an appearance of 'directness' (Bridges, 1994).

The variety of ecological models discussed above illustrates the number of situations in which adapted variants could be selected. It may be that the environmental phase of *V. cholerae* survival, particularly during starvation, acts as an engine to generate diversity and from this new pandemic strains arise. If this is the case, it would be equally true to say that the cholera cases behave as amplifiers of bacterial numbers and thus amplify the raw materials upon which such generators of diversity may act.

10.6 REFERENCES

Baker, R.M., Singleton, F.L. and Hood, M.A. (1983) Effects of nutrient deprivation on *Vibrio cholerae*. *Applied and Environmental Microbiology*, **46**, 930–40.

Barer, M.R., Gribbon, L.T., Harwood, C.R. and Nwoguh, C.E. (1993) The viable but non-culturable hypothesis and medical bacteriology. *Reviews in Medical Microbiology*, **4**, 183–91.

Bridges, B.A. (1994) Starvation-associated mutation in *Escherichia coli*: a spontaneous lesion hypothesis for 'directed' mutation. *Mutation Research*, **307**, 149–56.

Cairns, J., Overbaugh, J. and Miller, S. (1988) The origin of mutants. *Nature*, **335**, 142–5.

DiRita, V.J. (1992) Co-ordinate expression of virulence genes by ToxR in *Vibrio cholerae*. *Molecular Microbiology*, **6**(4), 451–8.

DiRita, V.J., Parsot, C., Jander, G. and Mekalanos, J.J. (1991) Regulatory cascade controls virulence in *Vibrio cholerae*. *Proceedings of the National Academy of Sciences, USA*, **88**, 5403–7.

Epstein, P.R., Ford, T.E. and Colwell, R.R. (1993) Marine ecosystems. *Lancet*, **342**, 1216–19.

Faruque, S.M., Roy, S.K., Abdul Alim, A.R.M., Siddique, A.K. and Albert, M.J. (1995) Molecular epidemiology of toxigenic *Vibrio cholerae* in Bangladesh studied by numerical analysis of rRNA gene restriction patterns. *Journal of Clinical Microbiology*, 33, 2833–8.

Glass, R.I., Claeson, M., Blake, P.A. *et al.* (1991) Cholera in Africa: lessons on transmission and control for Latin America. *Lancet*, **338**, 791–5.

Karaolis, D.K.R., Lan, R. and Reeves, P.R. (1995) The sixth and seventh cholera pandemics are due to independent clones separately derived from environmental, nontoxigenic, non-O1 *Vibrio cholerae*. *Journal of Bacteriology*, **177**, 3191–8.

Koblavi, S., Grimont, F. and Grimont, P.A.D. (1990) Clonal diversity of *Vibrio cholerae* O1 evidenced by rRNA gene restriction patterns. *Research in Microbiology*, **141**, 645–57.

Mekalanos, J.J. (1992) Environmental signals controlling expressions of virulence determinants in bacteria. *Journal of Bacteriology*, **174**, 1–7.

Miller, C.J., Drasar, B.S. and Feachem, R.G. (1982) Cholera and estuarine salinity in Calcutta and London. *Lancet*, **i**, 1216–18.

Miller, C.J., Drasar, B.S. and Feachem, R.G. (1984) Response to toxigenic *Vibrio cholerae* O1 to physico-chemical stress in aquatic environments. *Journal of Hygiene (Cambridge)*, **93**, 475–95.

Miller, C.J., Drasar, B.S., Feachem, R.G. and Hayes, R.G. (1986) The impact of physico-chemical stress on the toxigenicity of *Vibrio cholerae*. *Journal of Hygiene (Cambridge)*, **96**, 49–57.

Parsot, C. and Mekalanos, J.J. (1990) Expression of ToxR the transcriptical activator of virulence factors of *Vibrio cholera* O1 is modulated by heat shock response. *Proceedings of the National Academy of Sciences, USA*, **87**, 9898–902.

Pearson, G.D.N., Woods, A., Chiang, S.L. and Mekalanos, J.J. (1993) CTX genetic element encodes a site-specific recombination system and an intestinal colonization factor. *Proceedings of the National Academy of Sciences, USA*, **90**, 3750–4.

Ravel, J., Knight, I.T., Monahan, C.E., Hill, R.T. and Colwell, R.R. (1995) Temperature-induced recovery of *Vibrio cholerae* from the viable but nonculturable state: growth or resuscitation? *Microbiology*, **141**, 377–83.

Roszak, D.B. and Colwell, R.R. (1987) Survival strategies of bacteria in the natural environment. *Microbiological Reviews*, **51**, 365–79.

Siegele, D.A. and Kolter, R. (1992) Life after Log. *Journal of Bacteriology*, **174**, 345–8.

Singleton, F.L., Atwell, R., Jangi, S. and Colwell, R.R. (1982) Effects of temperature and salinity on *Vibrio cholerae* growth. *Applied and Environmental Microbiology*, **44**, 1047–58.

Tamplin, M.L. and Colwell, R.R. (1986) Effects of microcosm salinity and organic substrate concentration on production of *Vibrio cholerae* enterotoxin. *Applied and Environmental Microbiology*, **52**, 297–301.

Thaler, D.S. (1994) The evolution of genetic intelligence. *Science*, **264**, 224–5.

Xu, H.S., Roberts, N., Singleton, F.L., Attwell, R.W., Grimes, D.J. and Colwell, R.R. (1982) Survival and viability of non-culturable *Escherichia coli* and *Vibrio cholerae* in the estuarine and marine environment. *Microbiological Ecology*, **8**, 313–23.

Index

Accessory cholera enterotoxin (Ace) 142
Accessory Colonization Factor (ACF) 131
Activated sludge 308, 315
Adhesins 125
Aerobic systems 308, 314
Algae 319, 320
  Anabaena sp. 217, 219
  blooms 267
  blue-green 216, 217, 219, 320
  Chlorella 320, 321
  Euglena 320, 321
  green 215
  Laminaria longicruris 207
  Lemna minor 209, 211
Alkalinity 264, 267, 268
Altitude, role of 35, 36
Altona 49
Amino acids 77
Amoebae 244, 303
  Acanthamoeba palestinensis 244
  Acanthamoeba polyphaga 245
  Amoeba discoides 244
  Naegleria gruberi 245
Anabaena sp., 217, 218, 219
Anaerobic digesters 308
Angola 263
Animal models of cholera 128
Anti-diarrhoeal/anti-secretory agents 80
Antibiotics 78, 88
  resistance to 80
Antibody
  secreting cells (ASCs) 104, 105
  serum vibriocidal 158
Antigens
  lipopolysaccharide O-antigen 108, 109, 110
  outer membrane proteins (OMPs) 111
  pili/fimbrial 113
  protective surface of *V. cholerae* O1 158
Arcot 20
Attached growth 302
Australia 200

Bangladesh 12, 189, 201, 261, 271
Behavioural factors 257
Belgium 20
Bengal 34, 261
Bethlem Hospital 42
Black blood 54, 58
Bleeding 55, 57
Bombay 30
Brandy 57
Breast-feeding 276
Brixton 42
Broad Street 42, 45, 47
Budd, Dr William 31, 42

Calcutta 11
Canada 21
Carriers 191, 342
  animals 191
  humans 191
Causative mechanisms of cholera 255, 256
Chemotaxis 218
Chesapeake Bay 199
Chitinase 203
Chlorination 297
Chlorophyll 319

Cholera
  as a protective agent upon convalescent patients 96
  bacteria as causative agents 50
  carriers 191, 342
  causation of 29, 32, 50, 255
  cots 85
  diffusion of incidence 274
  disease 19
  disease distributions 255
  early approaches to management 58
  ecology 187, 333, 343
  endemicity of 190
  epidemics 7, 189
  epidemiology 188–202, 333–47
  geoecology 260
  geography 255–94
  history 18–53, 54–74, 187, 192
  management of 54
  mode of communication 37
  monitoring of 283
  pathogenesis 125
  pathology 37
  protection following clinical disease 96
  regulation 146, 148, 149
  rehydration 60, 68, 74
  seasonality 189, 270
  theories of 32
  toxin 14, 97, 99, 100, 101, 102, 104, 105, 106, 107, 111, 114
  toxin B subunit (CTB) 100, 101, 102, 107, 108, 114
  vaccine development 125
Cholera ships 28
  Devonshire 28
  HMS *Dreadnought* 26, 28, 33
  HMS *Grampus* 33
  Iphigenia 28
Cholera-fungus controversy 34
Climate, role of 29, 34
Clinical management 81
Cockles 236
Colonization factors 129
  Accessory Colonization Factor (ACF) 131
Comma-shaped bacillus 50
Commission of the European Community 297
Constantinople 30
Copepods 203, 205, 231
Core-encoded pilus 132
Crabs 199, 205, 234, 237
  blue crabs 241
ctx operon 140
Cuba 21
CVD 103-HgR 160, 162
  efficacy 164
  field trials of 165
  immunogenicity in populations in industrialised countries 163
  immunogenicity in populations in less-developed countries 163
  safety 161
CVD 111 167
CVD 112 168
Cyanobacteria 216, 217, 219, 320
Cytolysin 143
Czechoslovakia 12

Dams 262
Debt 280
Deficiency theory 32
Dehydration, assessment of 82
Dhaka 11
Diarrhoea watery (secretory) 54
Diet 84
  composition of 75
Diffusion of incidence of cholera 275
Disinfection of wastewater 316
Displaced people 264, 277
  forceably displaced people 277
  population displacement 277
  refugees/refugee camps 277, 279
DNA repair 344
Dormancy 261, 283
Dutch East India Company 19
  history 18
  theories 21

E. coli 300
Ecology 187, 333, 343
  aquatic 301
  models 335
  structure phenomenon 278
  systems 257
Education 259
Eighteenth Century 19
El Tor biotype 8, 51, 130, 132, 143, 211, 239, 276, 315, 319, 335
Electricity as a cause of cholera 32
Endemicity 190
  areas 8
  foci 228
England 23

Enterotoxin 135, 157
alternate mechanisms of action 137
cAMP 137
cellular response 137
ctx operon 140
enzymatic activity 135
genetics 139
structure 135
Environment 192, 257, 259
alkalinity changes 264, 267, 268
aquatic 260, 265
changed surface environment 264
devegetation 265
environmental hypothesis 192
estuarine 267
flooding 262, 268
global environmental change 267
localized environmental change 262
locational vulnerability 275
pH 267, 302, 313, 318
population density, high 268
salinity 264, 267, 268, 302
sunlight 272, 303
temperature 267, 272
Environmental transitions 336
Epidemics 8, 43, 44, 189
see also Pandemics
Epidemiology of cholera 7, 256, 333
see also Cholera
Evolution 301
Excreta disposal 297
Excreters 305
Exeter 23, 24

Faecal coliforms 298, 300, 317, 319, 324
Faecal indicators 300
Fauna 194, 230
aquatic 203
microhabitats 202, 228
Fimbriae 113, 115
First Pandemic 20
Fish 232, 267
Atlantic salmon 233
climbing perch 242
Crucian carp 242
Dover sole 233, 242
finless eel 242
goatfish 236
Hilsa ilisha 242
loach 242
murrel 242
snakehead fish 242
Tilapia 243
turbot 233
Fishermen 234
Flooding 268
Flora 194, 209, 214
aquatic 206, 207, 212
Foods 10
Fourth Pandemic 21
France 20
Fungoid theory 32

Ganges delta 20, 35, 203
Gastric
acid 260
hypoacidity 276
Geoecology 260
Geographic Health Information Systems (GHIS) 284
cross-sectional and spatio-temporal techniques 284
global positioning systems (GPS) 285
satellite imaging 285
Geography of cholera 255–332
ecology 270
landscape epidemiology 256
perspectives 258
Global warming 267
Glucose solutions 67
Glucose-facilitated intestinal absorption 74
Golden Square 42, 44
Great Britain 20
Gut-associated lymphoid tissue (GALT) 99

Haemolysin 142–3
Hamburg 23, 49
Heat-stable enterotoxin (ST) 144
History of cholera 18
Holism 257
Holland 20
Human mobility 276
Hybridomas 102, 109, 110
Hygiene
behaviour 257
personal hygiene 259
Hypochlorhydria 265

Immunity
antibacterial 157
infection-derived 154

Immunity (*contd*)
intestinal 95, 96
intestinal memory immune responses 104
local 100, 108
protective following natural infection 95, 99
Immunity, cellular
T-lymphocytes
CD4 105–6
CD8 105–6
cytokines
IL-4 106, 107
IL-5 107
Immunoglobulins
colostrum 111
coproantibody 99
IgA 96, 97, 99, 102, 105, 107, 109, 111
IgG 99, 100, 102, 107, 109, 111, 112
IgM 99, 100, 105, 109
polymeric IgA 100, 109
secretory IgA (SIgA) 96, 97, 99, 100, 112
serum immunoglobulins 97, 103, 100, 101, 112
Immunological memory 104, 105
India 196
Indicator bacteria 315
International Monetary Fund 280
Intestinal memory immune responses 104
Intravenous rehydration 60, 63, 64, 67, 69, 73
Irrigation 298, 299
Italy 21, 200

Lagoons 317
Lamina propria 98, 101
Latta 61, 65, 66
Lipopolysaccharide (LPS) 103, 107, 108, 111, 113
O-antigen 108, 109, 110
London Basin 36
London 25

M cells 99
Macromolecular scavenging 344
Macrophytes, marine 207
freshwater 209, 210
Madras 30
Malaysia 201, 235
Malnutrition 264, 280
Management of cholera 54
clinical 81
early approaches to 58
Mannose–fucose-resistant haemagglutinin 131
Mannose-sensitive haemagglutinin 132
Mecca 23
Medical service 259
Microcosm
studies 323
seawater 195
Microhabitats of *V. cholerae* 202, 228
Microphytes, marine 212
association with bacteria 213
freshwater 214
Mines 41
Monitoring cholera 283
in physical environment 283
surveillance 283
Monoclonal antibodies 102, 106, 109
Motility and flagella of *V. cholerae* 128
Mozambique 263
structural change in 280
Mucosal immunoglobulin 101
Mucus-dissolving enzymes 125
Mussels 126

New Cholera Toxin (NCT) 144
New York 49
Newcastle epidemic 44
Nineteenth Century, early 20
causation of cholera 29
Nitrous oxide 56
Non-culturable, viable 261
Norway 20
Nutrient intake during infection 81

Opium 57
Oral immunization 101, 102, 105, 106, 107, 109, 112, 115
Oral rehydration solution (ORS) 76, 77, 87
Oral rehydration therapy (ORT) 71, 47, 83, 87
complications of 84
Outer Membrane Proteins (OMPs) 103, 111, 112, 113, 114, 115, 133
antigens 111
Oxfam Sanitation Units 309
Oxygen 56
Oysters 205, 238, 243

Ozonic theory 32

Pandemics 8, 9, 19, 188, 263
  spread 342
  strains 343, 345
Pathogen indicators 295
Pathology and mode of communication of cholera (historical) 37
Peru 201, 261
Peru-14 167
Peyers patches (PPs) 97, 98, 99, 102, 106, 107, 109, 163
pH 267, 302, 313, 318
Philippines 235
Pilgrimages 22
Pili
  antigens 113
  core-encoded pilus 132
  Toxin Corregulated Pilus (TCP) 112, 125
  *tcp*A 113, 130
Plankton 230, 267
  phytoplankton 232, 303
  zooplankton 231, 242
Plasma cells 97
Ponds
  anaerobic 317, 318
  facultative 317, 318, 321
  macrophyte and reed beds 324
  maturation 317, 318, 321
Population density, high 268
Protozoa 321
Pump, the 45

Rainfall 271
  reduction 267
Recombinant vaccine candidates 159
Reed beds 324
Regulation of virulence 146
  in vivo regulation 149
  iron regulation 148
  toxR regulon 146
Rehydration
  complications of 84
  fluids 68
  intravenous 60, 63, 66, 67, 69, 71, 72
  oral 71, 74, 75, 76, 77, 83, 84, 87
  rectal 61, 70
  route of administration 68
Reservoirs 187, 191, 192, 202, 203, 206, 212, 214, 260
  alkaline aquatic 263
  animals 191, 202, 203, 205, 230
  aquatic 265
  of *V. cholerae* 187, 191, 193
  wastewater as 301
Resting phase 229
Rice 77
Risk reduction strategies 283
Royal Free Hospital, The 26
rRNA 300
Rwanda 279

Salinity 264, 268, 302, 303
Salmonella typhi Ty21a 110
Sanitary Conference, Rome 30
Sanitation 10, 259, 265
  standards in 278
Sea mullet 205
Seamen's Hospital Society 26
Seasonality 270
  behavioural factors 273
  epidemic curve 271
  estuarine cities 271
  fishing 274
  handwashing 274
  human mobility 271
  faecal contamination 271
  famine 274
  migration 274
  non-availability of water 274
  of the host 273
  poverty 273
  rainfall 271
  sewage lagoons 272
  socioeconomic effect 273
  sunlight 272
  temperature 272
Second Pandemic 20
Secondary transmission 276
Sedimentation 312, 314
Septic tanks 308
Sewage/sewers 296, 305, 306, 307
  treatment 296
Shellfish 234
Shiga-like toxin 144
Shrimp 235
Sicily 21
Snow, Dr John 31, 37, 39, 41, 255
Societal context 257
Socioeconomic
  factors 280
  status 257
Sociogeography 258

Soil 21
 elevation of 35
 role of 37
South/Central America 21, 239
Spain 21
Spatial perspective 255
 behaviour 256
 cause and effect 256
 causal mechanisms 255
 disease distributions 255
 ecological systems 257
 education 259
 geographical perspectives 258
 landscape epidemiology 256
 medical cartography 255
 medical service 259
 personal hygiene 259
 sanitation 259
 societal context 257
Squid 235
St. Petersburgh 23
Stomach acid barrier 125
Structural adjustment policies 279
 structural adjustment programmes 280
Sucrose 77
Sunderland 24
Sunlight 272, 303
Survival 211
 ecological models of 335
 experiments 193, 195
 inter-epidemic 334
 phase 338
 phenotypes 340
Suspended growth 302
Switzerland 21

Tanzania 271
Taxonomy of vibrios 1
Temperature 272
Third Pandemic 21
Toxin corregulated pili (TCP) , 125, 130
Toxins 134
 Accessory cholera enterotoxin (Ace) 142
 heat-stable enterotoxin (ST) 144
 enterotoxin 135, 157
 genetics of 139
 haemolysin/cytolysin 143
 New Cholera Toxin (NCT) 144
 role of additional toxins in disease 145
 Shiga-like toxin 144
 sodium-channel inhibitor 145
 *tcp*A 113, 130
 Zonula occludens toxin (Zot) 141
toxR 113, 125
 toxR regulon 146
toxR 125
Transudate from serum 100
Trickling filters 308, 315

Ultramicrobacteria 229
United Kingdom 198
USA 198, 237

V. cholerae, see Vibrio cholerae
Vaccination 11
 oral 97, 150
Vaccines 95
 attenuated *S. typhi* expressing *V. cholerae* O1 antigens 168
 attenuated *V. cholerae* O1 153
 attenuated *V. cholerae* O1 El Tor 165–6
 Bengal-15 169
 bivalent *Salmonella typhi* Ty21a 110
 combination cholera-toxin B-subunit-WCV (B-WCV) 102, 103, 104, 107, 109
 CVD-103HgR 108
 killed whole cells (WCV) 103, 104, 107
 live, attenuated 97, 108
 live, against *V. cholerae* O139 168
 non-living oral 150
 parenteral whole-cell inactivated 157
 rationale for live cholera vaccines 153
 recombinant vaccine candidates 159
 toxin-deletion mutants 108
Vellore 20
Viable, non-culturable cells (VNCs) 216, 232, 336, 303, 341
Vibrio cholerae El Tor 8, 15, 130, 132, 143, 211, 239, 276, 315, 319, 335
Vibrio cholerae non-O1 11, 12, 13, 14, 191, 229, 301, 315, 322
Vibrio cholerae O139 7, 11, 99, 101, 114, 246, 259, 262, 301
 Bangladesh strains 247
 dispersion 269
 immunity 270

Vibrio cholerae
biotypes 6
characteristics 1–17
colonization 129
ecology 187–254, 333–47
ecology, aquatic 301
epidemiology 7
identification 3
immunity to 95–124
isolation 197, 198, 199, 200, 201, 204
microhabitats 202, 228
motility/flagellae 128
non-culturable 217
pathogenesis 125–50
reservoirs 191, 192, 193, 202, 203
serotyping 5
survival 193, 195, 211, 334, 338, 340
taxonomy 1, 6, 194
vaccines 150–71
zoological microhabitats 228–54
Vibrio mimicus 3, 7, 11, 14, 301
Vibrios
biochemical characteristics 4
isolation 204
reservoirs 206, 212
taxonomy 1
Volunteer studies 126, 127

Waste stabilization pond (WSP) 297, 316
Wastewater 295–332
concentrations of *V. cholerae* in 304
disinfection of 316
effluent 322
treatment 295, 297, 299, 300, 303, 304, 308, 316
Water hyacinth 303
Water
barrages 262
brackish 268
drinking 41
groundwater 263
salt 265
storage 257
supply 280
Waterborne Theory 37
Whitehead, Rev. Henry 45, 48
World Bank 257, 280

Zonula occludens toxin (Zot) 141